AF797001

Nalco Water Fuel Field Manual

Nalco Water Fuel Field Manual

Kim B. Peyton

Nalco Water, an Ecolab Company

Third Edition

Mc Graw Hill

Cataloging-in-Publication Data is on file with the Library of Congress.

Nalco Water Fuel Field Manual, Third Edition

1 2 3 4 5 6 7 8 9 DOC 21 20 19 18 17 16

ISBN 978-0-07-184809-1
MHID 0-07-184809-6

Sponsoring Editor
Amanda Quinn

Editing Supervisor
Donna M. Martone

Acquisitions Coordinator
Lauren Rogers

Project Manager
Jyotsna Ojha,
Cenveo® Publisher Services

Copy Editor
Surendra N. Shivam,
Cenveo Publisher Services

Proofreader
Raghu Narayan,
Cenveo Publisher Services

Indexer
Vikas Makkar,
Cenveo Publisher Services

Production Supervisor
Lynn M. Messina

Composition
Cenveo Publisher Services

Art Director, Cover
Jeff Weeks

About the Author

Kim B. Peyton began his career in the petroleum industry in 1978. His early experience was in the development of synthetic fuels from coal and shale oil, the manufacture of automotive engine lubricants, and the formulation of oils for industrial applications.

He was an Industry Fellow at Nalco Champion in Sugar Land, Texas where he worked daily with the technical and engineering staff to provide products, technical service, and a diverse array of problem solving support to a variety of customers worldwide. This included petroleum refiners, independent fuel marketers, alternative fuel producers, and others in the petroleum industry.

He held a B.S. Degree in Chemistry from Marshall University and an M.S. Degree in Organic Chemistry from California State University, Fullerton.

Contents

Introduction

The *Nalco Water Fuel Field Manual,* Third Edition, contains a blend of original information taken from the first two editions of the book plus updated information on a wide range of topics used in fuel problem solving. Some examples of these topics include discussions on interactions between fuel additives that may impact fuel performance, contamination of fuel by sulfur and sulfur compounds and their effect on fuel quality, inhibiting peroxide formation in stored distillate fuel, and a review of new alternative fuel sources such as the production of cellulosic ethanol.

Discussion topics related to refining process including the production of commercial LP gases and fuel finishing process involving caustic treatment, salt drying, and clay filtration have been expanded. Also, further details describing the performance properties of hydroprocessed S15 diesel fuel oil, the characteristics of B100 FAME and FAME blends, and ethanol properties such as pHe are presented.

The continuing change taking place within today's petroleum refining and alternative fuel industry has not diminished the need to solve fuel-related problems. Even after fuel is processed, problems in handling, transportation, and storage of finished fuel can arise. For individuals providing technical support to the fuel industry, maintaining an up-to-date awareness of all possible causes of a specific problem will continue to be important.

This third edition is meant to serve as a guide to help identify the cause of a fuel problem and develop fitting solutions. The book can be used as a practical training manual or reference source for individuals wanting to understand why fuel problems occur. It can also be helpful for those in industry working to solve fuel problems as they develop.

Although a basic knowledge of chemistry, petroleum technology, and fuel refining makes reading through the text easier, it is not required. To gain information needed to help identify and solve problems, a chapter focusing on solving specific fuel problems and a separate problem-solving guide are included. Also, a detailed index has been developed to help locate information in the book by using a variety of keywords and phrases.

I wish to acknowledge and thank Terry Burleson, David Horsup, Jeff Montanye, Joseph Stark, and Jennifer Sorrells of Nalco Champion, plus Amanda Quinn of McGraw-Hill and Jyotsna Ojha of Cenveo Publisher Services for their support and guidance during production of the third edition of the *Nalco Champion Fuel Field Manual*.

Kim B. Peyton

Nalco Water Fuel Field Manual

CHAPTER 1

A Problem Solving Technique

Beginning the process of solving a problem can sometimes be the most difficult part of the process. In many circumstances, more information is needed than is available. The decision to bypass the tedious process of collecting information is sometimes easy to justify. The time required to look closely at all aspects of a problem before recommending a solution is often not available. Problem diagnosis and solutions, however, are often expected readily.

How then are many problems solved? Often they are solved by individuals with experience in facing similar situations. However, even if all of the information required becomes available, conclusions may still be difficult to reach. To help with problem solving, various techniques can be used that help provide focus on the most important components of a problem and to eliminate other factors.

Described in this chapter is a technique taken from the book written by Charles H. Kepner and Benjamin B. Tregoe titled *The New Rational Manager,* Princeton Research Press, 1981. This straightforward process of problem analysis provides steps to help focus on vital information needed to solve fuel problems.

The key elements of the Kepner-Tregoe problem analysis technique are presented in Table 1.1.

If answers to the above series of questions can be obtained, it is much easier to focus in on the actual cause and source of a problem. The strength of the Kepner-Tregoe technique is its ability to eliminate much of the "hearsay" information and fortify the relevant information.

The following example describing a gasoline contamination problem may be used to clarify the effectiveness of the Kepner-Tregoe technique.

Identity	Location	Timing	Magnitude
WHAT is observed?	WHERE is the problem observed?	WHEN is the problem observed?	HOW MANY occurrences are observed?
WHAT COULD BE but IS NOT observed?	WHERE COULD the problem be but IS NOT observed?	WHEN COULD the problem be but IS NOT observed?	HOW MANY occurrences COULD BE but ARE NOT observed?

Table 1.1 Key Elements of Kepner-Tregoe Problem Analysis

1.1 Problem Scenario

A retail marketer of gasoline received numerous complaints from customers regarding engine shutdown after fueling at the retailer's service stations. Complaints came from customers who fueled with premium-grade gasoline. Complaints were all received within a 2-week time frame in January. Below freezing overnight and daytime temperatures were common throughout this same 2-week time span.

The fuel retailer could not pinpoint the source of the problem since complaints were coming from customers who purchased premium-grade fuel at different stations throughout the area.

The problem began shortly after the retailer switched to a new gasoline additive for use in both regular- and premium-grade gasoline. Immediately, the gasoline additive supplier was contacted by the retailer to help resolve this problem. In the retailer's eyes, the customer complaints began immediately after initiating use of the new fuel additive. Therefore, the additive was probably the cause of the engine shutdown problem.

To resolve the problem, personnel from the gasoline additive supplier worked with the customer to collect service station gasoline samples, customer fuel filters, and samples of gasoline from the retailer's local service stations and from the fuel distribution terminal. All of these samples were extensively analyzed. Also, January fuel delivery records were closely audited.

From the information collected, the following observations were made:

WHAT is observed?

- Engine shutdown was observed in vehicles using premium unleaded gasoline.
- All gasoline was distributed to local service stations from a single fuel terminal.
- Water was found on the bottom of several service station tanks. Water contained high levels of iron (rust), magnesium, sodium, and calcium.

- The Mg:Ca ratios were close to 3:1. These Mg:Ca ratios are typical of seawater, not freshwater.
- Iron (rust), magnesium, sodium, and calcium were identified on customer fuel filters.

WHAT COULD BE but IS NOT observed?

- Gasoline additive was not observed on customer fuel filters.
- Problems in vehicles using regular unleaded gasoline were not observed.

WHERE IS the problem observed?

- Interviews with customers revealed that fuel purchases were from 15 service station locations.

WHERE COULD the problem be but IS NOT observed?

- Problems were not reported from customers who fueled from 140 of the retailer's remaining service stations.

WHEN IS the problem observed?

- Within a 2-week time period in January

WHEN COULD the problem be but IS NOT observed?

- Continuously

HOW MANY occurrences?

- 33

HOW MANY occurrences COULD BE but ARE NOT observed?

- More than 5000

1.2 Observations from Distribution Terminal Tank Records

An oceangoing barge delivery of premium unleaded gasoline was off-loaded into terminal storage 4 days before the first customer complaint was received. Storage tank records indicate that 1250 gal of water settled from the fuel after 24 hours in storage. Present tank records show a water level of zero gallons in the terminal storage tank.

Premium unleaded gasoline was delivered from this terminal to over 160 different service station locations during the 2-week time period that customer complaints were received. Two additional barge loads of fuel were off-loaded into the same terminal storage tank during this 2-week period.

From the information collected, the following conclusions were made:

- Engine shutdown occurred within a 2-week time period in January.
- Problem fuel was purchased from a limited number of stations.

- All fuel was distributed from a single terminal source.
- Fuel off-loaded from the barge contained significant levels of water.
- Significant levels of water are no longer observed in the terminal tank.
- Service station gasoline tanks contain water with a Mg:Ca chemical ion composition similar to seawater. Iron (rust) and sodium were also found.
- Metal contaminants found on customer fuel filters are similar to those found in service station tank water.

1.3 Probable Cause of the Engine Shutdown

Seawater and corrosion products were delivered with premium gasoline by barge to the fuel distribution terminal. The water settled to the bottom of the fuel distribution terminal storage tank and was drawn off with fuel and delivered to service station tanks. Customers fueling at these service station locations pumped premium gasoline containing seawater and corrosion products into their automobiles.

Fuel filters were partially blocked by rust and products carried by seawater. Below freezing temperatures possibly initiated the formation of ice within fuel lines containing water. The combination of ice and corrosion products blocked lines and filters. Gasoline flow to the engine was halted and engine shutdown followed.

Not all service stations receiving premium unleaded fuel from the distribution terminal were affected. Only a limited number of stations received fuel containing water and corrosion products. This explained the random nature and limited scope of the vehicle problems.

By using the Kepner-Tregoe approach to problem solving, information can be collected in a manner which seems to focus on only the critical details of a problem. Once collected, the information can be compiled into a format which enables one to develop logical conclusions and to determine the cause of a problem.

CHAPTER 2

Refining Processes Used in Fuel Production

The processes utilized in the refinery to produce finished fuels such as gasoline, jet fuel, diesel fuel, and heating oil can be quite complex. Constant monitoring of the refining process is required by engineers. This is because the characteristics of the various crude oils refined can frequently vary and also because the finished product volume demand can change often.

To fulfill finished product blend requirements, it is imperative that all processes within the refinery function and work together as a continuous system. An upset or problem with the operation of one unit in the refinery can sometimes dramatically affect the efficiency, output, and quality of the entire refining process.

A brief description of the various units found within the typical fuel refining operation is provided in this chapter.

2.1 Distillation

Fuels and other petroleum products are derived from crude oil through the use of a variety of different refining process techniques. Distillation, however, is the first significant processing step taken in crude oil refining. Both atmospheric and vacuum distillation can be utilized to process crude oil into fuels and other products. However, prior to distillation, metals and other water soluble impurities must be removed from crude oil. This process is termed *desalting*.

2.1.1 Desalting

Desalting of crude oil involves mixing the oil with water at a temperature of about 121°C (250°F) under enough pressure to prevent evaporation of both water and volatile crude oil components. The salts in the crude oil are dissolved and removed by water. Oil-water emulsions often form that also contain salts. The emulsions can be broken by the use of high-voltage electrostatic coalescers or by the use of demulsifying or emulsion breaking compounds.

Single stage and multistage desalting processes are utilized to remove as much of the salt as possible. Any remaining salts carried into the distillation unit could initiate corrosion.

From the desalting unit, crude oil passes through a series of heat exchangers to increase the temperature of the oil to about 288°C (550°F). From the exchangers, oil passes into a furnace or pipe still and is heated to approximately 343 to 371°C (650 to 700°F). The heated oil then enters the atmospheric distillation column.

Once in the column, volatile compounds vaporize and less volatile liquids drop to the bottom of the column. Reflux begins and oil fractions condense and collect in various trays throughout the column. Heating within the column can influence the rate of reflux and holdup within the column. Steam is introduced into the column at the level of the lower trays and strips the lighter compounds from the condensed fractions on these trays. As a result, the flash point of the fractions remaining on the lower trays increases.

2.1.2 Atmospheric Distillation

Most atmospheric distillation columns contain from 30 to 50 fractionation trays. For each sidestream desired, about five to eight trays are required, plus additional trays above and below the primary trays. The various sidestreams collected from the distillation column contain lighter boiling products that must be removed. Smaller reboiling units are used to remove lighter products and direct them back into the distillation column as vapor. Also, refluxing units are sometimes employed to condense and remove heavy end products from collected fractions. These condensed heavier products are reintroduced into the lower trays.

Typical distillation products and cut points obtained through atmospheric distillation are provided in Table 2.1. The products and cut points can vary depending upon the refinery operating parameters, crude source, and product demand. A schematic of an atmospheric distillation unit is provided in Fig. 2.1.

2.1.3 Vacuum Distillation

Removal of naphtha and distillate fuel fractions from the crude oil under atmospheric pressure distillation requires charge temperatures to be maintained below the cracking temperature of the crude oil components. The cracking temperature will vary but can typically range from 399 to 427°C (750 to 800°F). Above these temperatures,

Crude Tower Fractions	Distillation Temperature Range, °C (°F)
Butane and lighter gases	• <32 (<90)
Light naphtha or straight-run gasoline	• 32 to 104 (90 to 220)
Heavy straight-run gasoline or heavy naphtha	• 104 to 157 (220 to 315)
Kerosine or kerosene	• 157 to 232 (315 to 450)
Light gas oil	• 232 to 343 (450 to 650)
Heavy gas oil	• 343 to 427 (650 to 800)
Atmospheric residual oil	• >427 (>800)

Table 2.1 Atmospheric Tower Fractions and Cut Point Temperatures

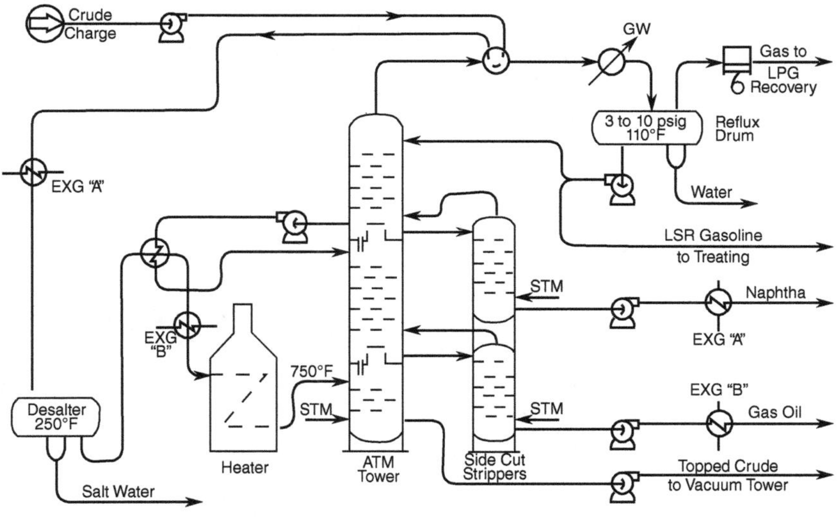

Figure 2.1 Atmospheric distillation unit.

crude oil components begin to thermally crack and foul processing equipment.

Vacuum distillation utilizes reduced pressures and temperatures as a means to remove remaining fuel components and higher boiling lubricant fractions from atmospheric tower bottoms. In the vacuum unit, atmospheric residual oil components are further distilled without being subjected to the high-temperature conditions that lead to cracking and degradation of oil components. Under absolute pressure levels ranging from 25 to 120 mmHg, vacuum distillation will remove higher boiling materials effectively.

In order to minimize the tendency of coke formation within the column, steam is utilized. Application of steam also helps to reduce the absolute pressure of the system to 10 mmHg or less and can aid in stabilizing vacuum levels. By operating at low vacuum pressures, the product yield will increase and the operating costs can often be reduced.

The vacuum-unit-charge heating temperatures are dependent on the operating vacuum pressure, initial boiling point (IBP), and end point (EP) of the crude oil charge. However, the unit operating temperatures are always maintained low enough to prevent thermal cracking of the charge. Upon removal of useful fuel and lubricant components from the vacuum unit, high molecular weight, high boiling point compounds remain as a vacuum residual fraction. The vacuum residual can vary in composition from refinery to refinery and sometimes from day to day within the same refinery if a variety of crude oil blends are processed. A vacuum distillation unit schematic is presented in Fig. 2.2. Also, a listing of typical products obtained from atmospheric and vacuum distillation of crude oil and their use are briefly described in Table 2.2.

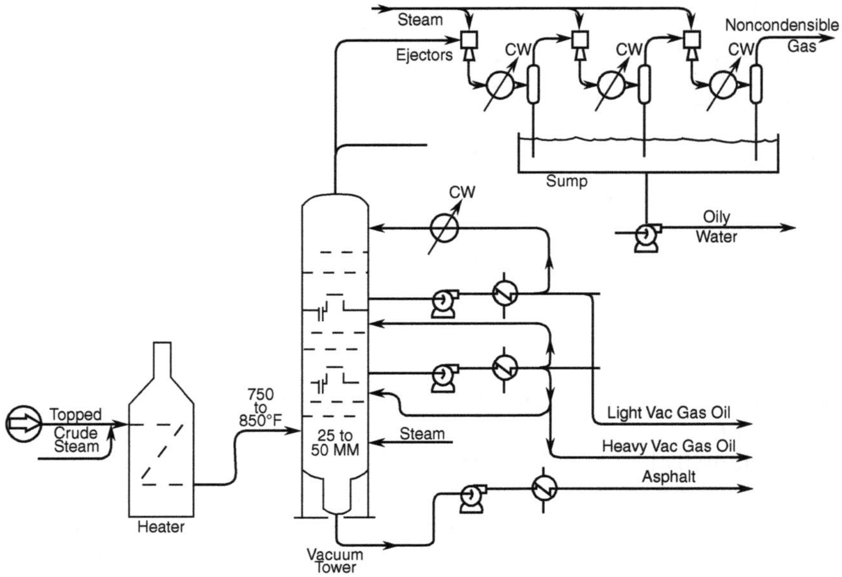

FIGURE 2.2 Vacuum distillation unit.

Product	Composition and Use
Flue gas	Primarily methane and ethane, sometimes propane, often referred to as "dry gas"
Wet gas	Primarily propane and butane, can also include methane and ethane; propane and butane are used in LPG; butane is used in gasoline blending
Light straight-run naphtha	After desulfurization, this stream is used in spark ignition fuel or gasoline blending
Heavy straight-run naphtha	Utilized primarily as catalytic reformer feedstock
Gas oils	Utilized as straight-run distillate after desulfurization; light atmospheric and vacuum gas oils are often hydrocracked or catalytically cracked to produce gasoline, jet fuel and diesel fuel fractions; heavy vacuum gas oils can be used to produce lubestocks or as fluid catalytic cracking (FCC) feedstock
Residual oil	Vacuum bottoms can be utilized as feedstock for the visbreaker, coker, or possibly the asphalt unit; can be used in No. 6 fuel oil or marine fuel blending

Table 2.2 Products from Atmospheric and Vacuum Distillation and Their Use

2.2 Thermal Cracking

By applying elevated temperatures, residual fuel and other high boiling point components of crude oil distillation can be converted into lighter fractions for use in naphtha and distillate fuel blending. Heating residual fuel oil to temperatures in the range of 427 to 538°C (800 to 1000°F) and at increased pressured, carbon-carbon bonds break to yield free radicals. This process of breaking carbon-carbon bonds is called *cracking* and is illustrated in Eq. (2.1).

$$C_{14}H_{30} \xrightarrow[\Delta]{} 2CH_3(CH_2)_5CH_2\bullet \quad (2.1)$$

At pressures between 250 and 1000 psi and temperatures in between 399 and 482°C (750 and 900°F), cracking reactions take place in the liquid phase. Naphtha and distillate type products are formed under these racking conditions.

At pressures less than 100 psi and temperatures more than 538°C (>1000°F), cracking reactions take place in the vapor phase. The formation of lower molecular weight, gaseous hydrocarbons is favored under these conditions.

Reaction	Description
Hydrogen transfer	Radical reacts with a hydrocarbon; hydrogen transfer occurs from a hydrocarbon to the radical forming a lower molecular weight compound and a new radical
Addition	Free radical reacts with an olefin to form a higher carbon number, sometimes branched, free radical; the radical may further crack or undergo hydrogen transfer
Dehydrogenation	Dehydrogenation to form a diene can occur under severe thermal cracking conditions

TABLE 2.3 Effect of Process Variables on Products of Thermal Cracking

Hydrogen transfer, addition, and dehydrogenation reactions occur among the produced radical species to yield new compounds. These reactions are summarized in Table 2.3.

Naphthenes may undergo ring cleavage or side chain removal when thermally cracked. Longer side chain naphthenes are the most susceptible to side chain removal by thermal cracking. Aromatic compounds are the most resistant to thermal cracking conditions. The ease by which various hydrocarbons are cracked thermally can be ranked as follows:

Easiest to crack	*n*-paraffins
↓	*iso*-paraffins
	cyclo-paraffins
↓	Aromatics
	Aromatic/naphthenes
Difficult to crack	Polynuclear aromatics

Because of the tendency of thermal cracking to produce coke as a by-product of the process, many refiners have replaced thermal cracking units with catalytic cracking processes.

2.3 Visbreaking

The process of visbreaking is used to reduce the viscosity of heavy residual fuel oils by heating the oil to temperatures of about 454 to 510°C (850 to 950°F) under pressures between 60 to 300 psi for only a few minutes. This modified version of thermal cracking yields a lower viscosity product by cracking heavier products to lighter components. Approximately 10 to 25 percent of the residual material is converted to lighter components. Refiners recover the cracked components for use as gasoline and diesel blending stocks. If these lighter components remain in the oil, the viscosity of the residual fuel is reduced.

Visbreaking severity is monitored to help minimize cracking and alteration of the nature of asphaltenes within the visbreaker feed. Paraffinic side chain cracking or destruction of the asphaltene resin complex may occur during visbreaking operations and may result in precipitation of asphaltenes from solution. Asphaltene precipitation has been seen especially when cracked material is blended with lighter viscosity paraffinic fuels.

2.4 Fluid Catalytic Cracking

Fluid catalytic cracking (FCC) is an extremely valuable refining process and is utilized extensively to upgrade lower quality petroleum fractions into valuable components, primarily gasoline grade stocks. The FCC process involves preheating a gas oil such as atmospheric gas oil (AGO), vacuum gas oil (VGO), or other fractions to temperatures between 315 to 399°C (600 to 750°F). The feed is then introduced into the reactor accompanied by steam. Average reactor temperatures can vary within a typical range of 454 to 538°C (850 to 1000°F). Temperatures are varied to control octane number and component distribution.

Feed to the FCC unit is mixed with hot catalyst and steam in a reactor line called a riser. The ratio of catalyst:oil can typically range from 4:1 to 9:1 by weight. Overall, FCC is an endothermic process. Heat provided by the hot, circulating catalyst is the prime source of energy driving the FCC process. In the riser, vaporized oil is cracked catalytically in less than 2 seconds. The vapors and catalyst flow out of the riser and into the reactor. At this point, most cracking reactions have occurred.

In the reactor, oil vapors and catalyst are separated by a deflector device or a cyclone. A cyclone facilitates separation by enabling catalyst to collect on its walls and drop from the cyclone to the stripping section at the bottom of the reactor. The vapors exit the reactor and flow into the FCC fractionating tower. A graphic depiction of an FCC reactor is presented in Fig. 2.3.

2.4.1 FCC Product Fractions

In the fractionating tower, the cracked oil is distilled into the following components:

2.4.1.1 Fuel Gas, C_3/C_4 Gases, and Light FCC Gasoline

These fractions are processed in the FCC gas plant. Compounds entering the FCC gas plant can include methane, ethane, ethylene,

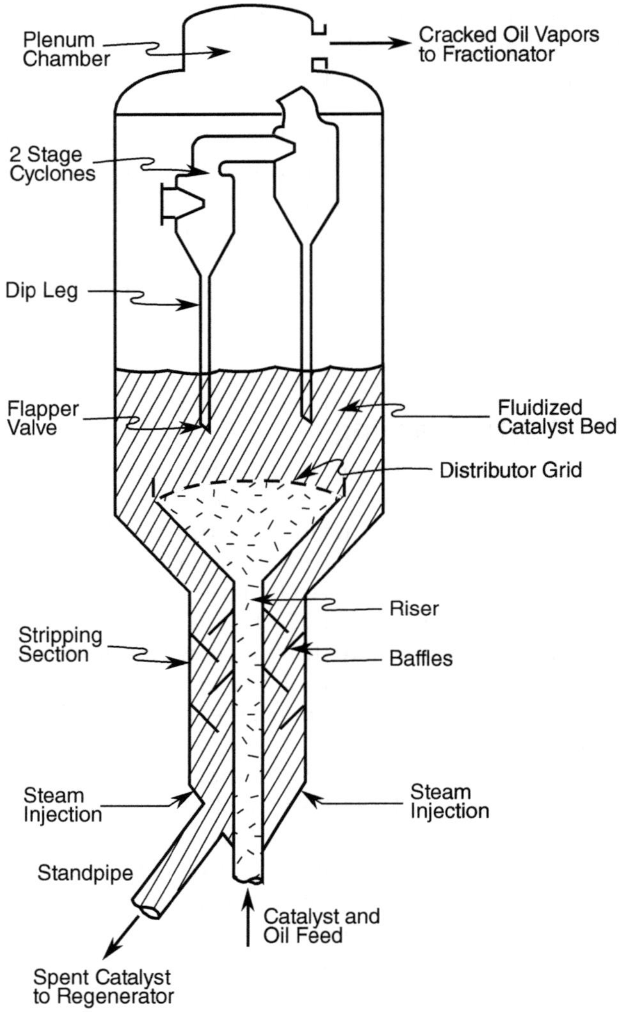

FIGURE 2.3 FCC reactor.

propane, propylene, *n*-butane, isobutane, butene isomers, unstable FCC gasoline, light mercaptans, hydrogen sulfide, cyanides, ammonia, and carbon dioxide.

Special absorbing, stripping, and splitting units are employed to separate these fractions. Once separated, it is necessary to further treat the fractions by washing with water, caustic, or an alkanolamine to remove impurities such as hydrogen sulfide, mercaptans, hydrogen cyanide, and ammonia. Separated fuel gas can be used to fire burners, the C_3 compounds are typically used as petrochemical feedstock, and the C_4 compounds as alkylate feed. Also butane can be used as a gasoline blend component.

2.4.1.2 FCC Gasoline

The produced light FCC gasoline typically contains a mixture of paraffins, olefins, and aromatic compounds in a ratio of around 5:3:2. This ratio will often vary depending upon the feedstock, catalyst quality, and reactor parameters. The research octane number (RON) of FCC gasoline will typically be much higher than the motor octane number (MON).

2.4.1.3 Cycle Oil

Heavier distillate range compounds formed during FCC processing can accumulate within the FCC fractionator. The primary fraction is called light cycle oil (LCO) and contains high percentages of monoaromatic and diaromatic compounds plus olefins and heavier, branched paraffins. Unhydrotreated LCO is often quite unstable and has a very low cetane number. For this reason LCO is blended into diesel fuel at limited amounts, or it is hydroprocessed to improve the color and storage stability. Heavy cycle oil and heavy naphtha are additional side cuts. These streams can be pumped around to remove heat from the fractionator, to supply heat to other refinery units, or used as a low-quality blendstock component.

2.4.1.4 FCC Residual/Slurry Cycle Oil/Decant Oil/Clarified Oil

This heavy product is primarily aromatic in nature and often has a density >1.0 g/mL. It can serve as a heavy fuel oil blendstock or possibly as a carbon black oil feedstock after catalyst fine removal.

During the cracking process, carbon deposits or coke can accumulate on the spent catalyst particles. These deposits can deactivate the catalyst performance and must be removed. This is typically accomplished in two stages:

1. The catalyst collected at the bottom of the FCC reactor is steam stripped to remove residual hydrocarbon.
2. The stripped catalyst then passes into the FCC regenerator and is heated with air to temperatures as high as 540 to 650°C (1000 to 1200°F). At these temperatures, coke burns off of the catalyst making it ready for reuse within the FCC unit.

A schematic of a typical gas processing plant is provided in Fig. 2.4. A depiction of the FCC regenerator is shown in Fig. 2.5.

2.4.2 Impact of Feedstock Characteristics on FCC Product Properties

Fluid catalytic cracking of paraffin, olefin, naphthene, and aromatic compounds in FCC feedstock can produce a wide variety of compounds. Compounds formed often depend on the original structure and characteristics of the feedstock molecules. A description of

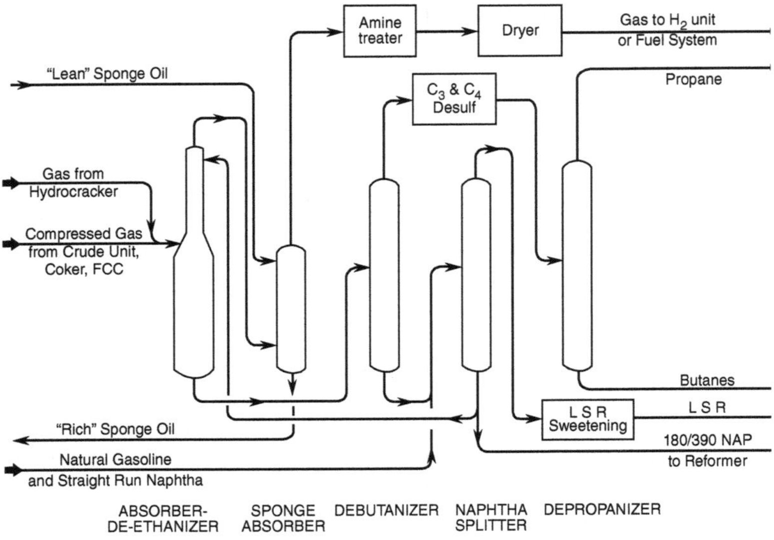

FIGURE 2.4 Gas processing plant.

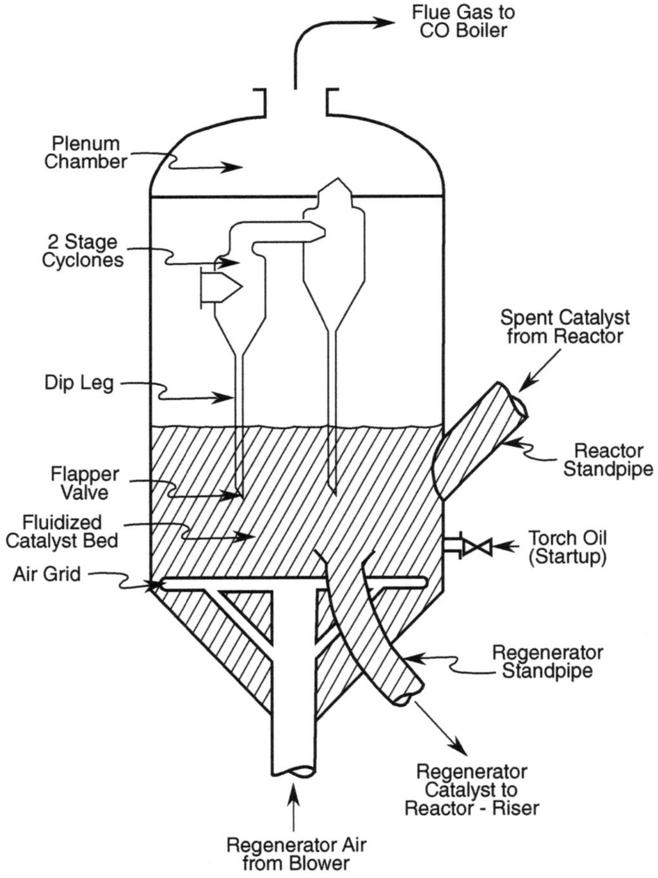

FIGURE 2.5 FCC catalyst regenerator.

Feedstock Component	Effect of Fluid Catalytic Cracking
Paraffin	High concentrations of C_3 and C_4 hydrocarbons are produced; olefins and aromatics are also produced; molecules containing greater than six carbon atoms are cracked more readily than molecules with fewer carbon atoms
Olefin	Crack at higher rates than paraffins; olefins isomerize and aromatize to yield higher octane components
Naphthene	Side chains are broken free of ring structure; dehydrogenation of naphthenic rings containing nine or more carbon atoms is common
Aromatic	Rings possessing three or fewer carbon atoms attached to the ring are quite resistant to cracking; long hydrocarbon chains are usually broken and freed from the aromatic ring

TABLE 2.4 Products Produced from Catalytic Cracking of Different Hydrocarbon Types

products produced from FCC of different hydrocarbon types is provided in Table 2.4.

2.4.3 FCC Cracking Reactions

Catalytic cracking proceeds through the formation of carbocations. The ease of formation depends on the structural nature and stability of the carbocation generated.

Easiest to form	Tertiary
↓	Secondary
	Primary
↓	Ethyl
Difficult to form	Methyl

Once formed, the carbocations undergo a variety of reactions including β-scission, isomerization, hydrogen transfer, and termination.

2.4.3.1 β-Scission

In β-scission reactions, carbocation cracking occurs at the C-C bond, which is β in position relative to the carbon possessing the positive charge. This type of cracking yields relatively large amounts of C_3, C_4,

C_5, and higher carbon number olefins. The carbocation which remains continues to crack, isomerize, and undergo further β-scission. The reaction is usually terminated at some point by hydrogen transfer from another molecule. The β-scission reaction is shown in Eq. (2.2).

$$CH_3CH_2CH_2\underset{\beta}{CH_2}\underset{\alpha}{CH_2}\overset{+}{C}H\text{–}CH_3 \longrightarrow CH_3CH_2CH_2\overset{+}{C}H_2 + CH_2{=}CHCH_3 \qquad (2.2)$$

2.4.3.2 Isomerization

Hydride or methyl group shift reactions to form the more stable carbocation species occur during catalytic cracking. For example, a methyl group shift to form a tertiary carbocation from a secondary ion species is favored. The presence of relatively high concentrations of branched species in FCC products can be explained by this mechanism. The carbocation isomerization reaction is provided in Eq. (2.3).

$$\underset{\text{Secondary carbocation}}{R\text{–}\overset{+}{C}H\text{–}CH_2\text{–}CH_3} \longrightarrow \underset{\text{Tertiary carbocation}}{R\text{–}\overset{+}{\underset{\substack{|\\CH_3}}{C}}\text{–}CH_3} \qquad (2.3)$$

2.5 Hydroprocessing

Hydroprocessing is used extensively to improve the quality and usefulness of petroleum products. Hydroprocessing can reduce unstable olefins to paraffins and can remove sulfur, nitrogen, and oxygen heteroatoms. Hydroprocessing can also effectively reduce aromatics to naphthenes or paraffins. Various hydroprocessing catalysts are available for each specific hydroprocessing need.

The process involves mixing oil with hydrogen under pressure and passing this mixture through a heated catalyst bed. Catalysts such as cobalt-molybdenum (CoMo) oxides on alumina and nickel-molybdenum (NiMo) compounds on alumina are commonly used. The CoMo catalysts are effective at reducing sulfur and are used throughout the refining industry to produce low-sulfur fuel products. The NiMo catalysts are effective at reducing both fuel sulfur and nitrogen compounds but often require careful application in the unit.

Upon hydroprocessing, organic compounds in fuel that contain sulfur and nitrogen are respectively converted to hydrogen sulfide and ammonia plus a hydrocarbon. Incomplete hydroprocessing of sulfur and nitrogen containing organics can yield mercaptans and amines. Although uncommon, carryover of these compounds into finished fuel can lead to copper corrosion, fuel color change, or odor problems.

$$\underset{\text{Mercaptan}}{\text{R-SH}} + H_2 \longrightarrow \text{R-H} + H_2S$$

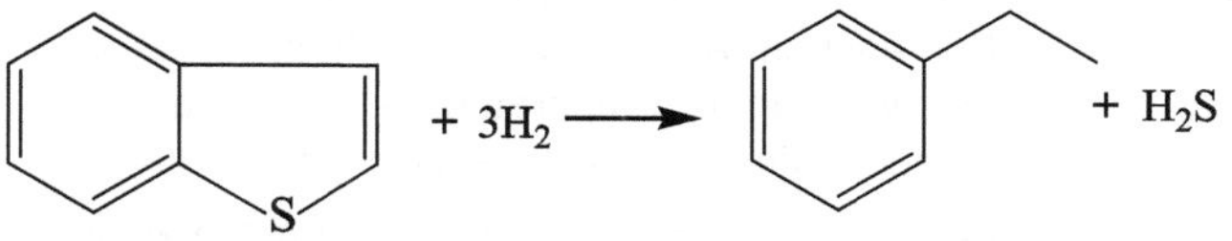

Benzothiophene

FIGURE 2.6 Desulfurization of fuel compounds through hydroprocessing.

Sulfur removal through hydroprocessing is used in the following refining applications:

- Production of low sulfur and ultralow-sulfur diesel fuel and kerosene
- Reduction of residual fuel viscosity and sulfur content to meet specifications
- Reformer feedstock sulfur reduction

Examples of desulfurization of fuel mercaptan and benzothiophene through hydroprocessing are provided in Fig. 2.6.

2.6 Catalytic Hydrocracking

The catalytic hydrocracking unit is typically used to produce naphtha and distillate components used in gasoline, jet fuel, and diesel production, including other components such as reformer feedstock. Feedstock to the hydrocracking unit consists of fractions such as cycle oil, coker distillates, and gas oils.

Most hydrocracking processes involve the use of single stage or dual stage, fixed bed reactors. The reaction occurs when feedstock and hydrogen are mixed at pressures of 1000 to 2000 psi and then passed into a 370 to 427°C (700 to 800°F) reactor heating zone. If the unit is a dual-stage unit, fractions boiling at temperatures less than 205°C (<400°F) are collected from the unit first. The heavier fractions then pass into the second-stage reactor and are hydrocracked to yield distillate grade jet fuel and diesel blendstock. There is no heavy fuel oil produced.

The hydrocracking catalysts are typically alumina-silica supporting a transition metal such as nickel, platinum, or palladium. Tin, a group IV metal, can also be utilized. Hydrocracking catalysts are poisoned by metals, oxygen, and certain organonitrogen and sulfur compounds. For this reason, feedstock is often hydrotreated prior to hydrocracking. Except for metals, this will result in removal of these hydrocracking catalyst poisons. Also, water removal from hydrocracker feedstock is essential to help ensure that the catalyst does not dissolve or collapse. Filtration through molecular sieve or silica gel

Process Variable	Primary Effect
Temperature	A 10°C (18°F) temperature increase doubles the reaction rate but does not affect conversion as significantly; temperature is typically used to compensate for loss in catalyst activity
Pressure	Increase in the partial pressure of hydrogen increases the rate of conversion
Space velocity	An increase in space velocity decreases conversion. To compensate for an increase in space velocity, temperature must be increased
Nitrogen content	Organic nitrogen decreases catalyst activity and decreases the conversion rate
Hydrogen sulfide	At low levels, hydrogen sulfide can inhibit aromatic ring saturation. This results in higher octane gasoline and low smoke point jet fuel. At high concentrations, cracking catalyst activity is adversely affected

Table 2.5 Effect of Process Variables on Hydrocracking Unit Operation

will help to remove water. Some hydrocracking processes, however, can tolerate up to 500 ppm of water in the feedstock. A description of the effect different hydrocracking parameters can have on the overall process is provided in Table 2.5.

2.7 Isomerization

The isomerization process is utilized to convert light paraffins such as butane, pentane, and hexane into higher octane isoparaffins. Isoparaffins have higher octane numbers than normal paraffins, *n*-paraffins, of the same carbon number. For example, *n*-pentane has a research octane number (RON) of 61, and isopentane has an octane number of 92.

The process involves first separating mixed butane compounds by distillation to isobutane and *n*-butane. The *n*-butane is then mixed with hydrogen, heated, and passed through a reactor containing a platinum catalyst or an HCl-activated aluminum chloride catalyst. The *n*-butane is isomerized to isobutene and separated.

Butane isomerization and pentane-hexane isomerization are the two most important isomerization processes. Isobutane is utilized primarily as alkylate feedstock. Isopentanes and isohexanes have become valuable, high-octane blending components in gasoline.

2.8 Catalytic Polymerization

In fuel refining, the polymerization process is utilized to produce high-octane gasoline components from three and four carbon olefins. It can also be used for the production of certain alcohols and aromatic compounds. Cumene and ethyl benzene can be produced through catalytic polymerization. Products from the catalytic polymerization process are primarily dimers, trimmers, tetramers, and oligomers rather than true polymers. However, the process name is still used throughout the refining industry.

The process involves reacting butenes and mixtures of propenes and butenes with either a phosphoric acid type catalyst or a nickel complex-alkyl aluminum type catalyst to produce hexene, heptene, and octene olefins. The reaction first proceeds through the formation of a carbocation which then combines with an olefin to form a new carbocation species. The acid proton donated to the olefin initially is then released and the new olefin forms. Hydrotreatment of the newly formed olefin species results in stable, high-octane blending components.

2.9 Catalytic Reforming

The catalytic reforming process is utilized for production of high octane number compounds from lower octane number fractions. It is an endothermic process that requires heat and a catalyst in order to maintain a constant reaction rate. Hydrogen is also produced in significant quantities through the reforming process.

Reformer feedstock is usually obtained from a straight-run naphtha fraction with a boiling point range from about 82 to 193°C (180 to 380°F). Higher boiling point compounds are typically not used due to a tendency to crack during the reforming process and to form coke on the catalyst. Lighter reformer feedstocks yield an excessive amount of butane and prove uneconomical. Examples of reforming reactions include the following:

- Dehydrogenation of cycloparaffins to yield aromatics and hydrogen
- Isomerization of *n*-paraffins to yield isoparaffins
- Isomerization of naphthenes
- Dehydrocyclization of paraffins to yield naphthenes and hydrogen
- Hydrocracking of higher carbon number paraffins to yield lower carbon number paraffins
- Dealkylation of alkyl aromatics to yield benzene and a paraffin

Naphtha with high cycloparaffin content provides the highest yield and is the easiest to process. Naphtha containing high levels of linear and branched paraffins requires more severe processing and generally produces a lower yield. The density, boiling point, and octane number values of the reformed compounds are higher than those of the reactants. Conversion of lower octane number cycloparaffins

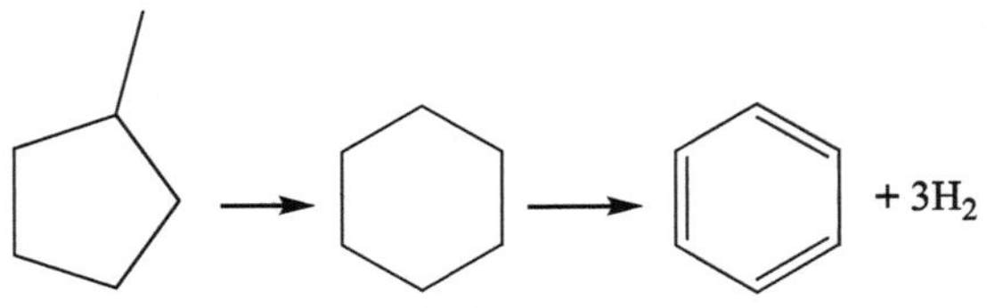

FIGURE 2.7 Reforming to produce high-octane compounds and hydrogen.

to higher octane number aromatics plus hydrogen is presented in Fig. 2.7.

Reforming catalysts are typically platinum on a silica-alumina support. The catalysts are deactivated by feedstock contaminants such as nitrogen and sulfur. Hydroprocessing and special absorption techniques are utilized to remove sulfur and nitrogen from the reformer. If not removed, feedstock sulfur will be converted to H_2S in the reformer. The H_2S will then serve as a poison to the platinum reformer catalyst. When present, H_2S can neutralize acid sites on the catalyst diminishing the ability of the catalyst to promote isomerization, dehydrocyclization, and hydrocracking reactions.

Metals such as arsenic, copper, and lead are severe reformer catalyst poisons. Only a few parts per billion are needed to poison the platinum catalyst. Water in the feed will promote hydrocracking reactions and lower reformer and produced hydrogen yield.

The reforming reaction involves preheating the feed to a temperature of approximately 510°C (950°F) and passing the feed through one or more reactors containing catalyst. A hydrogen pressure of 200 to 500 psig is typically maintained. Temperature drop throughout the process requires constant addition of heat. In reforming processes utilizing more than one reactor, the temperature drop in the final reactor is less severe. The liquid reformate product is separated from the produced hydrogen in a separation drum. Butane is then removed from the reformate by fractionation.

2.10 Alkylation

Products from the alkylation unit are utilized primarily as a high-octane gasoline blendstock. Octane numbers can range from MON 87–95 to RON 94–100. The process involves reacting an olefin with an isoparaffin possessing a tertiary carbon atom. The olefins and paraffins are usually low molecular weight molecules containing three or four carbon atoms. A five carbon atom olefin feedstock is rarely used. The isoparaffin:olefin ratio is kept high to minimize olefin polymerization. Isoparaffin:olefin ratios of 5:1 to 15:1 are common.

The alkylation process is highly endothermic and proceeds through the formation of tertiary butyl carbocation species. Alkylate

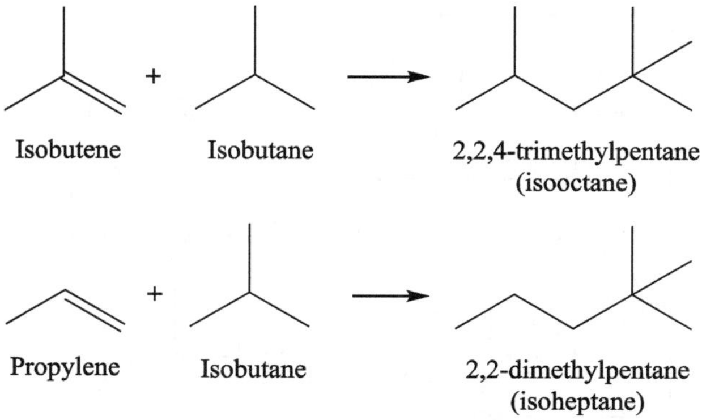

FIGURE 2.8 Typical alkylation reactions.

yields are usually quite high, especially at lower reaction temperatures. High octane number compounds such as isooctane and isoheptane are commonly produced. Reactions to produce isooctane and isoheptane are presented in Fig. 2.8.

Refinery alkylation processes utilize either sulfuric acid or hydrofluoric acid as reaction catalysts. The feedstock for both alkylation processes originates primarily from hydrocracking and FCC operations. Coker gas oils also serve as feedstock in some applications. The differences and similarities between sulfuric acid alkylation and hydrofluoric acid alkylation are reviewed in Table 2.6.

A sulfuric acid alkylation unit is illustrated in Fig. 2.9.

Sulfuric Acid Alkylation	Hydrofluoric Acid Alkylation
Reactions are performed at temperatures of approximately 10°C (50°F) or lower to minimize polymerization and SO_x formation; a costly refrigeration system is utilized to maintain low temperatures	Temperature is limited to <42.4°C (<110°F); reaction is cooled with water rather than by a refrigeration system
Catalyst volume is equal to the charge volume; acid strength is maintained at about 90 percent	Catalyst volume is equal to the charge volume; acid strength is maintained at about 85 to 90 percent
Requires vigorous mixing to prevent polymerization reactions	Requires less energy in mixing due to lower viscosity of HF and increased temperatures
Products are fractionated in the deisobutanizer and debutanizer to separate isobutane and butane from alkylate; HF-produced alkylate is fractionated in a similar manner	Octane number of HF produced alkylate is slightly lower than sulfuric acid alkylate

TABLE 2.6 Comparison of Sulfuric Acid and Hydrofluoric Acid (HF) Alkylation Processes

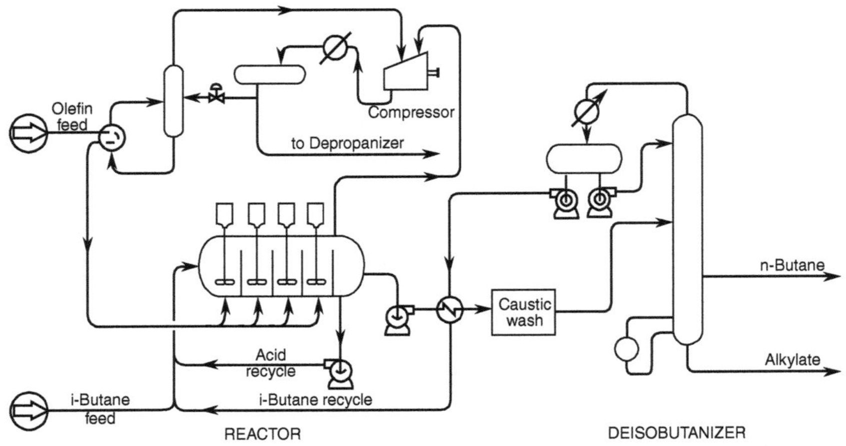

FIGURE 2.9 Sulfuric acid alkylation unit.

2.11 Coking

The coking process is used to produce light gases, naphtha, distillate fractions, heavy fuel oil, and petroleum coke by cracking heavy residual products such as atmospheric and vacuum residual oils. Both delayed coking and fluid coking processes are utilized.

2.11.1 Delayed Coking

The delayed coking process is a semi-batch type of process, whereby residual products are heated at temperatures in the range of 482 to 524°C (900 to 975°F) in a furnace and then transferred into a pressurized coking drum maintained at pressures ranging from about 10 to 100 psi. The volatile, thermally cracked product termed *coker gas oil* is removed from the coking drum along with the deposits, which formed onto the sides of the coke drum. To maintain somewhat of a continuous operation, an alternating cycle of filling and emptying of different coking drums is employed.

Characteristics of feedstock quality, recycle ratio, and drum pressure can affect the coke yield. Highly aromatic feedstock contains more carbon per feed volume and typically produces a high coke yield. Heavy coker gas oil can be recycled back into the coker feedstock to help improve the coke yield. Also, increasing the coking drum pressure tends to increase the coke yield. Typically, higher coke yield results in a reduced liquid product yield.

After residing in the coke drum for about 24 hours, the coke, referred to as green coke, is removed from the drum. Removal involves the following steps:

- Depressurizing the drum
- Filling the drum with water to cool the coke

- Draining the water
- Boring a hole into the coke bed
- Injecting high pressure water jets into the coke to break it free from the coking drum
- Once free, the coke and water are separated
- The entire process can take up to 48 hours

Coke which is low in sulfur and metal content is valued as a fuel, as a raw material for the manufacture of electrodes, and in graphite production. To produce high-purity coke, all traces of volatile matter must be removed from coke. A calcination process is utilized for this purpose. This requires the coke to be heated to temperatures of 1093°C (2000°F) or higher. The product is a high quality, pure carbon source.

2.11.2 Fluid Coking

When compared to the delayed coking process, higher yields of liquid products are typically produced by fluid coking. This continuous process utilizes a fluidized reaction zone of hot coke particles held in motion by steam. The coke particles are first heated in a burner to temperatures ranging from 593 to 649°C (1100 to 1200°F). The hot coke particles are then blown into the reactor by steam. The residual fuel is fed into the reactor and cracks on the hot surface of the fluidized coke particles.

Vaporized products pass through cyclones, out of the reactor and into a scrubbing unit. Coke particles pass downward to the base of the unit and are steam stripped to remove residual vapors. The coke then passes into the burner, and is again heated and recycled back into the fluidized reactor. Large particles of coke are removed from the process in the burner. The smaller particles are recycled. Coke yield ranges from approximately 5 to 25 percent or more. Naphtha yields can range from 15 to 25 percent, and distillate can range from approximately 50 to 75 percent. Light gases are also produced.

2.11.3 Flexicoking

This more recent process utilizes elevated temperatures to react coke with air and steam to yield CO, CO_2, H_2, N_2, H_2O, H_2S, and minute residual solids. After removal of residual solids and H_2S, the remaining gas can be used in burners and processing furnaces.

2.12 Finishing Processes

A number of processes exist that can be utilized to improve the quality of the refined and blended fuels. These processes are often an essential part of ensuring that finished fuels are free of contamination and meet required specifications. Some of the common fuel contaminants

Fuel Contaminant	Source of Contaminant
Ammonia	Results from cracking and hydrotreating of nitrogen containing compounds, typically heterocyclic compounds
Asphaltenes	Usually constituents of heavy fuel fractions; can form during cracking processes
Carbon dioxide	Typically formed during fluid catalytic cracking processing or other cracking processes
Color bodies	Sulfur, nitrogen, phenolic, and conjugated compounds formed during processing; the result of oxidation and polymerization
Gums	Conjugated diolefins and other olefinic compounds formed during catalytic and thermal cracking processes; heterocyclic compounds present in fuel can also initiate gum formation
Hydrogen sulfide and mercaptans	Sour crudes; formed by decomposition of sulfur compounds during distillation, cracking, reforming, and hydroprocessing
Naphthenic and phenolic acids	A component of certain crude oils; formed during cracking processes
Water	Crude oil; water washing; caustic washing; condensate water and contamination during storage

Table 2.7 Common Fuel Contaminants Resulting from Refining Processes

and constituents that can cause fuel storage, handling, and performance problems are provided in Table 2.7.

2.12.1 Caustic Treatment

Various processes are used in refining to remove objectionable odors and contaminants such as mercaptans, phenols, hydrogen sulfide, and naphthenic acids from fuels. Many of these compounds are low in molecular weight and appear in the naphtha and light distillate fuel fractions. Commercial processes incorporating caustic sodium hydroxide are the most commonly utilized in removal of these contaminants and odors. Caustic solutions of various strengths can be used for most applications and usually consist of a 5 to 10 percent solution applied at 10 to 20 volume percent of the charge.

Mercaptan removal processes can involve sparging a caustic and fuel mixture with air in the presence of a catalyst such as a metal chelate or a phenylenediamine compound. Water soluble mercaptide salts are formed in this process and are removed. Some remaining mercaptans react to form disulfides, which are fuel soluble and noncorrosive.

A countercurrent treatment method involves feeding fuel into the bottom of a vertical reactor and adding air, caustic, and a catalyst from the top of the column. Mercaptans are extracted into the caustic and react to form disulfides. Removal of these compounds from fuel can improve the color, odor, stability, demulsibility, and corrosive nature of the fuel. Reactions taking place during caustic treatment of mercaptans are presented in Eq. (2.4).

$$\underset{\text{Mercaptan}}{\text{R-SH}} + \text{NaOH} \longrightarrow \text{NaS-R} + H_2O$$

$$2\text{NaS-R} + 0.5O_2 + H_2O \longrightarrow \underset{\text{Disulfide}}{\text{R-SS-R}} + 2\text{NaOH} \tag{2.4}$$

Thermal and catalytically cracked gasoline fractions can contain significant concentrations of phenols, low molecular weight organic acids, and alkyl aryl mercaptans. All of these compounds can initiate gum formation in gasoline. Caustic treatment readily removes these compounds.

Naphthenic acids occur primarily in heavy naphtha, light distillate, and some heavy fuel fractions. Typically, caustic treatment effectively removes these compounds. However, even after caustic treatment, some alkali salts of heavier naphthenic acids may remain in the fuel as an emulsion or at the fuel-water interface in tank storage.

2.12.2 Water and Particulate Matter

Water and particulates are the sources of fuel quality problems such as corrosion, fuel filter plugging, and system component fouling. Water can be removed through salt drying, coalescence, filtration, and good housekeeping.

2.12.2.1 Salt Drying

The salt drying process is typically used to remove water from distillate fuel and is effective at reducing the final concentration of water to about 50 to 100 ppm. During salt drying, fuel is passed up through a heated column or drum packed with sodium chloride (rock salt). The unit is maintained at a temperature ranging from 32 to 38°C (90 to 100°F). Mined rock salt of 2.5 to 6 mesh is recommended and much less prone to channeling than solar salt of 8 to 10 mesh. Free water or insoluble water is removed by rock salt in the drier, but dissolved or soluble water is usually not removed. If fuel contains high levels of soluble water, a calcium chloride drier is typically required for effective removal.

Calcium Chloride Hydration	Melting Point, °C (°F)
$Ca(Cl)_2$ anhydrous	772 (1422)
$Ca(Cl)_2 \cdot H_2O$	260 (500)
$Ca(Cl)_2 \cdot 2H_2O$	176 (349)
$Ca(Cl)_2 \cdot 4H_2O$	45.6 (114)
$Ca(Cl)_2 \cdot 6H_2O$	30 (86)

TABLE 2.8 Calcium Chloride Salt Melting Temperatures

There are various grades of calcium chloride, each with different levels of hydration and melting points. Care must be taken to ensure the temperature of a calcium chloride salt drier does not exceed the melting point of the specific salt. If the calcium chloride drier is heated above the melting point of the salt hydration level, liquid calcium chloride can travel with the heated fuel and recrystallize downstream when the fuel cools below the melting point of calcium chloride salt. Table 2.8 provides information on the melting point temperatures of various hydrated calcium chloride salts.

The possibility of bridging or fusing of $Ca(Cl)_2$ in the salt drier can occur in calcium chloride driers. This can result in uneven flow through channels that develop in the unit. This phenomenon is termed *channeling*, and it can develop using calcium chloride in either flake form or pellet form. Newer forms of calcium chloride have been developed that resist channeling. Figure 2.10 depicts a salt drying unit with both rock salt and calcium chloride drying drums.

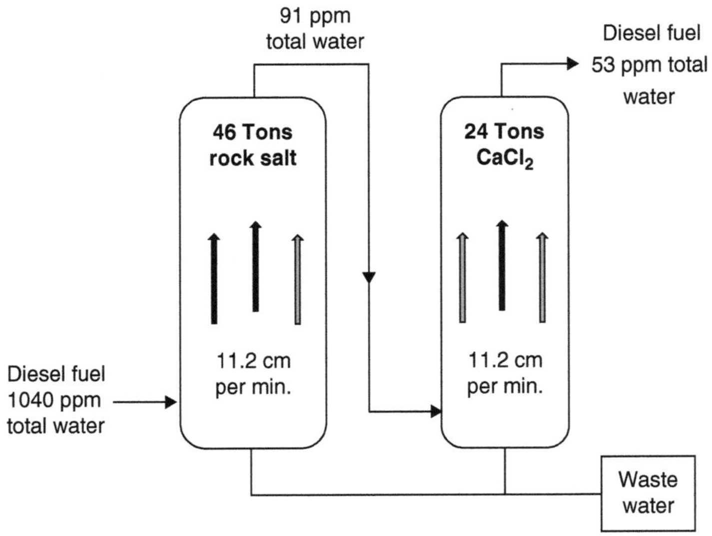

FIGURE 2.10 Distillate fuel salt drying unit.

2.12.2.2 Clay Treatment of Aviation Fuel

As jet fuel is refined, processed, and transported through pipelines and by common carrier, the fuel may pick up particulates, color bodies, additives, and other surfactants. Clay treating vessels are utilized at refineries, fuel terminals, and airports to remove these contaminants.

The clay used in these vessels is typically a high grade of Attapulgus clay composed of granules having a sand-like appearance and possessing numerous, crystalline pores that can retain fuel contaminants. The surface area of the clay crystals is quite extensive with an area of approximately 24,000 m^2/kg or 13 acres/lb.

Cylinder-like, cartridge filters for jet fuel filtration that are tightly packed with clay are available for use in jet fuel filtration. These cartridges are typically aligned vertically in a large vessel with fuel flowing from the outside of the filter cartridge, into a central collection core. Upon collection, the filtered jet fuel flows and exits into a main pipe leading to a fuel storage tank. Vessel sizes can differ in capacity to hold filter cartridges of various quantities. As an example, one manufacturer of clay treating hardware recommends a maximum flow rate of 340 gal/min of filtered jet through a 51-cartridge clay treating vessel. Clay filtered jet fuel is typically lighter in Saybolt color, lower in particulate matter, and yields higher MSEP ratings than nonfiltered fuel.

2.12.2.3 Coalescence

Electrostatic coalescers can be utilized for removal of free water from distillates and some solvents. During coalescence, fuel is heated to a temperature of about 38°C (100°F) and passed upward through a column containing electrodes and a desiccant. Free water coalesces onto the electrodes and the dried fuel then passes through a desiccant for additional water removal.

2.12.2.4 Good Housekeeping Practices

The presence of water and oxygen in a fuel storage tank can lead to problems with finished fuel quality. Microbial growth at the fuel-water interface and the accumulation of rust and corrosion products in tank water bottoms can develop in water contaminated storage tanks. The turbulence which results during filling, blending, and pumping operations can cause these contaminants to mix with fuel. Hazy fuel can result. Regular removal of these water bottoms from contaminated storage tanks can help ensure that finished fuel remains bright and clear.

2.12.3 Molecular Sieves

Molecular sieves can be used in a variety of fuel processing applications. Uses include drying and water removal from fuel, product purification, hydrocarbon separation, and catalysis. Molecular sieves are composed of sodium and calcium aluminosilicate crystals which

Pore Diameter, Å	Common Applications
3	Water removal from olefins, methyl and ethyl alcohols, and refrigerants
4	Water removal
5	Sweetening of light hydrocarbon streams; *n*-paraffin separation; improving jet fuel quality; purification of aromatics
10	Catalysis and filtration

TABLE 2.9 Uses of Molecular Sieves in Fuel Processing

have been produced from natural or synthetic zeolite compounds. The crystals are dehydrated through heating and are processed to ensure that pore sizes are tightly controlled.

Molecular sieves function by both absorption and separation mechanisms. Through absorption, polar compounds such as water and hydrogen sulfide are absorbed. Through separation, molecules which are held within the pores of the sieve are retained. Pore sizes are typically 3, 4, 5, and 10 Angstroms in diameter. Common uses for each of the pore sizes are presented in Table 2.9.

The various pore size beads are also produced in different sphere diameters. Larger diameter beads are utilized primarily in liquid phase hydrocarbon separation processes, and smaller diameter beads are used primarily in vapor phase and certain liquid phase processes.

CHAPTER 3

Critical Properties of Crude Oil and Common Hydrocarbon Fuels

3.1 Crude Oil

Crude oil is composed primarily of hydrocarbon compounds. Organic and inorganic species containing sulfur, oxygen, and nitrogen are also found in crude oil. Additionally, water, vanadium, nickel, iron, sodium, and other metals may be present. The hydrocarbons present are complex but are principally paraffinic, naphthenic, and aromatic in composition. The complexity of the compounds present increases with the boiling range of the crude oil.

No generally accepted method of classifying crude oil exists. What is termed ultimate crude oil analysis describes the composition as a percentage of carbon, hydrogen, nitrogen, and sulfur. Also, a chemical analysis is performed that describes the composition as a percentage of paraffinic, naphthenic, and aromatic compounds present. These analyses are valuable in order to develop a general idea of the usefulness of the crude oil in producing various refined products. However, the results do not yield much information about the variety of final product grades that can be obtained. To gain this information, an additional evaluation termed the crude oil assay is required.

3.1.1 Crude Oil Assay

A crude oil assay is based on a number of standard tests. Some of the more important tests are listed as follows:

3.1.1.1 Fractional Distillation

Fractional distillation of crude determines the initial boiling point, volume percent of materials boiling at specified temperature ranges, end point, residue, and loss. A procedure termed *15:5 distillation* is frequently used to assist in crude oil distillation. This procedure utilizes a column possessing the equivalent of 15 theoretical plates and a reflux ratio of 5:1. Fractional distillation results can help determine the volume of naphtha, distillate, and residual fractions the crude oil will yield.

3.1.1.2 Gravity

Either American Petroleum Institute degrees (°API) or specific gravity provides information on the paraffinic or aromatic nature of crude. The API gravity of most crude oil falls between the ranges of 20 to 45°API.

3.1.1.3 Sulfur Content

High-sulfur crudes, especially those with a sulfur content more than 0.5 weight percent, complicate refining and are also more expensive to refine. Sulfur and some sulfur-containing compounds are corrosive toward metals and can decrease the pH of the crude oil. The sulfur content of most crude oil varies from less than 0.1 to more than 5 weight percent.

The terms *sweet* and *sour* are often used to characterize crude oils. These terms were originally coined to describe the characteristic hydrogen sulfide and mercaptan odor of the crude. They are now commonly used to distinguish crude oil sulfur level. In general, those crudes containing approximately 0.5 weight percent sulfur or less are considered sweet and those containing more than 0.5 weight percent sulfur are considered sour.

3.1.1.4 Nitrogen Content

Nitrogen is a severe refining catalyst poison. Crude oils containing more than 0.25 weight percent nitrogen are difficult to refine.

3.1.1.5 Salt Content

Salts form deposits and scale that can foul the internal surfaces of crude oil refining equipment. Under refining temperatures and conditions, salts can break down to liberate ions that produce acids during processing. If the NaCl expressed salt content of crude oil is more than 30 parts per million (ppm), the crude oil must be desalted before processing.

3.1.1.6 Water and Sediment

Water and salts can lead to major problems in refining equipment such as corrosion, uneven heat distribution, and equipment fouling.

3.1.1.7 Viscosity and Pour Point

The viscosity and pour point values of crude oil are important for establishing proper pumping, transportation, and storage parameters.

3.1.1.8 Metals

The metal content can range from only a few ppm to more than 1000 ppm. Trace elements such as iron, sodium, vanadium, lead, and arsenic can corrode metallic parts and damage heating equipment. Low levels of nickel, vanadium, and copper are known to deactivate refining catalysts.

3.1.2 Hydrocarbons in Crude Oil

Linear, branched, and cyclic hydrocarbons are all found in crude oil. The concentrations, however, vary from crude source to crude source and even from well to well. The physical and chemical properties of the crude oil hydrocarbons can influence the way refineries process the crude and can also impact the performance of refined fuels. A brief description of the properties of various crude oil hydrocarbon type is presented below.

3.1.2.1 Alkanes or Paraffins

The carbon number of paraffins in crude oil typically ranges from C_1 to C_{40}. Higher carbon number paraffin species have been identified in crude oil, but their percentage is not great. Straight-chain *n*-alkanes or linear paraffins with a carbon number C_{20} and higher are responsible for wax-related problems in crude oil and fuels. Wax deposition and high pour point values are typically caused by paraffins.

Branched paraffins are commonly found in crude oil. Compounds with one or more methyl, ethyl, and isopropyl groups have been identified. Ten carbon number branched paraffins are especially common.

3.1.2.2 Cycloalkanes or Cycloparaffins

Cyclic, ring-like, saturated hydrocarbons in crude oil are called *naphthenes*. About one-half of the total weight of compounds present in crude oil can be naphthenes. Methyl derivatives can be abundant such as methylcyclopentane and methylcyclohexane shown in Fig. 3.1. Single ring naphthenics possessing either five or six carbon

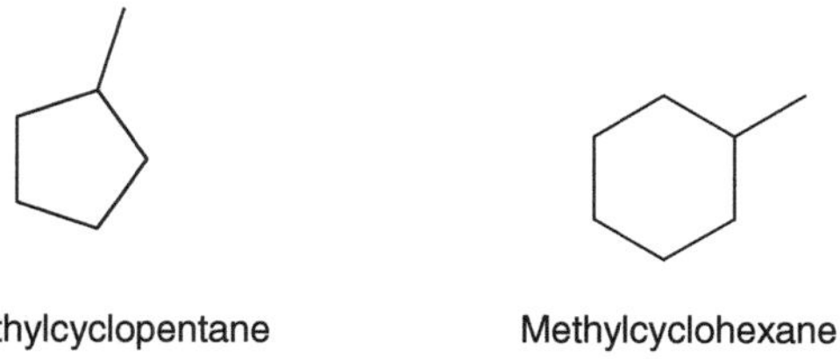

Figure 3.1 Methylcyclopentane and methylcyclohexane.

atoms in the ring structure are the most common. Higher carbon number or lower carbon number rings are less prevalent.

Bicyclic naphthenes are common in kerosene and diesel fuel. Both bicyclic and polycyclic naphthenes comprise a large portion of the mass of heavy distillates and residual oils. These polycyclic higher molecular weight naphthenes possess four or five rings similar in structure to naturally occurring steroids and higher-molecular weight terpenoids. These molecules can also be present in naphthenic lubricating oil fractions.

3.1.2.3 Alkenes or Olefins

Alkenes or cycloalkenes are commonly termed *olefins* in the refining industry. These double-bonded, reactive compounds are rarely found in crude oil. Olefins appear in finished products as a result of cracking reactions and other refining processes.

3.1.2.4 Arenes or Aromatics

Aromatic compounds are typically no more than 15 percent of the total weight of most crude oil. The branched or alkylated aromatic derivatives are more prevalent than the nonalkylated aromatics. For example, toluene and xylene are found in higher percentages than benzene. Also, monoaromatic compounds such as toluene and xylene are some of the most common aromatics in crude.

Polynuclear aromatics can also be found. Naphthalene compounds possessing two aromatic rings and phenanthrene compounds possessing three aromatic rings are among the most common. Ring structures possessing four or more aromatic rings are less prevalent. The structures of biphenyl and naphthalene, both diaromatic compounds, are provided in Fig. 3.2. Biphenyl has a noncondensed or isolated ring structure while naphthalene has a condensed ring structure.

Residual fuel oils, heavy gas oil fractions, and vacuum residual fractions can all contain higher molecular weight polynuclear aromatics. These compounds are known to contribute to deposit formation and fouling in refining equipment and fuel combustion furnaces. Also, some polynuclear aromatics found in crude oil have been determined to be human carcinogens.

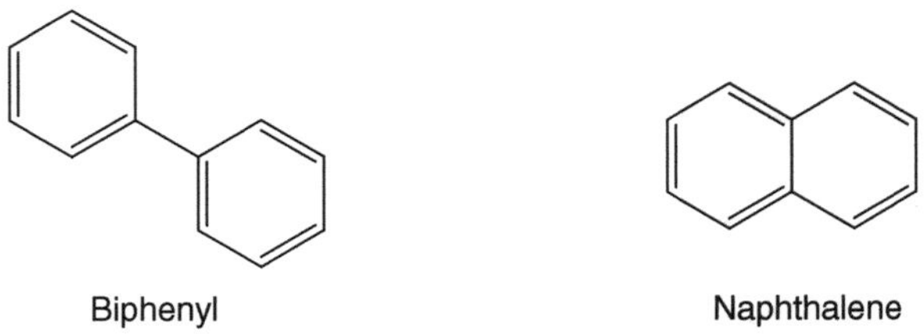

FIGURE 3.2 Biphenyl and naphthalene.

Indane Tetralin

Figure 3.3 Indane and tetralin.

3.1.2.5 Naphthenic-Aromatics or Mixed Polycyclic Hydrocarbons

Naphthenic-aromatic compounds are formed by the condensation of an aromatic ring with a cycloparaffin. These compounds are common components in distillate fuel and light gas oil fractions. Indane and tetralin are examples of two mixed polycyclic hydrocarbons and are shown in Fig. 3.3.

3.1.3 Other Compounds in Crude Oil

Asphaltenes, heterocycles, and particulates can also be present in crude oil. These compounds are present in various distilled crude oil fractions but seem to concentrate in heavier distillate and residual oil fractions. The effect of these materials on processing equipment, refining catalysts, and finished product quality can be dramatic. Corrosion, catalyst poisoning, and fuel stability problems can all be due to the effect of these compounds.

3.1.3.1 Asphaltenes, Maltenes, and Resins

Very high molecular weight, high-boiling point compounds composed of fused ring and waxy paraffin structures termed *asphaltenes*, *maltenes*, and *resins* are found in crude oil. Asphaltenes appear to exist within a colloidal dispersion in crude oil and are surrounded by maltenes and resins. The asphaltenes are amorphous, dark-colored solids. Maltenes contain resins and higher molecular-weight naphthenic and paraffinic waxes and oils. Resins are aromatic components of maltenes and are typically found in higher concentrations than asphaltenes. They are lighter in color and appear as viscous fluids rather than amorphous solids. Also, maltenes and resins are lower in molecular weight than asphaltenes and possess fewer condensed aromatic rings.

Asphaltenes are present as sheet-like structures composed of as many as 20 condensed aromatic rings. These rings can contain nitrogen, sulfur, or oxygen heteroatom species and will typically possess naphthenic and paraffinic side chains. Molecular weights of individual asphaltene compounds can range from approximately 3000 to 10,000. In micellular form, asphaltene molecular weight values can be significantly higher. Also, it is not unusual for heavy metals found in crude oil, such as vanadium and nickel, to be complexed with asphaltenes.

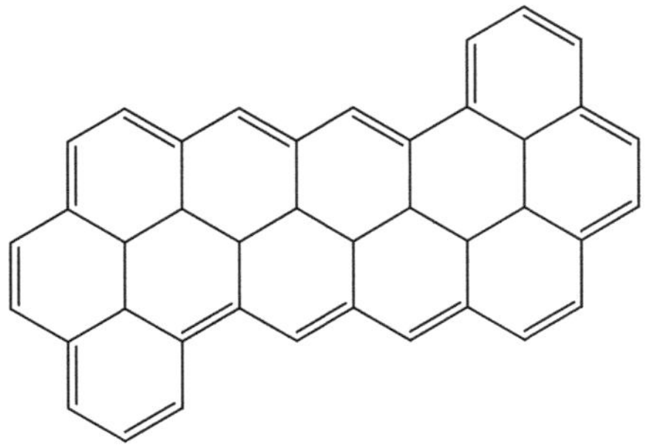

Peri-condensation of aromatic rings in asphaltenes

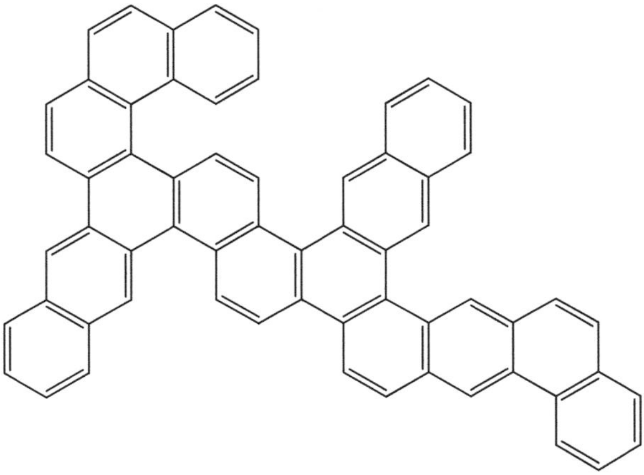

Kati-condensation of aromatic rings in asphaltenes

FIGURE 3.4 Peri-condensed and kati-condensed aromatic rings in asphaltenes.

Asphaltenes can exist in various structural forms such as peri-condensed and kati-condensed rings, and these structures are presented in Fig. 3.4. Upon heating, peri-condensed asphaltenes in crude oil have been shown to rearrange into kati-condensed asphaltenes.

A significant portion of the polycyclic aromatic and sulfur and nitrogen-containing species in crude oil can be found in heterocyclic form within asphaltene, maltene, and resin compounds. Oxygen-containing compounds may also be present. Examples of these compounds are provided in Fig. 3.5.

3.1.3.2 Organosulfur Compounds

Following carbon and hydrogen, sulfur is the third most abundant element found in crude oil. The average sulfur content of crude oil is

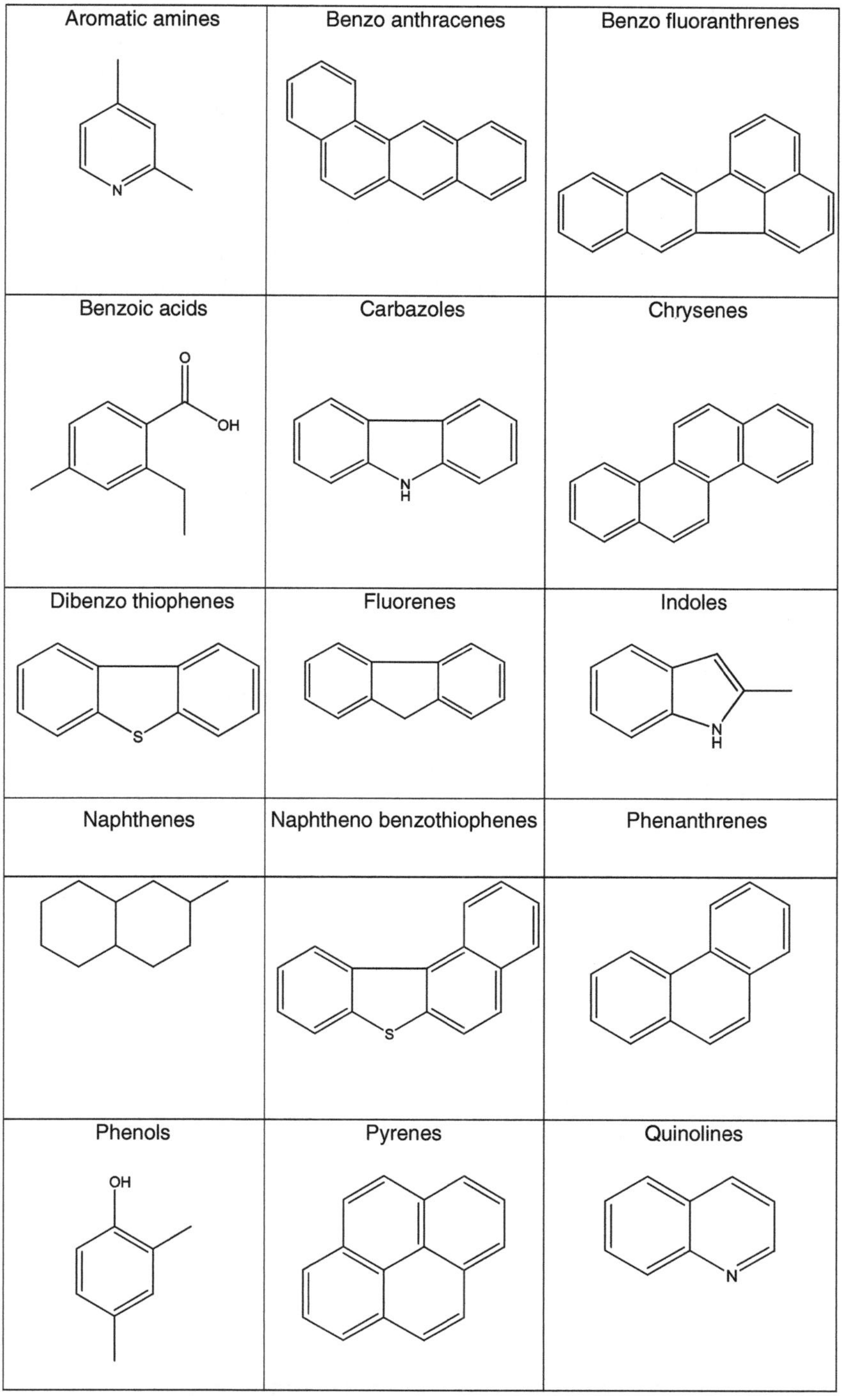

Figure 3.5 High molecular weight aromatics, maltenes, and resins in crude oil.

approximately 6500 ppm but can be identified at concentrations more than 5 weight percent. Sulfur can be found in most distilled petroleum fractions. In the high boiling fractions, sulfur may be bound in asphaltene and resin matrices in combination with nitrogen and oxygen. The organosulfur compounds present in most fuels are either thiols, sulfides, or thiophenes.

3.1.3.2.1 Thiols or Mercaptans Also known as mercaptans, thiols possess very unpleasant odors and are most commonly found in the naphtha fraction. These low boiling point compounds typically contain from one to eight carbon atoms and can be linear, branched, or cyclic. Most mercaptans are removed from crude oil and petroleum products through hydroprocessing or chemical sweetening. Cyclopentyl mercaptan is presented in Fig. 3.6.

3.1.3.2.2 Sulfides Diethyl sulfide and diethyl disulfide are examples of sulfides found in crude oil. Sulfide compounds are low in concentration and in molecular weight. Diethyl disulfide is presented in Fig. 3.7.

3.1.3.2.3 Thiophenes Thiophene as a compound is not found in high concentration in crude oil. However, benzothiophenes and other aromatic thiophenes are critical components of high-sulfur crude oil. Thiophenes are present in relatively high concentration in oils containing increased percentages of aromatics, asphaltenes, and resins. Figure 3.8 presents the structure of 2-ethyl-thiophene and 2-methylbenzothiophene.

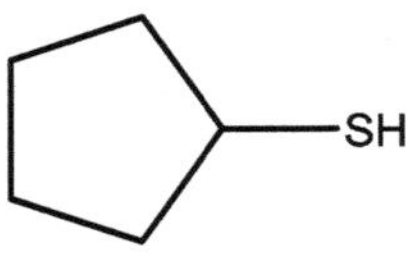

FIGURE 3.6 Cyclopentyl mercaptan.

$CH_3CH_2\text{–}S\text{–}S\text{–}CH_2\text{–}CH_3$

FIGURE 3.7 Diethyl disulfide.

2-ethyl thiophene

2-methyl benzothiophene

FIGURE 3.8 Aromatic thiophenes 2-ethyl-thiophene and 2-methylbenzothiophene.

Figure 3.9 Organonitrogen compounds quinoline and indole.

3.1.3.3 Organonitrogen Compounds

The nitrogen content is relatively low in most crude oils and typically less than 2000 ppm. The average concentration is approximately 950 ppm. The nitrogen content of distilled crude oil fractions increases with increasing boiling point and are typically quite stable throughout the refining process. Naphtha and kerosene contain little nitrogen, whereby higher boiling distillates and residual oils can contain higher concentrations of organonitrogen compounds such as quinolone and indole shown in Fig. 3.9.

The nitrogen-containing porphyrin ring species can be found in most asphaltic crude oils. The porphyrin ring structure in crude oil usually holds a metal ion such as nickel or vanadium. These metals are known as powerful poisons for petroleum refining catalysts.

3.1.3.4 Oxygen-Containing Compounds

The oxygen content of crude oil is typically low and usually in the form of naphthenic acid compounds. Asphaltic crudes contain significantly more of these acids than paraffinic crudes. Also, the concentration of oxygen increases with the boiling point of the crude oil fraction. Most of these compounds concentrate in the heavy naphtha, kerosene, fuel oil, diesel fuel, and lighter lubricant fractions. The naphthenic acids in crude oil are usually monocarboxylic acids possessing an alkylated cyclopentane single ring structure. Fused ring, branched aliphatic and dicarboxylic acid compounds are found in lower numbers. Most species contain 10 carbon atoms, but 20-carbon atom species have been identified. A 10-carbon, monocyclic naphthenic acid is presented in Fig. 3.10.

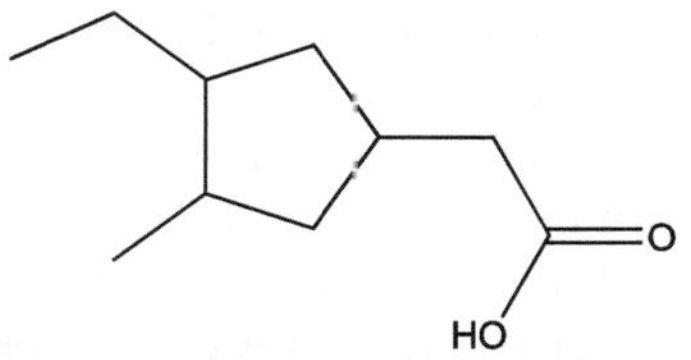

Figure 3.10 Monocyclic naphthenic acid.

3.1.3.5 Metals and Metal Complexes

Nickel and vanadium are the most abundant metals found in crude oil. Other metals, metalloids, and nonmetals including aluminum, arsenic, barium, calcium, chromium, copper, gold, iron, lead, magnesium, manganese, phosphorus, silicon, silver, and titanium have also been identified. These elements exist in a variety of forms and concentrations. Metal complexes incorporating asphaltenes, porphyrins, and other polar organics can stabilize metal solubility throughout the crude oil matrix. Concentrations of metals can range from less than 1 ppm to 1500 ppm or more.

3.2 Gaseous Fuel Products

The abiogenic theory suggests that very large quantities of natural gas exist deep within the earth and migrate upward through cracks and fissures. Deep gas is believed to lie at depths of 15,000 to 30,000 ft. Natural gas is a mixture of hydrocarbon and non-hydrocarbon gases found in porous, underground formations. Raw natural gas may be subclassified as follows:

- *Nonassociated gas*—free gas not in contact with significant amounts of crude oil in the reservoir.
- *Associated gas*—free gas in contact with crude oil in the reservoir.
- *Dissolved gas*—gas in solution with crude oil in the reservoir.

Methane is the main constituent of natural gas with ethane, carbon dioxide, nitrogen, hydrogen, water vapor, hydrogen sulfide, helium, argon, and natural gas liquids comprising the remainder. Some produced natural gases contain as much as 50 percent H_2S and/or CO_2. After removal of water vapor, hydrogen sulfide, and natural gas liquids, the methane gas is ready for shipment and use. Methane is primarily used as a fuel for home and industrial heating and in power generation. Liquefied natural gas (LNG) and compressed natural gas (CNG) are also produced for long distance and short distance transportation of the fuel.

The natural gas liquids separated from methane are termed *natural gas liquids* (NGLs) or *condensates*. The NGL/condensate may be sold as a raw product or can be fractionated into commercial products such as LP gas, natural gasoline, or separate gases such as ethane, propane, *n*-butane, or isobutane. The LP gas consists primarily of propane, *n*-butane, and isobutane.

Refining of crude oil also yields gaseous products. During crude oil distillation, paraffinic gases are released. Crude oil cracking reactions such as fluid catalytic cracking and coking yield both light paraffinic and olefinic gases such as propane, propene, butane, isobutane, butene, and isobutene.

A brief description of commercial LP gas products and selected critical properties for gases meeting requirements under ASTM D1835 Standard Specification for Liquefied Petroleum (LP) Gases are outlined.

3.2.1 Commercial LP Gases (LPG)

- *Commercial propane:* A hydrocarbon product where high volatility is required. Commercial propane is suitable for certain low severity internal combustion engine applications.
- *Commercial butane:* A hydrocarbon product for use where low volatility is required.
- *Commercial PB mixtures:* Mixtures of propane and butane for use where intermediate volatility is required.
- *Special duty propane:* A product composed primarily of propane that exhibits superior antiknock qualities and was specifically developed for use as a fuel in spark ignition internal combustion engines. Equivalent to HD-5 propane.

3.2.2 Selected Properties of LPG

3.2.2.1 Carbonyl Sulfide

If present in LPG, carbonyl sulfide itself is not corrosive toward copper. However, in contact with water, carbonyl sulfide can hydrolyze to release H_2S. This is the primary reason the LPG copper corrosion test vessel is water wetted prior to corrosion testing. Rates of COS in LPG that lead to copper corrosion vary widely. Rates of COS greater than 100 ppm are believed to increase the risk for copper corrosion, while other reports state that corrosion can occur with as little as 3 to 4 ppm of COS.

3.2.2.2 Residue

Most LPG is produced without the presence of heavier hydrocarbons and other low volatility compounds. However, during tank transfer, shipment by pipeline, or contact with incompatible materials, residue can be picked up by LPG. A residue specification is in place for LPG to help ensure low volatility compounds do not lead to deposits in LPG powered engines and equipment. Residue can be determined by method ASTM D2158 which involves evaporation of volatile gases and measurement of the amount of residual material remaining. Two gas chromatographic methods, ASTM D7756 and ASTM D7828, are suggested as methods to help identify the composition of components in the residue.

3.2.2.3 Copper Corrosion

Hardware composed of copper, brass, or bronze is commonly used in the construction of equipment to handle, store, and utilize LPG in the industrial and consumer market. Because of this, compounds present in LPG that may lead to corrosion of copper and copper alloys must be closely monitored. Elemental sulfur and H_2S are typically the most common corrosive agents in LPG. At rates less than 5 ppm, together elemental sulfur and H_2S can be highly corrosive.

3.3 Automotive Spark Ignition Fuel

Nicholas Otto designed and developed the first liquid fuel powered, four-cycle, internal combustion engine in Germany in 1876. The first Model T automobile introduced by Henry Ford in the United States in 1908 was equipped with a four-cycle, internal combustion engine. The Model T engine was designed to operate on kerosene, ethanol, or gasoline. At the time, kerosene was the most dominant fuel produced from crude oil for use as lamp oil alongside of coal oil and whale oil. The natural gasoline in crude oil was a surplus product of refining at the time, but within a few years, due to the popularity of the automobile and the increased engine power provided by gasoline, the production of gasoline surpassed kerosene as the dominant product of crude oil refining in the United States.

The ASTM D4814 Standard Specification for Automotive Spark-Ignition Fuel provides a description of various fuel grades acceptable for use in a wide range of vehicle operating conditions. Differing vapor pressure classes and related distillation property limits during seasonal climactic changes are examples of fuel characteristics listed in ASTM D4814. Automotive spark-ignition fuel, commonly described in various worldwide regions as gasoline, gasolina, motor gasoline, petrol, benzin, or essence, is essentially a blend of hydrocarbons derived from petroleum and is used to fuel internal combustion, spark-ignition engines. Gasoline contains hundreds of individual hydrocarbons which range from *n*-butane to C_{11} hydrocarbons such as methyl naphthalene. Heterocycles from crude oil processing, ethanol, and a variety of fuel performance additives can also be found in gasoline.

The typical boiling point range for gasoline is 29.4 to 226.7°C (85 to 440°F). Properties of commercial gasolines are influenced by both refinery practice and by the nature of the crude oils being refined. Gasoline contains low-molecular weight, linear and branched paraffins, naphthenes, and aromatics. Olefins present in gasoline originate from olefin-containing blend streams produced by cracking reactions taking place in the refinery. In a finished gasoline blend, addition of these cracked streams may add from 5 to 20 percent olefins to the final blend. The final blend may be hydroprocessed to saturate olefins to yield paraffins.

Commercial motor gasoline is usually a blend of several refined components such as straight-run gasoline, FCC gasoline, alkylate, reformate, isomerate, and other available fractions. Many of the typical components used in gasoline blending are listed in Table 3.1. Individual refineries may not produce all blend components listed in

Gasoline Component	Characteristics
Normal butane	Blended into gasoline to increase the vapor pressure for improved cold weather startability
Light straight-run gasoline	Composed primarily of components $>C_5$ with a cut point typically between 82.2 to 93.3°C (180 to 200°F); it is separated from heavy straight-run gasoline and is caustic washed, hydroprocessed, or sweetened; the octane number is sometimes improved through isomerization
Heavy straight-run gasoline	Contains compounds with a boiling range from approximately 82.2 to 193°C (180 to 380°F); this fraction is typically blended directly into gasoline or used as reformer feedstock
Cracked gasoline; FCC gasoline	Composed of paraffinic, olefinic, and aromatic compounds; branched compounds are present in a relatively high amount; typically has a higher RON than MON; high olefin content gasoline can lead to gum formation and fuel color degradation
Hydrocracked gasoline	Relatively olefin-free, low aromatic gasoline blendstock
Alkylate	Composed of C_7 and C_8 branched paraffins; this is a high octane number blend stock with MON values ranging from approximately 87 to 95 and RON values from 94 to 100
Isomerate	Contains C_4 to C_5 branched paraffins; used to improve startability and warmup properties
Reformate	Composed of higher octane number aromatics, branched paraffins, and cycloparaffins; high boiling point reformate can improve power
Oxygenate	Ethanol, butanol, various tertiary butyl ethers are added to fuel to help lower CO and exhaust HC emissions and improve octane number

Table 3.1 Typical Components Used to Prepare Finished Gasoline Blends

Table 3.1. Also, from day-to-day individual blend component percentages may vary depending upon availability.

3.3.1 Antiknock Properties

Engine knock is a physical and chemical combustion phenomena in gasoline engines related to engine design and operating conditions. Resistance to knock is used to rate gasoline into various octane number grades posted on fuel pumps such as 87, 89, and 93. Fuel aromatics and isoparaffins help to suppress and limit knock in gasoline engines during combustion.

Engine knock is measured by two ASTM engine test methods, ASTM D2699 or D2885 and D2700 or D2885. Method ASTM D2699 is identified as the research octane number (RON) method and correlates with low-speed, low-temperature engine operation. Method D2700 is identified as the motor octane number (MON) method and predicts relative performance under high-speed, higher temperature engine operating conditions. The primary differences between these two test methods are summarized in Table 3.2. ASTM D2885 is titled as a Test Method for Determination of Octane Number of Spark-Ignition Fuels by On-Line Direct Comparison Technique.

The reported octane number for gasoline is a calculated average of the RON and MON. This value is called by various names such as the antiknock index (AKI), octane rating, posted octane, and (R + M/2). Using antiknock index, the value is determined as follows:

$$\text{Antiknock index (AKI)} = (\text{RON} + \text{MON})/2$$

Isooctane and *n*-heptane are utilized as primary reference fuels for octane number determinations. Isooctane has an octane rating of 100, and *n*-heptane has an octane number rating of zero. Therefore, if gasoline has a reported octane number of 90, its performance is equal to that of a blend containing 90 percent isooctane and 10 percent *n*-heptane.

The term *octane number requirement* (ONR) is used to describe the octane number of fuel that will provide initial knock or trace knock

Test Parameter	RON	MON
Test result value	Usually higher than MON	Usually lower than RON
Engine speed, rpm	600	900
Air temperature	15.6 to 51.7°C (60 to 125°F) Air intake temperature will vary with barometric pressure; temperature can be as high as 59.4°C (139°F)	37.8°C (100°F)
Lube oil	SAE 30	SAE 30
Usefulness	Predicts relative performance under low-speed, mild knocking conditions; best correlates with engine run-on tendency	Correlates with high-speed, high-temperature operating conditions

TABLE 3.2 Research Octane Number (RON) and Motor Octane Number (MON) Parameters

under high-speed and high-load conditions. Engines vary in their ONR. This is due to both engine design and environmental operating conditions. The following conditions can affect ONR:

ONR may increase if:

- The compression ratio is increased.
- The coolant temperature increases.
- Combustion chamber deposits form.
- The ambient temperature increases.

ONR may be minimized if:

- The absolute humidity decreases.
- The altitude increases.

In some circumstances, the octane number of gasoline used in an engine must be increased to prevent knock cause by accumulated deposits that build on the piston walls and crown. Deposits act to increase the compression ratio and temperature of the cylinder environment. This phenomenon is termed *octane requirement increase* (ORI) (Fig. 3.11).

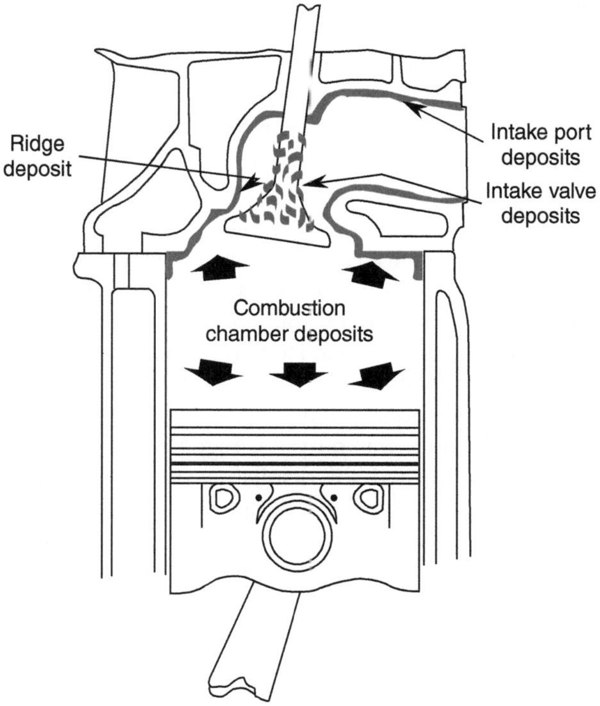

FIGURE 3.11 Engine deposit locations impacting octane requirement increase.

Fuel Component	RON	MON	Sensitivity
n-pentane	62	62	0
Isopentane	92	90	2
Toluene	120	115	5
Cyclohexane	83	77	6
2,3,4-trimethylpentane	103	96	6
Ethylbenzene	107	98	9
Isopropylbenzene	113	99	14
1,4-cyclohexadiene	75	40	35

Table 3.3 Sensitivity of Various Gasoline Components

The term *sensitivity* is used to describe the difference between the RON and MON of fuel. High-sensitivity fuels have a greater difference between RON and MON values, 12 numbers, for example. Low sensitivity fuels will have RON and MON only a few numbers apart. Sensitivity value represents the effect changes in engine operating conditions will have on the knock properties of fuel.

The hydrocarbon composition of fuel can influence sensitivity. Fuels containing higher percentages of unsaturated and aromatic compounds typically have high sensitivity. Linear paraffins in fuel have low sensitivity while highly branched paraffins have higher sensitivity. Sensitivity values for various gasoline components are presented in Table 3.3.

3.3.2 Volatility

Gasoline volatility is established to help ensure engine operation under a range of ambient temperature conditions. Gasoline volatility must be high enough to ignite in extreme cold conditions, but upper volatility limits of gasoline must be controlled to minimize evaporative loss of fuel and help prevent vapor lock in fuel lines. Different vapor pressure/distillation classes and vapor lock protection (vapor:liquid ratio or V/L) classes are established for this purpose. Together, gasoline vapor pressure, V/L and T10, T50, and T90 distillation properties impact vehicle startability, fuel economy, lubricant dilution, and overall vehicle driveability. These properties are controlled by the physical and chemical characteristic of the refined components used in fuel blending.

3.3.3 Other Gasoline Parameters

3.3.3.1 Phosphorus Content

Gasoline blend components or additives used in gasoline must not contain phosphorus. Metals used in the catalytic converter to reduce hydrocarbon exhaust emissions react with phosphorus degrading the efficiency of the converter.

3.3.3.2 Sulfur Content

Sulfur levels in fuel are continuously falling due to environmental regulations limiting SO_x and particulates in fuel exhaust emissions. This trend will likely continue. A sulfur limit of 30 ppm is targeted for automotive spark ignition fuel.

3.3.3.3 Existent Gum and Stability

Limits on the level of existent gum and the storage lifetime of gasoline are in place to help ensure that gums and residue do not deposit and accumulate during the transport and storage of fuel. Gums may also contribute to deposits onto carburetors, intake manifolds, valves and valve guides and fuel injectors.

3.3.3.4 Gravity

The °API of gasoline typically falls between 48 and 69°API.

3.3.3.5 Rust and Corrosion

Internal corrosion in storage tanks, transfer lines, and underground pipelines can occur if fuel becomes contaminated with water. Copper and silver-containing components in fuel systems may also corrode when in contact with elemental sulfur and sulfur-containing compounds such as hydrogen sulfide. Corrosion inhibitors are available to help contend with water and sulfur contamination of fuel.

3.3.3.6 Hydrocarbon Composition

Breakdown of the percentage of paraffins, olefins, naphthenes, and aromatics in gasoline can provide valuable information on gasoline quality. Olefin content, for example, is monitored because of the tendency of high olefin gasoline to form deposits on fuel injectors and contribute to gums and color in stored fuel.

3.3.3.7 Oxygenates and Performance Additives

The use of oxygenated compounds and additives are permitted in gasoline. One of the primary requirements described in the United States Environmental Protection Agency (EPA) Clean Air Act is the substantially similar rule. This rule prohibits the use of any fuel or fuel additive, which is not substantially similar to any fuel or fuel additive used for emission certification of any model year 1975 or subsequent model year unless a waiver is obtained from the EPA. The rule states the following:

- The fuel must contain C, H, O, N, or S, or combination of these elements. The fuel may contain a combination of the following:
  - Hydrocarbons
  - Aliphatic ethers

- Aliphatic alcohols other than methanol:
    - Up to 0.3 volume percent methanol
- Up to 2.75 volume percent methanol with an equal volume of butanol or higher molecular weight alcohol.
- The fuel must not contain greater than 2.0 mass percent of oxygen unless it contains aliphatic ethers and alcohols other than methanol, in which case it must not contain greater than 2.7 mass percent oxygen.
- A performance additive for fuel must contain only C, H, O, N, or S. An EPA approval is required for variance from this limitation.
- No more than 0.25 mass percent of a fuel additive may be used. If a fuel additive contains more than 15 ppm sulfur, an explanation must be provided describing the maximum rate of fuel additive addition that will contribute 15 ppm sulfur to the fuel.
- Other exceptions must be reviewed by the EPA.

See Table 3.4 for a listing of common spark ignition fuel performance additives and their function.

3.3.3.8 Ethanol Blending

Denatured ethanol was blended into gasoline during the 1970s as a high-octane blend component to help offset the need for petroleum-based fuel during these years of energy shortages and supply uncertainty. Since this time, ethanol has become the primary oxygenate used in gasoline blending, essentially displacing ethers such as methyltertiarybutylether (MTBE), ethyltertiarybutyl ether (ETBE), and tertiaryamylmethyl ether (TAME).

Although ethanol can be produced by synthetic routes such as the ethylene sulfonation-hydrolysis process and by direct catalytic hydration of ethylene, fermentation of biomass yields the greatest supply of fuel grade ethanol today. After processing and before commercial sale of ethanol for use in gasoline blending, ethanol is denatured with low volumes of natural gasoline, gasoline blendstock, or unleaded gasoline. A typical blending range for denaturant is specified by customers, such as 2.5 to 4.0 volume percent. Specification ASTM D4806 Standard Specification for Denatured Fuel Ethanol for Blending with Gasolines for Use as Automotive Spark-Ignition Engine Fuel describes the requirements for fuel-grade ethanol used at a typical blending rate of 1 to 15 volume percent in automotive gasoline. A volume of 10 percent denatured ethanol blended into 90 percent gasoline is described as E10. Some important properties

Additive	Function
Antioxidant	Helps prevent oxidative degradation and gum formation in fuel
Corrosion inhibitor	Two types of inhibitors could be used: 1. Copper and silver corrosion inhibitors to prevent sulfur, hydrogen sulfide, mercaptan, and other chemical attack on copper and silver 2. Ferrous metal corrosion inhibitor to prevent water/oxygen initiated corrosion of iron and steel components
Metal deactivator	Chelate metals which may act to catalyze fuel oxidation reactions
Demulsifier	Usually added to help prevent fuel detergent additive from emulsifying with water
Detergent/deposit control additive	Prevents deposit buildup on fuel system parts such as carburetors, fuel injectors, intake valves, valve seats, valve guides, and combustion chambers; aids in removal of existing deposits
Dyes	Used to identify gasoline grades and types
Knock control compounds	Help control predetonation and ignition
Oxygenate	An oxygen-containing, ashless organic compound, such as an alcohol or ether, which can be used as a fuel or fuel supplement

Table 3.4 Function of Spark Ignition Fuel Performance Additives

of E10 compared to gasoline without 10 percent ethanol are presented in Table 3.5.

Worldwide, the ethanol blend volume in motor gasoline can vary from E5 in the EU to E22 or higher in South America. Additionally, ASTM D5798 Standard Specification for Ethanol Fuel Blends for Flexible-Fuel Automotive Spark Ignition Engines describes fuel blends containing 51 to 83 volume percent ethanol.

3.4 Aviation Gasoline

Aviation gasoline is possibly the most complex fuel produced in the refinery. ASTM D910 Standard Specification for Leaded Aviation Gasolines details stringent requirements for properties such

Property	Effect of E10
Antiknock index	Normally increases up to three numbers
Vapor pressure	May increase 1.0 to 5.0 psi
Copper corrosion	Typically no change
Induction time	Typically no change
Existent gum	Typically no change
Water tolerance	Phase separation can occur if fuel contains over 0.9 wt% water. Water tolerance can be improved by the addition of high molecular weight alcohols such as butyl alcohol
Plastics and elastomers	Can degrade chlorinated polyethylene, neoprene, nitrile rubber, polyurethane

Table 3.5 Effect of E10 on Gasoline Properties

as volatility, calorific value, and antiknock ratings. Aviation gasoline, commonly termed *AvGas,* contains a high concentration of alkylate and aromatics, typically those of C_7 and C_8 due to their high calorific value. High octane paraffinic components such as isohexane and isoheptane, can be obtained from careful processing and fractionation of straight-run naphtha and are valuable blending components. A listing of selected properties for AvGas grades 80, 91, 100VLL, 100LL, and Grade 100 is provided in Table 3.6.

Property	Grade 80	Grade 91	Grade 100VLL	Grade 100LL	Grade 100
Net heat of combustion, MJ/kg	43.5	43.5	43.5	43.5	43.5
Color	Red	Brown	Blue	Blue	Green
Copper corrosion, 2 hours @100°C, max	No. 1	No. 1	No. 1	No. 1	No. 1
Electrical conductivity, pS/m, max	450	450	450	450	450
Oxidation stability, mg/100 mL, 5 hours aging					
Potential gum, max	6	6	6	6	6
Lead precipitate, max	3	3	3	3	3

Table 3.6 Selected Properties for Aviation Gasoline

3.4.1 Aviation Gasoline (AvGas) Quality Criteria

3.4.1.1 Antiknock Properties

Various AvGas grades are formulated to ensure knock-free operation. Knocking combustion can damage an aviation engine and can result in serious power loss if allowed to persist. AvGas octane ratings are summarized below:

3.4.1.1.1 Lean Rating The lean rating is termed the *aviation octane rating* (F-3), and is used to simulate lean fuel/air conditions experienced during aircraft cruise operation. The number is determined by the ASTM D2700 motor octane number procedure and converted to the aviation rating.

3.4.1.1.2 Rich Rating The rich rating is termed the *supercharged octane rating* (F-4) and is used to simulate takeoff conditions. This rating is typically higher than the lean rating.

3.4.1.1.3 Performance Number The performance number is primarily used to rate the antiknock values of AvGas with octane number values over 100. This value is defined as the maximum knock-free power output obtained from AvGas expressed as a percentage of the power obtainable from isooctane. The relationship between octane number and performance number is presented in Eq. (3.1).

$$\text{Octane number} = 100 + (\text{performance number} - 100)/3 \qquad (3.1)$$

3.4.1.2 Volatility

Both low-end and high-end vapor pressure vapor limits are established. Low-end volatility control ensures that fuel will vaporize adequately for engine starting and flow throughout the combustion system. High-end volatility limits ensure that vapor-lock or fuel line icing will not occur.

3.4.1.3 Freezing Point

The freezing point temperature is important to ensure that wax crystals do not form and interfere with the flow and filtration of fuel at high altitudes and during ground operations in arctic environments.

3.4.1.4 Storage Stability

Gums and fuel degradation product accumulation in AvGas could lead to deposits on induction manifolds, carburetors, valves, injectors, and other fuel intake system components. Strict potential gum limits are established for all grades of AvGas.

3.4.1.5 Sulfur Compounds

Sulfur is limited to 0.05 weight percent. Higher levels of sulfur compounds in AvGas may lead to degradation of the efficiency of antiknock compounds and may lead to metal corrosion.

3.4.1.6 Water Reaction

The original intent of this specification was to prevent the addition of high octane, water soluble components such as alcohols and diisopropyl ether to AvGas. ASTM D910 specifications require fuel, water, and interface ratings. The specification limits the volume change in the aqueous layer to ±2 mL.

3.5 Aviation Turbine Fuel, Aviation Kerosene, or Jet Fuel

Grades of commercial and military jet fuel are essentially identical worldwide and differ primarily in the types of additives permitted. There are some differences in testing methodology used to qualify Russian jet fuels compared with fuels produced in Western countries. Efforts are underway to harmonize these differences. A listing of some of the major specifications governing jet fuel production worldwide are listed in Table 3.7.

Jet fuels are typically prepared from conventional straight-run or hydrocracked kerosene or from wide-cut kerosene/naphtha blends. Synthetic, hydroprocessed components that comply with ASTM D7566 requirements can be blended with conventional Jet A or Jet A-1 fuel components. However, the percentage of synthetic component permitted is limited to a specific volume and is also limited by the impact on Jet A and Jet A-1 physical properties.

Specification	Grade	Fuel Type
Defence Standard 91–91	AVTUR, NATO F-35	British; NATO civil kerosene
GOST 10227	TS-1	Russian civil aviation kerosene
ASTM D1655	Jet A, Jet A-1	U.S. civil aviation kerosene
GB 1778	No. 2	China civil aviation kerosene
DCSEA 134	Jet A-1	France civil aviation kerosene
ASTM D6615	Jet B	U.S. wide cut civil turbine fuel
JIS K 2209	Jet A-1	Japan civil aviation kerosene

Table 3.7 Various Worldwide Jet Fuel Specifications

Synthetic components permitted in jet fuel blending include the following:

- Fischer-Tropsch synthesized paraffinic kerosene (SPK) up to 50 percent by volume but limited by the impact on jet fuel physical properties.
- SPK from hydroprocessed esters and fatty acids up to 50 percent by volume but limited by the impact on jet fuel physical properties.
- Synthesized iso-paraffins (SIP) from hydroprocessed fermented sugars up to 10 percent by volume but limited by the impact on jet fuel physical properties.

3.5.1 Aviation Turbine Fuel, Aviation Kerosene, Jet Fuel Quality Criteria

Stringent quality requirements have been established through specifications to ensure jet fuel performance. Selected properties for civil jet fuels Jet A, Jet A-1, and Jet B and military jet fuel JP-8 are presented in Table 3.8. Some specifications are not listed but are based upon established agreements between the fuel producer and purchaser.

A discussion of several of the aviation turbine fuel properties listed in Table 3.9 is presented further.

3.5.1.1 Chemical Properties and Combustion

Jet fuels are blended primarily from straight-run distillate and hydrocracked distillate components that contain virtually no olefins. The aromatic content of jet fuel is also limited since aromatics greater than approximately 25 weight percent can lead to smoke formation during combustion and the possibility of carbon deposition in engines. A higher aromatic content may lead to elastomer deterioration and the potential for fuel leakage.

3.5.1.2 Sulfur Content

Jet fuel sulfur normally ranges from 0.15 to 0.30 weight percent. Metal corrosion may also result when high-sulfur fuel is burned. Mercaptan sulfur is limited to 30 ppm. Odor, elastomer degradation, and metal corrosion can result from high mercaptan levels.

3.5.1.3 Oxygenated Organics

Oxygen containing impurities such as phenols and naphthenic acids can adversely affect water separation properties and initiate gum formation in fuel. No limit presently exists to control the amount of these compounds in jet fuel. However, tests for existent gums, neutralization number, and water separation properties indirectly limit the presence of these oxygen containing compounds in jet fuel.

Property	Jet A	Jet A-1	Jet B	JP-8
Freeze point, °C (°F), max	−40 (−40)	−52.6 (−47)	−58 (−50)	−52.6 (−47)
Sulfur, total mass %	0.30	0.30	0.30	0.30
Sulfur, mercaptan, mass % max	0.003	0.003	0.003	0.003
Thermal stability, 2 hours at 260°C, min				
Filter pressure drop, mm Hg	25	25	25	25
Tube deposits less than	3	3	3	3
Electrical conductivity, pS/m, max (if additive is specified by purchaser)	50 to 600	50 to 600	50 to 600	150 to 600
Water reaction interface rating	—	—	—	1b

Table 3.8 Selected Properties for Civil Jet Fuel Grades and JP-8

3.5.1.4 Volatility

The 10, 50, and 90 percent distillation points, the final boiling point, distillation residue, and distillation loss are specified to ensure that a properly balanced fuel is produced.

3.5.1.5 Low-Temperature Properties

Jet fuel must flow during long periods at high altitudes. Fuel must not solidify and block lines, filters, and nozzles. Low-temperature viscosity must also be controlled to ensure adequate fuel flow and pressures are maintained. Fuel viscosity can also significantly impact fuel pump service life.

3.5.1.6 Combustion Quality

The combustion properties are primarily based on the paraffin, naphthene, and aromatic content of jet fuel. The smoke point, together with the naphthalene content, help to characterize fuel combustion quality.

3.5.1.7 Thermal Stability

Fuel must not produce lacquer or form deposits which can adversely affect jet engine operation. The Jet Fuel Thermal Oxidation Test (JFTOT) helps predict the tendency of fuels to form deposits.

3.5.1.8 Contaminants

Clay filtration can be used at refineries, terminals, and airports to remove water, solids, and unwanted compounds from jet fuel prior to use in jet engines.

3.5.1.9 Water Retention and Separation

Measurements such as fuel viscosity, density, and existent gum work together to help determine whether jet fuel will have a tendency to retain water or fine particulate matter. Small traces of free water or particulates could result in ice formation or adversely impact fuel filtration and flow to the engine.

3.5.1.10 Electrical Conductivity

The ability of fuel to conduct electric charge must be maintained to a minimum and maximum level. Adequate conductivity helps ensure that static charge does not build in the fuel. However, if fuel is too highly conductive, some capacitance type aircraft fuel gauges can be disrupted.

3.5.1.11 Aviation Fuel Additives

Additives are introduced into aviation fuel for a variety of reasons. Primarily, additives are added to improve fuel performance or to prevent undesirable effects during fuel use. Additives may be added into aviation fuel under the following limitations:

Mandatory: The additive must be present within minimum and maximum limits defined.

Permitted: The additive may be added by the fuel producer up to a maximum limit.

Optional: The additive may be added with user/purchaser agreement.

Not allowed: Additives not listed are not allowed.

A brief description of the various additives approved for use in AvGas and aviation turbine fuel is provided below:

3.5.1.11.1 Antioxidant It is used to prevent gum formation in fuel. The approved additives are generally phenolic based. Use is optional in AvGas. Antioxidants are permitted in civil and military jet fuels. Antioxidants are mandatory in hydroprocessed British and U.S. military fuels as well as in international civil Jet A-1.

3.5.1.11.2 Metal Deactivator Metal deactivator chelates metal ions, primarily copper. Copper catalyzes the oxidation and degradation of jet fuel. Use is not permitted in AvGas. Metal deactivator is permitted in civil and military jet fuels.

3.5.1.11.3 Corrosion Inhibitor It helps prevent rusting of metal in fuel storage and handling system components. Corrosion inhibitor also provides a protective film on metal surfaces to aid in improving the lubricating properties of jet fuel. Use is optional in AvGas and civil jet fuels but mandatory in military jet fuel grades.

3.5.1.11.4 Fuel System Icing Inhibitor This inhibitor prevents ice crystal formation in jet fuel systems. Ethylene glycol monomethyl ether (EGME) and diethylene glycol monomethyl ether (DEGME) are approved anti-icing additives. The products are optional in AvGas and civil jet fuels. Use is mandatory in military fuels. Products are known by the U.S. military as Grade FSII.

3.5.1.11.5 Antistatic Additive This dissipates static charge in jet fuel. Static charge buildup can result in unwanted ignition of jet fuel/air mixtures. Use is optional in AvGas and civil jet fuel and mandatory in military jet fuel.

3.5.1.11.6 Microbiocides These are used to prevent microbial growth in jet fuel storage and handling systems. Biobor JF and Kathon FP 1.5 are currently approved biocides for jet fuel. Consult the aircraft maintenance manual regarding biocide application and use procedures.

3.5.1.11.7 Lubricity Additive This additive is used to help prevent wear of the high-pressure fuel injection equipment. Use is optional in AvGas and civil jet fuel but mandatory in military jet fuel grades. Typically, the corrosion inhibitor also functions in providing adequate fuel lubricity performance.

3.5.1.11.8 Tetraethyl Lead Tetraethyl lead is used to improve the octane number (antiknock) properties of gasoline. Use is mandatory in AvGas. Tetraethyl lead is not permitted in civil and military jet fuel.

3.5.1.11.9 Antismoke Additives All are ash containing and are not approved for flight. Used only in engine testing.

3.6 Diesel Fuel and Fuel Oil

Fuels used to power internal combustion engines used in automotive, over-the-road transport trucks and lorries, railroad engines, marine vessel power plants, and stationary industrial diesel engines fall into the general category of diesel fuel. Fuels used to fire industrial burners and heaters, power turbines and generators, stoves, and lamps fall into the general category of fuel oil.

Today's diesel fuel and fuel oil grades are typically blends that may contain straight-run distillate, heavy cracked distillate, various gas oils, cycle oil, hydroprocessed natural oils, and fatty acid methyl ester (FAME). Kerosene or jet fuel may also be formulated into the distillate blend to improve the low-temperature viscosity and handling properties of the fuel.

Throughout the world, both low-sulfur and high-sulfur grades of fuel are being refined. Because fuel sulfur level has been identified as a primary component in fuel combustion emissions and the formation of "acid rain," fuel sulfur reduction has been mandated and implemented in several parts of the world. Fuels classed as high-sulfur diesel and fuel oil, termed *S5000*, can contain up to 5000 μg/g of sulfur. Low-sulfur fuels, termed *S500*, can contain up to 500 μg/g of sulfur, and ultralow-sulfur fuels, termed *S15*, can contain up to 15 μg/g of sulfur.

3.6.1 Grade S15 or Ultralow-Sulfur Diesel Fuel

Hydroprocessed ultralow-sulfur diesel fuel, ULSD or S15, can differ in properties and performance from higher-sulfur fuels. Some important differences are presented here.

3.6.1.1 Corrosion Properties

Polar organic compounds and heterocyclic compounds present in fuel can provide the fuel with various properties such the ability to dissipate static charge buildup in transportation pipelines and to provide a film-like layer on the iron and steel surfaces of fuel storage and transportation equipment.

Removal of sulfur heteroatoms from diesel fuel by hydroprocessing to produce S15 diesel results in fuel that has little ability to prevent water initiated corrosion of steel surfaces. The naturally occurring, sulfur containing, film-forming inhibitors have been removed enabling water to reach the metal surface to initiate corrosion. The addition of a film-forming corrosion inhibitor to fuel is often necessary to restore corrosion protection.

3.6.1.2 Lubricity

In high-pressure fuel injection systems, such as those found in jet, diesel, and gasoline direct injected (GDI) engines, the fuel serves to lubricate moving pump parts as they deliver fuel at high pressure to fuel injectors. Polar organic compounds and certain aromatics help to provide a protective, lubricating film on metal system parts as they move past one another. Sulfur is part of many polar organic compounds and heterocyclic aromatics found in diesel fuel.

Hydroprocessed S15 diesel has poorer lubricity properties than higher sulfur content fuels of a similar viscosity grade. Many of the compounds that contribute to providing adequate lubricity are removed by hydroprocessing. Metal wear and loss of fuel pump pressure can occur if fuel with poor lubricity is used. Sulfur-free lubricity improvers can be added to restore fuel lubricating properties. Product chemistries and treating rates vary.

3.6.1.3 Electrical Conductivity

The ability of fuel to conduct electrical charge is of benefit during pipeline transport and during high-velocity loading and off-loading operations. Fuel movement in pipelines and transfer lines creates a static electrical charge. If polar compounds are present in fuel, then this charge will be conducted throughout the fuel matrix and to a ground source.

Hydroprocessing required to produce S15 removes most of the polar organic compounds from fuel. The resulting product is virtually unable to dissipate static charge as it builds in fuel moving through a pipeline. Instead, electrical discharge into pockets of gas within the pipeline or into air and vapors as fuel is being dispensed can occur. An explosion may result. Low levels of fuel-soluble electrical conductivity improver into S15 can help restore fuel electrical conductivity to safe levels.

3.6.1.4 Cetane Number

Conversion of heterocyclic aromatic compounds or other aromatic fuel components to paraffins by hydroprocessing can help increase the cetane number of diesel fuel. Fuel paraffins have significantly higher cetane number values than fuel aromatics. For this reason, the cetane number of S15 will typically be one to five numbers greater than higher sulfur, higher aromatic content fuel grades.

3.6.1.5 Heating Value

The energy available from fuel influences the efficiency of systems used to generate heat and produce power such as boilers, furnaces, turbines, and internal combustion engines. Hydroprocessing yields hydrogen-rich fuel that will tend to burn more efficiently producing lower hydrocarbon emissions. This same fuel, however, can have less potential energy per unit volume than most high sulfur, higher aromatic content fuels due to displacement of carbon volume by hydrogen. As hydrogen adds to carbon-carbon double bonds during hydroprocessing, the total carbon content of a given volume of fuel will be reduced. Consequently, heavily hydroprocessed S15 will tend to be lower in potential heating value per unit volume than higher sulfur, higher aromatic content diesel fuel.

3.6.1.6 Oxidative Stability

As hydroprocessing may remove many of the naturally occurring antioxidants contained in fuel, other less stable, more reactive components will also be reduced. As processed, S15 will tend to be much less prone to color degradation and deposit formation than non-hydroprocessed diesel fuel.

3.6.1.7 Elastomer Compatibility

Although the issue of ensuring S15 compatibility with elastomers used in fuel systems has essentially been resolved, a discussion of this problem may still be of value when switching from a higher sulfur content fuel to hydroprocessed ultralow-sulfur fuel.

Aromatic fuel components tend to cause swelling of some fuel system elastomers to create a tight seal between metal parts and prevent fuel leakage. Hydroprocessing to produce S15 can remove a portion of the aromatic fuel components that contribute to elastomer swelling and seal formation. The introduction of a more paraffinic, less aromatic ULSD into the same system may tend to shrink and crack elastomers due to leaching of aromatics from the swollen elastomer. Problems such as leakage of fuel from pumps now in service could occur. A good preventive step to take when switching to ULSD from a higher sulfur content fuel, would be to check fuel pumps carefully for signs of leakage and replace old gaskets or elastomers if necessary.

3.6.2 Diesel Fuel and Fuel Oil Classifications and Quality Criteria

3.6.2.1 Fuel Grades

The various grades of diesel fuel are based primarily upon the viscosity of the fuel, sulfur content, and the intended application. Both diesel fuel and fuel oil grades exist for this purpose. The terms ultralow-sulfur diesel, low-sulfur diesel, and high-sulfur diesel have been replaced with the terms S15, S500, and S5000 to better define the fuel according to maximum sulfur content. In the United States, high-sulfur S500 and S5000 fuels are dyed red to distinguish them from ultralow-sulfur S15 fuel. The current diesel fuel sulfur grades and their intended applications are listed in Table 3.9.

Diesel fuel grades No. 1-D and No. 2-D are commonly used in long-haul trucks, lorries, heavy construction equipment, railroad engines, and some stationary engines. Grade No. 4-D fuels are used in marine and certain industrial diesel applications.

Fuel oil grades described in specification ASTM D396 are formulated for use in a variety of applications other than internal combustion engines such as burners, stoves, and illumination. Also, cetane number and cetane index are not specifications for fuel oil grades. This significant difference contrasts with diesel fuel grades whereby

ASTM D975 Diesel Fuel Grade	Application
S15 No. 1-D	Very low-sulfur, low-viscosity fuel for use in high-speed engines in on-highway service requiring ultralow-sulfur diesel; also for low-sulfur kerosene applications
S500 No. 1-D	A low-sulfur, low-viscosity fuel for use in high-speed engines in off-highway service
S5000 No. 1-D	A high-sulfur, low-viscosity fuel for use in high-speed engines in off-highway service
S15 No. 2-D	Ultralow-sulfur fuel for use in high-speed engines for on-highway use
S500 No. 2-D	A low-sulfur fuel for use in high-speed engines for off-highway use
S5000 No. 2-D	A high-sulfur fuel for use in high-speed engines for off-highway use
No. 4-D	High-viscosity, high-sulfur content fuel for use in medium- and low-speed engines utilized in sustained load, constant speed applications

TABLE 3.9 Diesel Fuels Grades and Applications

cetane number and cetane index are key specifications. Fuel oil grades and applications are listed in Table 3.10.

3.6.2.2 Cetane Number

This number is a measure of diesel fuel combustibility using a variable compression ratio cetane engine. A typical diesel fuel for use in internal combustion engines has a cetane number ranging from 40 to 55. High cetane number fuels generally allow for an easier start in cold weather, provide more complete combustion of the fuel, and yield an overall increase in engine efficiency, primarily power.

3.6.2.3 Cetane Index

The cetane index is an estimation of the cetane number utilizing fuel physical properties such as the T-10, T-50, and T-90 boiling points and the °API.

3.6.2.4 Gravity °API

Diesel fuels usually have °API values between 23 and 49°API.

3.6.2.5 Distillation

The distillation profile can be used to predict the following:

- *Ease of engine starting*: Poor if T-10 is too high.
- *Engine warmup time*: Increased if boiling range is wide between T-10 and T-50.

ASTM D396 Fuel Oil Grade	Application
No. 1 S500	Vaporizing-type burners and low-smoke lamps
No. 1 S5000	Vaporizing-type burners and lamps
No. 2 S500	Atomizing-type domestic and industrial burners
No 2. S5000	Atomizing-type domestic and industrial burners
Grade No. 4 Light	Low-viscosity, heavy distillate fuel for pressure-atomizing-type burners
No. 4	Heavy distillate fuel for pressure-atomizing-type burners
No. 5 Light	Intermediate viscosity fuel for residual fuel oil burners
No. 5 Heavy	Residual fuel oil burners
No. 6	(Also termed *Bunker C*) High-viscosity oil usually requiring heating before use in residual fuel oil burners

Table 3.10 Applications of Fuel Oil Grades

- *Exhaust smoke and odor*: Minimized if T-50 is low.
- *Carbon residue*: Increased if T-90 is high.

3.6.2.6 Viscosity

The shape of the fuel spray from a fuel injector is related to viscosity. High viscosities cause poor atomization and a solid-stream jet spray pattern. Poor combustion and low power result. Low viscosities can result in a soft, nonpenetrating fuel spray, leakage of fuel past the injector plunger, and possible wear of fuel system components.

3.6.2.7 Heat of Combustion

The heat of combustion can be determined from the °API of the fuel. Low °API fuels have higher British thermal unit (Btu) ratings per fuel volume than high °API gravity fuels. See Appendix 1 for °API versus Btu nomograph.

3.6.2.8 Cloud Point

The cloud point measurement is used to predict the temperature at which wax in fuel may begin to create operating problems such as filter plugging in engines and blockage of lines in fuel storage and transportation equipment.

3.6.2.9 Pour Point

The pour point is used to predict the temperature at which cold fuel will gel and no longer flow.

3.6.2.10 Flash Point

This temperature is an important fuel transportation, handling, and storage safety property. The flash point can also be used to help identify whether distillate fuels and lubricating oils have been contaminated with more volatile compounds such as gasoline or solvents.

3.6.2.11 Sulfur Level

There is worldwide concern about exhaust emissions resulting when fuel sulfur combusts to form sulfur oxides, or SO_x. These gaseous products further react to form environmental pollutants such as sulfuric acid and metal sulfate particulates. Under low temperature engine operating conditions, these same compounds can accumulate within the combustion chamber to initiate corrosion and deposit formation.

3.6.2.12 Carbon Residue

This value helps predict the deposit forming tendency of fuel during combustion. Deposits in oil burner systems can form "hot spots" on metal surfaces that can lead to stress, distortion, and cracking of system components.

3.6.2.13 Ash Content

These are the nonburnable components, typically metals and metalloids, found in fuel. Depending upon size, these particles can contribute to fuel system wear and plugging of fuel filters and injection nozzles.

3.6.2.14 Neutralization Number

This value is a measure of the acidity or basicity of fuel. The presence of acidic compounds can indicate a fuel instability or oxidative degradation problem.

3.6.2.15 Stability

Fuel stability is a property describing the tendency of fuel to remain free of gum, and deposit formation during storage. If fuel degrades in storage, problems such as color change, filter plugging, combustion chamber deposits, and sticking of pump and injection system components can occur.

3.6.2.16 Water and Sediment

Water and sediment in fuel usually result from contamination during fuel handling and storage. Water and sediments can enter via fuel transportation systems, atmospheric air contamination, and blending operations. Corrosion, filter blocking, injection system wear, and deposit formation can result.

3.6.2.17 Fatty Acid Methyl Ester (FAME) Content

As described in specifications ASTM 396 and D975, up to 5 volume percent B100 FAME biodiesel meeting ASTM D6751 requirements can be blended into diesel fuel grades No. 1-D and No. 2-D. Rates higher than 5 volume percent FAME are not included in the ASTM D396 and D975 diesel fuel specifications. Specification ASTM D7467 describes properties of biodiesel blends ranging from 6 to 20 volume percent biodiesel.

3.7 Marine Fuels

Fuels used in marine applications are quite diverse in their properties and performance. Low-viscosity distillate fuels and high-viscosity marine residual fuels are both utilized. The applications can include direct injected marine diesel engines, boilers, and gas/oil-fired turbines.

Worldwide, marine fuels are mostly defined under International Standards Organization specification ISO 8217. Changes in marine engine technology, fuel refining processes, and environmental regulations directly impact ISO 8217 specifications. Critical factors such as maximum fuel density, particulate contamination, and total sediment limits are of prime concern in the marine fuel industry. Examples of marine fuels are listed in Table 3.11.

3.7.1 Marine Fuel Properties and Quality Criteria

Some of the more critical properties of marine fuels include ash content, carbon residue, carbon aromaticity index, density, sulfur content and sulfur emissions, total sediment, and viscosity. A description of these properties is presented below.

3.7.1.1 Ash Content

Residual fuel oils, FCC clarified oils, and other heavy petroleum fractions used to blend marine residual oils may contain catalyst fines and other metals. If delivered with fuel, these metals may score pistons and piston liners to the extent that engine performance is seriously impacted. For this reason, an aluminum + silicon specification limit has been established for all marine residual fuel oil grades. The current aluminum + silicon limit for various marine fuel grades is provided in Table 3.12.

3.7.1.2 Carbon Residue

Fuel asphaltenes, resins, and other higher boiling point compounds can build up as residue on engine components after evaporation and burning away of the more volatile fuel components. These residues can accumulate as piston deposits and interfere with heat transfer, lubrication, and efficient fuel combustion.

Fuel Type	Description
Marine gas oil (MGO) • Grade DMA and Grade DMZ • Grade DMX	Grade DMA and DMZ are blended from straight-run and cracked gas oils. The fuels can sometimes be similar to No. 2-D diesel fuel oils used in high-speed automotive diesel engines with a minimum cetane index of 40. Grade DMX is a light distillate used for emergency generators and small engines with a minimum cetane index of 45. Not for high-speed diesel engines.
Marine diesel oil (MDO) • Grade DMB • Grade DMC (reclassified as RMA 10)	Grade DMB is blended from gas oils and a trace amount of residual oil. The fuel has a minimum cetane index of 35 and a maximum viscosity of 11 cSt @ 40°C.
Marine residual fuels	Residual marine fuel grades RMA, RMB, RMD, RME, RMG, and RMK currently exist in viscosity ranges from 10 to 700 mm^2/s. Grades contain varying percentages of low-grade distillate and cracked components, FCC clarified oil, atmospheric residual, vacuum residual, and visbreaker oils.
Naval distillate fuel • NATO F76	The physical and chemical properties of F-76 fuel are listed in specification MIL-F-16884J. The physical properties are similar to MGO Grade DMA. Restrictions on additive addition into fuel apply.

Table 3.11 Fuels Used in Marine Applications

Fuel Grade	Aluminum + Silicon Limit, mg/kg
RMA	25
RMB and RMD	40
RME	50
RMG and RMK	60

Table 3.12 Aluminum + Silicon Limits for Marine Fuel Grades

3.7.1.3 Calculated Carbon Aromaticity Index and Calculated Ignition Index

Because of their high viscosity and content of higher molecular weight compounds, it is impractical to determine the cetane number of residual fuels in the ASTM D613 cetane engine. Because of this, the

Calculated Carbon Aromaticity Index (CCAI) and the Calculated Ignition Index (CII) were, respectively, developed by Shell and BP. These values can be determined from the following equations where d = density in kg/m^3 @ 15°C and v = viscosity in cSt @ 50°C.

3.7.1.3.1 Calculated Carbon Aromaticity Index

$$\text{CCAI} = d - 140.7 \log \log (v + 0.85) - 80.6$$

3.7.1.3.2 Calculated Ignition Index

$$\text{CII} = 275.985 - (0.254565)d +- 23.708 \log \log (v - 0.7)$$

Marine residual fuels with CCAI values less than 860 are considered to be of better quality than fuels with values greater than 860. A CCAI value less than 800 is often preferred. Fuels with CII values less than 30 are considered to be of poor quality.

3.7.1.4 Density

Marine fuel density is reported in values of kg/m^3 or g/mL. Typical densities for marine residual fuels range from 900 to 990 kg/m^3. As fuel density approaches 1000 kg/m^3, water separation properties will be impaired and ignition properties will be poorer than that of lower density fuel.

3.7.1.5 Sulfur Content

Marine residual fuel sulfur content is based on statutory limits for SO_x emissions in control areas and on purchaser defined agreements for sulfur content. Sulfur values can range from 1.5 to 4.5 percent (m/m) for fuel grades RMA 10 through RMK 700. For ships at berth and upon inland waterways, the sulfur limit is 0.1 percent (m/m).

3.7.1.6 Hydrogen Sulfide

The hydrogen sulfide content maximum limit for marine residual fuel is 2.0 mg/kg.

3.7.1.7 Used Lubricating Oil

Used lubrication oil (ULO) is not permitted as a blending component in marine residual fuels. A marine residual fuel is considered to contain ULO when either one of the following is detected in the fuel:

- Calcium greater than 30 mg/kg and zinc greater than 15 mg/kg
- Calcium greater than 30 mg/kg and phosphorus greater than 15 mg/kg

3.7.2 Marine Engines

Engines used in marine applications are classified based on speed and horsepower and are defined in Table 3.13.

Marine Engine Type	Engine Properties
High-speed engines	These engines operate at over 1000 rpm and are found in tug boats, lifeboats, small pleasure boats, and portable power generators.
Medium-speed engines	These engines operate at speeds of 300 to 1000 rpm and can develop 100 to 200 hp (74.6 to 149.1 kw) per cylinder with cylinder bore diameters up to 12 in (30.5 cm). Small stationary and marine power plants can be medium speed engines. Also, railroad engines can be of this type.
Slow-speed engines	These engines operate up to 300 rpm and can develop 1000 hp (745.7 kw) per cylinder at speeds of 80 to 150 rpm. Cylinder bore diameter measurements can be up to 36 in (91.5 cm). These engines are limited primarily to large stationary and marine power plants.

Table 3.13 Marine Engine Properties

3.8 Burner Fuels

A wide range of fuels are used in burners, furnaces, and ovens. Examples of fuel types frequently used include No. 1 fuel oil, kerosene, range oil, No. 2 fuel oil, heating oil, and liquefied petroleum gas (LPG). These fuels must have several important properties to be considered as a good quality burner fuel. The qualities include high heat content, low smoking tendency, good low temperature handling characteristics, good ignitability, and low deposit forming tendency. Desired burner fuel performance characteristics and fuel properties which affect fuel performance are provided in Table 3.14.

Selected properties from the ASTM D3699 Standard Specification for Kerosene are provided in Table 3.15.

Other important burning properties of kerosene determined by test method ASTM D187 include a minimum burning time, a specific burning rate, chimney deposit color rating, and flame characteristics.

3.8.1 Low-Pressure Oil Burners

The most common types of equipment utilizing No. 1 fuel oil, kerosene, and range oil include the following:

3.8.1.1 Wick Vaporizing Burner

This type of burner is found in lamps and portable stoves used for cooking and heating.

Desired Performance	Fuel Property Affecting Performance
Ignites easily and safely	The flash point, initial boiling point, and viscosity values affect ignitability. If the values are too high, the fuel may not ignite readily.
Good flame height	If the fuel viscosity is too high, it may not travel readily through a wick to burn effectively. High fuel viscosity may be a reflection of high initial boiling point (IBP) and/or endpoint (EP) temperatures. Also, water can interfere with the fuel burning quality.
Low smoke and soot	Fuel paraffins burn with less smoke than aromatic compounds and can provide a higher flame height without smoking. High fuel density indicates the presence of greater concentrations of fuel aromatics. Also certain sulfur containing compounds can burn to form lamp deposits.
Low deposit forming tendency	High boiling point components do not volatilize and burn completely and can contribute to deposits. High distillation residue, EP, and carbon number values may be linked to formation of deposits. More paraffinic, lower boiling point fuel tends to form less deposit.
Good heating value	Higher density fuel has more potential heating value or Btu/kWh per volume than lower density fuel. High Btu/kWh fuel has a greater concentration of aromatic compounds per fuel volume than low Btu/kWh fuel. See App. 1 for fuel heating value determination.
Low odor	Low fuel sulfur content helps to minimize combustion vapor odor.
Remains liquid at low temperatures	The pour point is an indication of the low-temperature handling properties of fuel. Low IBP, low EP, low viscosity fuels will typically have good low-temperature handling properties.

Table 3.14 Burner Fuel Performance Properties

3.8.1.2 Perforated Sleeve Vaporizing Burner

This type of burner is used on ranges. Fuel burns with a silent, blue flame maintained by a natural air draft.

3.8.1.3 Vaporizing Pot Burner

These burners are used in room space heaters, water heaters, and central heating systems. Both natural air draft and forced air draft versions are available.

Property	ASTM Test Method	Limit
Flash point, °C, min	D56	38
Sulfur, % mass	D2622	
– No. 1-K, max		0.04
– No. 2-K, max		0.30
Kinematic Vis at 40°C, mm^2s	D445	
min		1.0
max		1.9
Mercaptan sulfur, % mass, max	D3227	0.003
Copper strip corrosion, max 3 hours at 100°C	D130	No. 3
Freezing point, °C max	D2386	–30
Saybolt color, min	D156	+16

Table 3.15 Selected Properties of Kerosene

3.8.2 High-Pressure Burners

Described below are common high-pressure systems, such as home furnaces and boilers, fitted with atomizers that utilize No. 2 fuel oil and heating oil:

3.8.2.1 High-Pressure Atomizing Gun Burner

This burner utilizes a high-pressure fuel atomizer to spray and meter heating oil into the firebox. Combustion air is supplied independently. An electronic ignition system is used to ignite the air-fuel mixture.

3.8.2.2 Low-Pressure Air Atomizing Burner

Both air and fuel are injected through the same atomizing nozzle in this burner. Fuel is injected at low pressure while air is injected at high velocity, resulting in a finer spray than is provided by a high-pressure atomizing gun burner. A secondary, external air supply is also provided. This type of atomizer is less susceptible to nozzle plugging.

3.9 Residual Fuel Oil

The terms *residual fuel and "resid"* are sometimes used to describe high boiling fractions obtained from crude oil distillation and processing. The following petroleum fractions are often described as residual oils:

- Atmospheric tower bottoms or reduced crude oil
- Visbreaker bottoms
- Heavy vacuum gas oil
- Vacuum tower bottoms

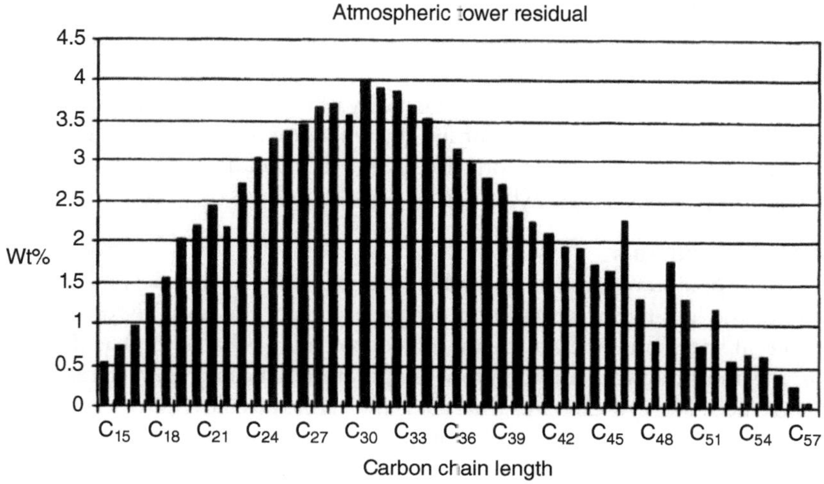

FIGURE 3.12 Paraffin carbon number analysis of a typical atmospheric tower residual oil.

- FCC bottoms; clarified oil; slurry oil
- Heavy coker gas oil

An atmospheric tower residual oil carbon number analysis is presented in Fig. 3.12.

Residual oil fractions are usually quite high in their Btu/kWh rating and are components of fuel blends such as No. 6 fuel oil, bunker fuels, and marine fuels. Blends of these fractions are also used to fire industrial, commercial, and residential heating systems. The fuels are used to heat water to produce steam for use in radiant heating systems and in power generation. Energy for production of steam to power turbines often comes from residual fuel-fired systems. These petroleum fractions are used to produce fuels for direct injected marine engines and other medium- and low-speed engines. The high flash point, lower cost, and good potential energy value make residual fuel a relatively safe and economical energy source.

3.9.1 Typical Residual Fuel Oil Quality Criteria

3.9.1.1 Specific Gravity

This value provides an indication of the potential heating value provided by the fuel. See App. 1 for heating value determination.

3.9.1.2 Flash Point

The flash point identifies the minimum temperature whereby fuel vapors will ignite. In residual fuel applications, this is helpful in determining whether fuel may be contaminated with lower flash point materials.

3.9.1.3 Viscosity

This is one of the more important properties of residual fuel oil. It is an indication of both the pumpability characteristics of the fuel and the fuel atomization potential. The viscosity of residual fuel decreases rapidly with increasing temperature. If preheating is available, residual fuels atomize and burn more effectively. If preheating is not available, it may be necessary to burn lower viscosity fuels.

3.9.1.4 Pour Point

The pour point temperature of residual fuel is usually not the best measure of the low-temperature handling properties of the fuel. Viscosity values at various temperatures are considered more reliable in providing meaningful information on low-temperature handling characteristics of fuels. Nevertheless, residual fuels are classed as "low pour" and "high pour" fuels to define their handling properties. Low pour point fuels have a maximum pour point of 15.5°C (60°F). There is no maximum pour point temperature specified for high pour point residual fuels. In marine residual fuels maximum pour point temperatures range from 0°C (32°F) for RMA 10 to 30°C (86°F) for RMB through RMK.

3.9.1.5 Ash Content

Catalyst fines, metal particles, rust, sand, and other material can be present in residual fuel. These compounds arise from the crude oil, processing catalyst, water contamination, transportation vessels, and storage of the fuel. If the total ash content is greater than 0.20 weight percent, deposits can form in burner systems and corrosion may occur in high-temperature burners.

Alumina, silica, iron, calcium, sodium, nickel, and vanadium are examples of compounds which can be found in residual fuel ash. If the vanadium content of residual fuel is high, severe pitting type corrosion can occur on steel components such as piston heads of residual fuel-fired direct injected engines and exhaust blades of power generating turbines burning residual fuel. Depending on the application, vanadium enhanced corrosion can occur at temperatures ranging from 450 to greater than 650°C (842 to greater than 1200°F).

3.9.1.5.1 Vanadium/Sodium Enhanced Corrosion Vanadium/sodium enhanced corrosion of metal occurs when molten slag containing vanadium compounds forms on metal system parts. The steps in the corrosion sequence are listed below:

1. Vanadium and sodium compounds present in the fuel are oxidized during fuel combustion to V_2O_5 and Na_2O.
2. The V_2O_5 and Na_2O adhere to and react with the ferrous metal surface and form a molten liquid or eutectic with the metal.
3. The liquid dissolves the magnetite or Fe_3O_4 protective layer on the metal surface. This exposes the underlying metal layer to rapid oxidation, most likely catalyzed by vanadium.

4. The oxidation which occurs in the hot environment rapidly reduces the thickness of the metal components and leads to component failure. Corrosion rates as high as 0.1 to 0.2 in/year have been demonstrated at 704°C (1500°F) operating temperatures.

Various compounds of magnesium can be added to fuel containing vanadium to help combat its corrosive effects. Typical addition rates range from 2:1 Mg:V at a 650°C (1200°F) operating temperature and 3:1 Mg:V at an 815.5°C (1500°F) operating temperature. However, total ash content contributed to fuel by magnesium addition must still be monitored.

3.9.1.6 Carbon Residue

Carbon residue measurement serves as an indication of the tendency of fuel to form deposits in a system where available air supply is limited. This value is less meaningful for home heating units burning No. 2 fuel oil and fed with an ample supply of air.

3.9.1.7 Water and Sediment

Filter plugging and fuel injector deposit problems can be caused by the presence of water-insoluble sediment and waterborne solids in residual fuel. Specifications on bottoms, sediment, and water, or BS&W, are set from 0.5 to 2.0 volume percent maximum for fuel oils blended with residual fuels. For RMA marine residual fuel, a maximum water content of 0.30 volume percent and a total sediment content of 0.10 volume percent is specified. For marine residual fuel grades RMB through RMK, a maximum water content of 0.5 volume percent and a total sediment content of 0.10 volume percent is specified.

3.9.1.8 Sulfur Content

Residual fuel oils are typically higher in sulfur content than lower viscosity oils. The sulfur content of residual oil is a function of the crude oils used in its production. Values for sulfur can range from 1.0 to 2.0 weight percent or higher in residual oil products. Burning of these higher sulfur content fuels may lead to deposit formation and eventual corrosion problems within transportation and burner systems.

Combustion of fuel sulfur produces SO_x, which can lead to both atmospheric air emissions and burner system corrosion. About 1 to 5 percent of the fuel sulfur burned is converted to SO_3, and the remainder is converted to SO_2. If a system operates below its dew point, the SO_3 can react with water to produce sulfuric acid. Much work is being done through hydrodesulfurization, neutralization, and engineering to reduce the amount of sulfur oxides produced through the burning of residual fuel oil.

3.9.1.9 Distillation

The distillation temperatures are typically not specified for residual fuels. End points may be over 540°C (1004°F).

CHAPTER 4

Sources of Fuel Performance Problems

4.1 Environmental Sources

Elements of the environment continually impact fuel performance. The effect of water, cold temperatures, heat, air, light, and external contamination can initiate a variety of problems. The first place to begin looking for the cause of a fuel handling or performance problem would be environmental sources.

4.1.1 Water

Water is one of the most abundant, unique, and utilized compounds on earth. It functions as a powerful solvent, provides heating and cooling, is a common reagent in most chemical reactions, and comprises most of our body weight.

Water is unique because of its powerful hydrogen bonding character and its distinct polarity. However, in fuel systems, these characteristics of water make it a source for a variety of problems. Corrosion of fuel system components, emulsification with fuel performance additives, and ice formation in fuel lines are some of the problems directly related to the presence of water in fuel.

4.1.1.1 Solvent Properties of Water

The ability of water to function as a solvent for many compounds also makes it useful in the fuel refining and processing industry. It is utilized to wash and remove salts from crude oil prior to distillation, to wash accumulated salts from the overhead of distillation towers, and to remove H_2S and mercaptans from refined fuel fractions as caustic water.

Water can dissolve a variety of both inorganic and organic compounds. The presence of these compounds can increase the ability of

Conductivity, pS	TDS, mg/L
1	0.5
25	13.4
50	27.4
100	56.3
500	303.0
1,000	630.0

Table 4.1 Effect of Total Dissolved Solids on the Electrical Conductivity of Water

water to conduct electrical change. Because of this, it is possible to estimate the total dissolved solids (TDS) in water by measuring a change in the ability of fuel to conduct electrical charge. The relationship between electrical conductivity and TDS in water is shown in Table 4.1.

Organic compounds present in water can be natural compounds or contaminants. Microorganisms and plant decomposition products can occur naturally in water. Alcohols, organophosphates, glycols, and organohalides are examples of organic contaminants found in water. Analytical methods including the biological oxygen demand (BOD), chemical oxygen demand (COD), total organic carbon (TOC), total organic halide (TOX), and oil and grease measurements are utilized to measure the quality and concentration of organic compounds in water. These methods are summarized in Table 4.2.

4.1.1.2 Water in Fuel Systems

Described below are some of the more common problems caused by the presence of water in fuel systems:

4.1.1.2.1 Ferrous Metal Corrosion Most fuel system storage tanks, transfer lines, and underground pipelines are composed of 1018/1020 carbon steel. These system components are all susceptible to internal corrosion whenever fuel containing water is introduced. Other factors that can enhance fuel storage and transportation system corrosion include:

- Acid carryover from fuel processing.
- Products of microbial growth.
- Seawater contamination.
- Water from salt drying operations.

Whenever deposits from fuel systems are analyzed and found to contain high levels of iron, corrosion is probably occurring somewhere within the fuel system. If water pH is less than 6, iron

Method	Summary
Biological oxygen demand (BOD)	Five day incubation is utilized to determine the amount of oxygen consumed when bacteria digest organic matter in water. High BOD may indicate an oxygen shortage in water for fish and other aquatic life.
Chemical oxygen demand (COD)	Utilizes potassium dichromate to oxidize and measure the amount of oxygen consumed by biodegradable and non-biodegradable organic material and oxidizable inorganic compounds.
Total organic carbon (TOC)	The total amount of organic carbon is measured in water by complete oxidation to CO_2 and H_2O. Oxidation methods utilized for measurement include combustion-IR (infrared) and persulfate UV (ultraviolet).
Total organic halide (TOX)	Organohalides are separated from water by adsorption onto a charcoal packed column. The charcoal is pyrolyzed releasing HX. The HX gases are then analyzed by coulometry.
Oil and grease	Mineral oil, lipids, dyes, detergents, and elemental sulfur are extracted into hexane or onto a solid phase silica-hydrocarbon medium and separated.

Table 4.2 Methods Utilized to Measure the Concentration of Organic Compounds in Water

corrosion and the formation of corrosion products such as gel-like ferrous hydroxide or colloidal ferric hydroxide can result. Ferrous hydroxide gel is insoluble in fuel and may plug filters or accumulate on the bottom of a fuel tank. Colloidal ferric hydroxide, however, is only about 1 μm in diameter and difficult to detect and remove by filtration. Fuel containing these particles appears bright and clear, but the particles can pass through filters and deposit onto fuel system components. Once exposed to air, colloidal ferric hydroxide forms visible rust.

4.1.1.2.2 Caustic Water Carryover Processing of fuel sometimes requires the use of a caustic sodium hydroxide solution. Sweetening of sour fuel streams and the neutralization of the acidic alkylation catalyst are both accomplished by utilizing caustic.

The carryover of caustic into a finished fuel blend usually has minimal effect on the corrosion of ferrous metal. However, in fuel blends containing a conventional dimer-trimer fatty acid based corrosion inhibitor, a monomeric diacid inhibitor, or a partially esterified inhibitor, caustic can react with the inhibitor to negate its effect. As a calcium or sodium salt, these inhibitors will no longer function effectively as an oil-soluble, fuel corrosion inhibitor. The reaction of an organic acid with sodium hydroxide to form an organic salt is presented in Eq. (4.1).

$$\text{R-COOH} + \text{NaOH} \rightarrow \text{R-COO}^- \text{Na}^+ + \text{H}_2\text{O} \tag{4.1}$$

As salts, these inhibitors can readily interact with water and fuel to form emulsions. The ability of the inhibitor to effectively prevent fuel system corrosion is lost. Once emulsified, corrosion inhibitors can initiate other problems such as filter plugging and sticking of moving parts in fuel pumping and injection equipment. Failure of sensitive particulate contamination tests, such as, ASTM D2276 or ASTM D5452 for aviation fuel or ASTM D6217 for middle distillate fuel may result if emulsions carry particulate or dissolved solids.

4.1.1.2.3 Acid Carryover/Low pH Water Sulfuric acid and sometimes hydrofluoric acid are used as catalysts in the alkylation process to produce high octane fuel blend components such as iso-octane and iso-heptane. These compounds, termed alkylate, are separated from the acidic reaction catalysts and washed with water or a dilute caustic solution to remove residual acid catalyst before use in fuel blending.

Although uncommon, the possibility of residual acid catalyst carryover into finished fuel exists. When this occurs, ferrous and nonferrous metal components in fuel transportation and storage systems may corrode.

4.1.1.2.4 Water Initiated Emulsification during Fuel Processing In refining and processing of crude oil to produce fuels, various chemical compounds and processing steps are utilized to prevent corrosion and deposit formation in the processing units. During fuel processing, these compounds typically do not initiate finished fuel quality problems. On occasion, these additives may aid in suspending water in the fuel or may participate in creating an emulsion with fuel. If this occurs, the finished fuel may appear hazy or may produce an emulsion that settles from the fuel. This type of emulsion typically settles as a film like layer on the bottom of the fuel storage tank, or if water is present in the tank bottom, the emulsion can exist as a scum at the fuel-water interface. The result may be as follows:

- Particulate contamination of jet fuel.
- Site for development of microbial growth plaques.
- Development of a corrosion cell.

4.1.1.2.5 Water Initiated Emulsification of Fuel Deposit Control Additives Deposit control additives are added to gasoline and diesel fuel to help prevent and remove deposits within the fuel intake system and combustion chamber of a vehicle engine. These additives are composed of highly polar organic compounds such as succinimides and polyether derivatives. Because of their polarity, water can have an affinity for most deposit control additives.

If fuel containing a deposit control additive is contaminated by water or comes in contact with water that resides on the bottom of a fuel storage tank, emulsification of the additive with water may occur. The result could be suspension of the additive as a water:oil emulsion within the fuel creating haze or may result in settling of the water:oil emulsion as a gel-like mass from the fuel. As a consequence of emulsification, the deposit control additive will no longer provide performance in keeping fuel injectors, intake valves, and combustion chambers free of deposits.

Demulsifiers can be formulated into deposit control additive formulations to prevent water from interacting with the additive. Specialized research and testing is required to identify a demulsifier for this application.

4.1.2 Heat

4.1.2.1 Change in Bulk Volume

Petroleum products expand when heated. The coefficient of expansion for most petroleum products will vary and is dependent upon the density or °API of the product. A listing of expansion coefficients for petroleum products is provided in Table 4.3.

An example of how to use the information in Table 4.3 is described as follows:

1. A 6000 gal load of kerosene at 21.1°C (70°F) leaves a fuel terminal by tank truck.
2. Upon arrival at its destination, the fuel has cooled to a temperature of 4.4°C (40°F).
3. The °API of the kerosene @ 15.6°C (60°F) is 44.0.
4. What will be the volume of kerosene at the new temperature of 4.4°C (40°F)?

°API @ 60°F	Expansion Coefficient per °F @ 60°F	Effective °API Range
6.0	0.00035	0 to 14.9
22.0	0.00040	15.0 to 34.9
44.0	0.00050	35.0 to 50.9
58.0	0.00060	51.0 to 63.9
72.0	0.00070	64.0 to 78.9
86.0	0.00080	79.0 to 88.9

Table 4.3 Expansion Coefficient for Petroleum Products

Calculation: 70°F – 40°F = 30°F
30°F × 0.00050 = 0.015
6000 gal × 0.015 = 90 gal
6000 gal – 90 gal = 5910 gal volume @ 4.4°C (40°F)

This 90-gal reduction in kerosene volume is the result of an actual contraction or shrinking of kerosene as it cools. Likewise, fuel volumes will expand comparatively when heated.

The expansion coefficient values per °F listed in Table 4.3 are for the °API ratings shown. Expansion coefficient values for measurements that fall within the effective °API range limits can be extrapolated. A more detailed volume conversion table appears in App. 2.

4.1.2.2 Air Temperature—Combustion Efficiency Relationship

As air temperature increases, air density decreases. This results in a reduction in the amount of oxygen available per unit volume of air to participate in the combustion of fuel. The information in Fig. 4.1 demonstrates the loss in fuel combustion efficiency as ambient air temperature increases above 32.2°C (90°F). A similar reduction in combustion efficiency occurs with increasing altitude.

4.1.2.3 Change in Rate of Reaction

Heat increases reaction rates. It is common knowledge that chemical reaction rates double whenever the temperature of a system increases by 10°C . This also means that reaction rates can slow by one-half whenever the temperature decreases by the same amount.

Therefore, in environments where high temperatures are maintained for long periods of time, the processes of oxidation and corrosion could be expected to accelerate. The graph in Fig. 4.2 demonstrates how corrosion of mild steel increases as water

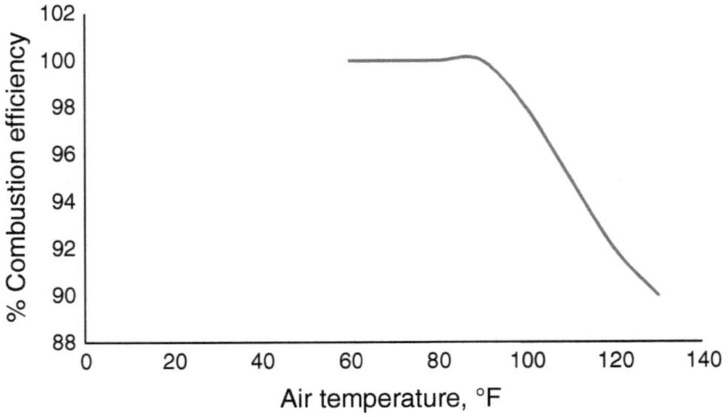

FIGURE 4.1 Loss of fuel combustion efficiency with increasing air temperature.

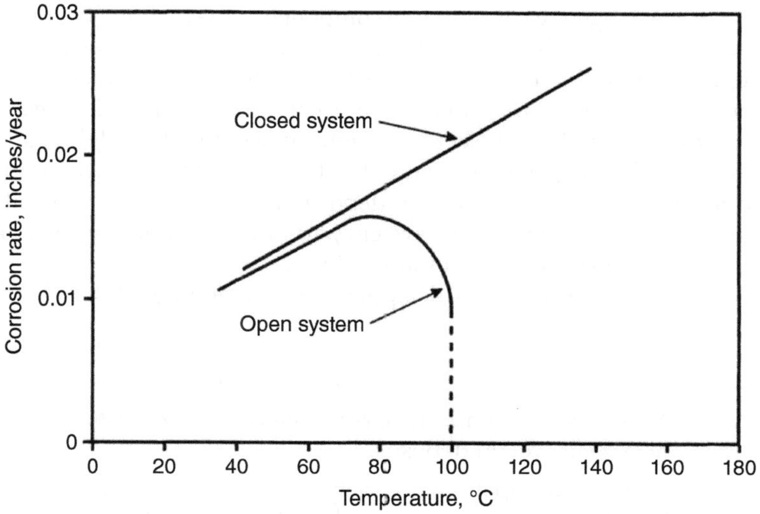

Figure 4.2 Effect of temperature on the corrosion of carbon steel in water.

temperature increases. Also shown is the reduction in corrosion rate that occurs when oxygen is free to escape from an open system.

4.1.2.4 Fuel Problems Enhanced by or Related to Heat

Examples of other problems related to heat are listed below. A detailed description of each of these problems can be found in later chapters of this book.

- Evaporation and loss of fuel components upon storage; resultant loss of fuel performance.
- Increase in gum formation of stored gasoline and diesel fuel.
- Degradation of diesel fuel color and accumulation of insoluble deposits.
- Possibility for enhancement of microbial growth between 32.2°C (90°F) and 37.8°C (100°F).

4.1.3 Air Contact and Deposit Formation

Air contains approximately 21 percent oxygen, 78 percent nitrogen, and less than 1 percent of other elements such as carbon dioxide and argon. Since oxygen is known to react with numerous materials, it can be a source of fuel problems. Fuel hydrocarbons which have reacted with oxygen from air can yield a variety of compounds such as peroxides, organic acids, and gums.

Most of the deposits formed in fuel and oil systems are rich in oxygen content. Oxygen in these deposits is not naturally occurring in the petroleum product, but typically comes from atmospheric oxygen.

When oxygen reacts with fuel components, the fuel first begins to darken in color. This is especially true for fuel components possessing conjugated double bonds. The oxidized hydrocarbons which result tend to further react to form polymeric compounds. These polymeric materials are high in molecular weight, usually fuel insoluble, and possess a slightly negative charge. The tendency of these compounds to form sludge and varnish-like deposits is more. Formed deposits can stick tightly to metal fuel system parts such as bulk storage tank surfaces, pipeline walls, and transfer lines.

The following general rule applies to the oxidation of petroleum hydrocarbons:

- Paraffins oxidize to form acidic compounds.
- Naphthenic and aromatic compounds oxidize to form high molecular weight deposits.
- Olefins participate in a variety of oxidation reactions which lead to the formation of carbonyl compounds, gums, and high molecular weight deposits.

4.1.4 Cold Temperature Conditions

Low temperatures can be the source of a large variety of fuel problems. When fuels and oils are cold, the following can result:

- Gelation.
- Significant increase in viscosity.
- Precipitation of wax.
- Become cloudy or hazy.
- Loss of volatility.
- Exposure to condensed water.
- Reduction in volume.

All of these phenomena can result in operating problems with pipelines, engines, pumps, and other equipment involved with the handling and use of fuels and oils.

4.1.4.1 Gelation

Some fuels and oils, especially those containing higher molecular weight compounds such as waxes, do not crystallize rapidly into a hard solid when cooled. Instead, they form a gel-like network throughout the fuel matrix. This network can begin forming well above the pour point temperature and may render the product nonpumpable.

It is recommended that any higher-viscosity product such as residual oil or heavy distillate fuel be evaluated for changes in low-temperature handling properties over time. Methods to evaluate this property include the British Admiralty Pour Point Reversion Test, the

Shell Amsterdam Reversion Test, and ASTM D5853 Procedure A for maximum (upper) pour point. Also viscosity increase verses temperature decrease measurements for products stored at low temperatures for an extended period of time may be beneficial.

4.1.4.2 Significant Increase in Viscosity

As temperature decreases, fuel viscosity increases. This fact can lead to a variety of problems related to pumping, mixing, and, utilization of fuels in cold environments. Pump cavitation and starvation can occur if cold, viscous fuel does not flow to the pump. Vane pumps, gear pumps, and piston pumps all rely on the material being pumped to serve as both a pump lubricant and as an agent for removing frictional heat. If fuel does not reach a pump due to its high viscosity, pumps can be severely damaged.

4.1.4.3 Precipitation of Wax

Fuels such as diesel fuel and heating oil are sometimes stored in large tanks for extended periods of time. At temperatures below the cloud point of the fuel, wax can form and fall from solution. Accumulated wax within fuel systems can deposit onto component parts and settle into areas of low turbulence. Problems such as filter plugging and flow limitations can be due to accumulated wax.

4.1.4.4 Become Cloudy or Hazy

A low level of water, up to approximately 50 to 75 ppm, is typically soluble in most middle distillate fuel at ambient temperature without causing the fuel to appear hazy. However as fuel cools, the dissolved water begins to appear as visible haze. Upon reheating, the fuel haze begins to clear. Figure 4.3 charts the effect of temperature on the solubility of water in middle distillate fuel at increasing concentrations.

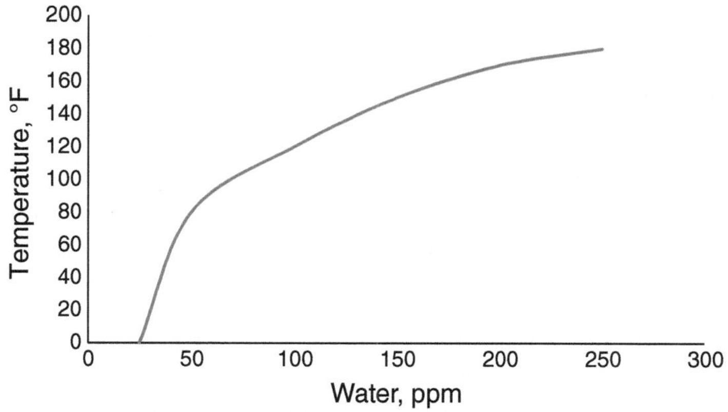

FIGURE 4.3 Effect of temperature on the solubility of water in middle distillate fuel.

4.1.4.5 Loss of Volatility

Fuel volatility is an extremely important factor related to fuel combustion and burning efficiency. Evaporation, vaporization, and vapor pressure of fuel are all reduced in cold environments. Poor starting and warmup of engines can be directly related to fuel volatility. Also, fuels that rely on volatility as a means to maintain complete combustion, such as wick fed kerosene lamps, will not vaporize and burn as effectively under cold temperature conditions.

Two complications of cold temperature volatility on fuel behavior in an operating engine are described below:

- *White smoke:* Upon starting a cold engine in cold weather, it is not unusual for exhaust to appear as white smoke. This is because combustion chamber and exhaust system temperatures are not hot enough to enable water of combustion and unburned fuel to exhaust as an invisible vapor. Instead, both appear as a condensed, white smoke.
- *Lubricating oil dilution by fuel:* Higher boiling, less volatile components in gasoline may not completely vaporize during cold weather startup of an engine and can remain as a liquid. In the combustion chamber, the still liquid fuel components may flow past the piston and piston rings to eventually mix with lubricating oil as it returns to the engine's oil pan. The result is a dilution and reduction in viscosity of the lubricating oil. This type of dilution by fuel can interfere with the ability of the lubricant to prevent engine wear in cold weather startup.

4.1.4.6 Exposure to Condensed Water

As water vapor in the air cools and reaches its dew point, it begins to condense and fall from the air. Condensation and accumulation of water vapor on the inside of fuel storage and transportation equipment can lead to unexpected water-related problems such as emulsification, bacterial growth, and corrosion.

- *Carburetor Icing:* In equipment using carbureted fuel intake systems, the phenomenon of carburetor icing may develop. Ice formation on carburetors occurs when ambient temperature is slightly above freezing and humidity is high. As fuel pumped into the carburetor vaporizes, heat is removed from carburetor parts by evaporating fuel. Water vapor in the intake air then freezes as ice around the venturi and other carburetor parts. As ice builds, the flow of air and fuel through the carburetor is restricted and the engine stops. Once radiant heat from warmer engine parts melts away the ice, the carburetor is again ready for use.

4.1.4.7 Reduction in Volume

As reviewed in Sec. 4.1.2 of this chapter, the coefficient of expansion and contraction for petroleum products is dependent on specific gravity. As fuels cool, the volume will be reduced. Gelled or partially solidified fuel, such as diesel fuel and residual oil, may contract or collapse slightly, pulling away from the sidewalls of the tank or container.

4.1.5 Light and Its Effect

Fuel performance problems initiated by light are not common. However, fuel quality can be affected. The primary concern related to light exposure is darkening of fuel color and the possible formation of high molecular weight compounds due to free-radical initiated polymerization of fuel components.

Light activated shift of double bonds in organic compounds can promote the reaction of oxygen with some molecules in fuel. Once oxygen has bonded to a reactive compound in fuel, organic compounds such as alcohols, aldehydes, esters, ethers, and acids can form. These compounds can then continue to react with other fuel components to form color bodies, gums, and insoluble deposits.

Light is an excellent catalyst for promotion of free radical reactions. An example is provided in Eq. (4.2) showing the reaction of an olefin with a mercaptan to form an organosulfide.

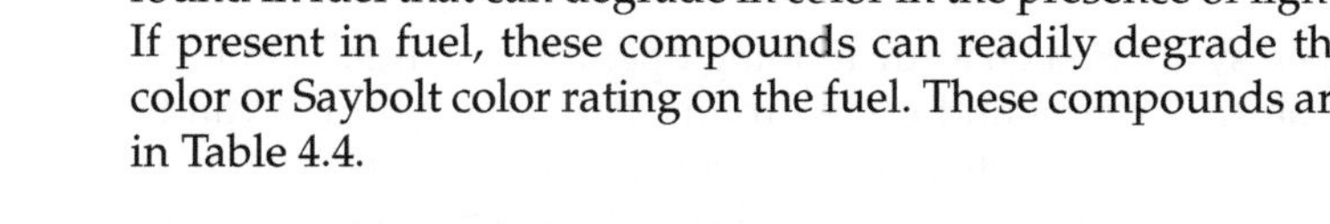

(4.2)

Phenols, cresols, and quinolines are examples of compounds found in fuel that can degrade in color in the presence of light and air. If present in fuel, these compounds can readily degrade the ASTM color or Saybolt color rating on the fuel. These compounds are shown in Table 4.4.

4.1.6 Rust, Dirt, and Debris

Once crude oil reaches the refinery, it is typically held within a closed system throughout most of the refining process. Outside of the refinery, finished fuels can be openly exposed to the environment. Upon loading into barges, tankers, and storage vessels, fuels may be exposed to airborne debris. After sale, fuel may be stored for various lengths of time. During storage, exposure to environmental contaminants is quite possible.

Within the closed system of a refinery, rust, metal salts, and catalyst fines exist. While being refined, processed, and blended, fuels may pick up some of these materials and carry them throughout the fuel distribution system. At some later time, the contaminants may lead to fuel darkening, combustion system deposits, or filter plugging.

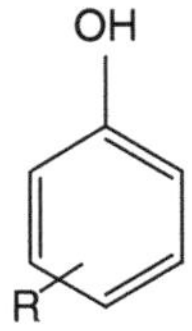 Alkyl phenol $R>C_1$	Phenols turn pink or red if not perfectly pure when exposed to light. This change is hastened in an alkaline environment.
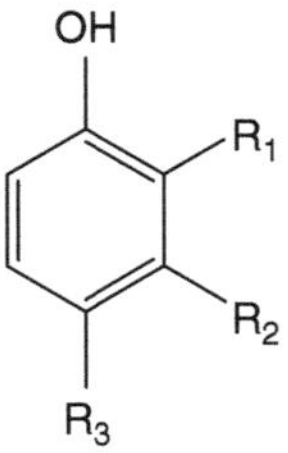 Cresols $R_{1,2,3}$ = methyl R_1 = ortho; R_2 = meta; R_3 = para	Cresols become dark, yellow-brown or red in color on exposure to light and air. *O - Cresol* – colorless or yellowish liquid *M- Cresol* – crystals or liquid becoming dark with age and exposure to air and light
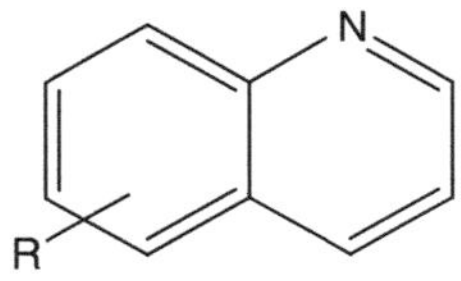	Quinolines are hygroscopic and darken in color when exposed to light.

TABLE 4.4 Effect of Light on Fuel Components

In open systems such as drumming operations, barge hold loading, and above ground and underground tank storage, fuels can be exposed to a range of potential contaminants. Rust, dirt, sand, filter debris, metal fines, elastomer degradation products, and water are some of the contaminants that can be found. These materials can result in fuel problems such as pump wear, emulsion formation, color degradation, combustion system deposits, and filter plugging.

Although difficult to predict, fuel problems related to rust, dirt, and debris are usually the easiest to identify.

4.2 Wax in Petroleum Products

Wax-related problems are common throughout the petroleum product industry. Fuels and lubricants contain wax at various concentrations. Filter plugging, line blockage, viscosity increase, and product haziness are all symptoms of wax formation within a fuel or oil.

A wax can be defined as a linear, branched, or cyclic hydrocarbon typically containing from 17 to 60 carbon atoms. Low carbon number waxes are found in middle distillate fuels and typically constitute a low percentage of the paraffins found in distillate fuel. Higher carbon number waxes can be found in residual fuel and lubricating oil. The percentage of wax in residual fuel oils can vary widely depending upon the refining process utilized. Processing factors that can influence the wax-related character of fuels and oils can include the following:

- Composition and character of the crude oils refined.
- Ninety percent distillation temperature.
- Hydroprocessing and hydrocracking aromatics and asphaltenes to paraffins.

4.2.1 Wax Classifications

Three basic classifications of wax in petroleum products are defined below and are depicted in Fig. 4.4.

4.2.1.1 Plate

Plate wax structure is the most common and consists primarily of linear *n*-paraffin molecules. The linearity of the wax molecules enables the wax to accumulate in the form of a uniform crystal lattice. The wax molecules build or stack together into a geometric form resembling a flat plate.

4.2.1.2 Mal-Shaped

Mal-shaped wax formations are not as symmetrical as plate crystals. Mal-shaped wax is composed of both linear and branched paraffin molecules. The branched paraffin molecules change the conformation of the wax so that stacking and alignment common in plate wax formation does not occur. The result is a wax crystal that is smaller than plate crystal wax. The temperature of mal-shaped wax formation is usually lower than plate crystal formation temperature.

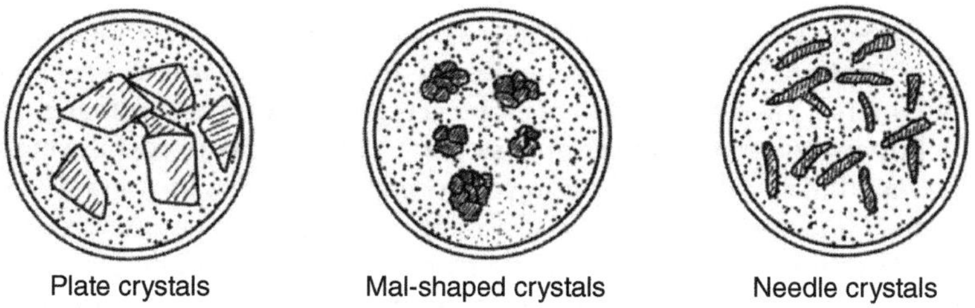

Figure 4.4 Wax crystal classifications.

4.2.1.3 Needle

The shape of needle wax crystals resembles a hollow cylinder or cone. Needle wax crystals form at the lowest temperature of all wax crystals and are quite small. Cycloparaffins and branched paraffins comprise the needle wax crystal structure.

4.2.1.4 Macrocrystalline and Microcrystalline Wax

- Macrocrystalline wax is larger in size and typically contains waxes from 17 to approximately 30 carbons in length. Plate structure wax is macrocrystalline and can be found in low concentration in some distillate fuel. Heavy distillates and residual oils contain higher concentrations of macrocrystalline wax.
- The term *microcrystalline wax* is commonly used to describe wax that is either mal-shaped or needle-like. This wax typically contains molecules more than 30 carbons in length. The wax can be present in various high boiling point fractions such as residual fuel oil or lubricant fractions.

Within given molecular weight limits, wax components having higher melting points crystalize in plates. Low melting point components crystalize in needles. Others crystalize in mal-shaped conformations. This process is independent of the fuel or oil where the crystals form.

4.2.2 Diesel Fuel Wax-Related Problems

At low temperatures, wax in most distillate fuel can cause severe problems with fuel pumping and filtration. As diesel fuel cools, wax begins to crystalize and become visible. The temperature at which crystallized wax becomes visible is known as the *cloud point*.

Further cooling of the fuel leads to wax crystal formation throughout the fuel matrix. The growing wax crystals develop into a larger lattice-like network encompassing the bulk fuel volume. This network of wax eventually causes the fuel to become highly viscous and to eventually gel into a semisolid mass. The lowest temperature whereby fuel remains in the liquid state just prior to gelation is termed the *pour point*.

Diesel fuel pumped from storage tanks, through pipelines, and within internal combustion engines passes through various types of fuel filters. When cold fuel containing crystallized wax accumulates on fuel filters, the trapped wax may slow or halt the flow of fuel through the fuel filter. Fuel filter plugging can occur. Wax may also accumulate in small lines and on orifices in fuel distribution equipment. As a result, fuel flow through a dispenser filter or a primary diesel engine filter may stop. Common test methods to determine the filter plugging tendency of diesel fuel and B100 biodiesel are listed in Table 4.5.

Test Method	Purpose
ASTM D4539 Filterability of diesel fuels by low temperature flow test (LTFT)	Determines the operability or lowest filterability temperature of diesel fuel when slowly cooled and passed through a 17-micron filter screen
ASTM D6371 Cold filter plugging point of diesel and heating fuels	A rapid cooling method for use in estimating the lowest filtration temperature for diesel fuel and heating oil
ASTM D7501 Determination of fuel filter blocking potential of biodiesel (B100) blend stock by cold soak filtration test (CSFT)	An accelerated method for determining the presence of substances in biodiesel that may plug filters when B100 biodiesel is cooled

Table 4.5 Testing Methods to Predict Filter Plugging Tendency of Fuel

4.2.2.1 Kerosene Blending

By diluting No. 2 distillate fuels with lower viscosity, lower wax content streams, the cold temperature handling and filtration properties of the fuels can be minimized, but not eliminated. It is a common practice to use kerosene, jet fuel, and No. 1 diesel as a blending component to reduce the cloud point, pour point, and cold filtration temperature of No. 2 diesel, No. 2 fuel oil, and other distillate fuel blends. If the No. 2 distillate fuel is an ultra-low sulfur fuel, care must be taken to ensure that the kerosene or No.1 diesel used as a blend component is also an ultra-low sulfur product.

4.2.2.1.1 Cloud Point Reduction During winter months and other low temperature operating environments, fuel cannot be effectively filtered at temperatures below its cloud point unless diluted with kerosene or treated with a wax crystal modifier. As a general rule, a reduction in No. 2 distillate fuel cloud point by 1 to 2°C (approximately 2 to 4°F) is typical for each 10 percent of kerosene added. This effect on cloud point reduction is presented in Fig. 4.5.

4.2.2.1.2 Pour Point Reduction Kerosene can effectively reduce the pour point of most distillate fuel. Dilution limits are often based upon whether kerosene will negatively impact fuel properties such as viscosity, distillation temperature limits, sulfur content, or cetane number. As a general rule, the pour point of diesel fuel can usually be reduced by 3 to 5°C (approximately 5 to 10°F) for each 10 percent of kerosene added.

4.2.2.1.3 Cold Filter Plugging Point Reduction The effect of kerosene blending on the reduction in No. 2 distillate cold filter plugging point

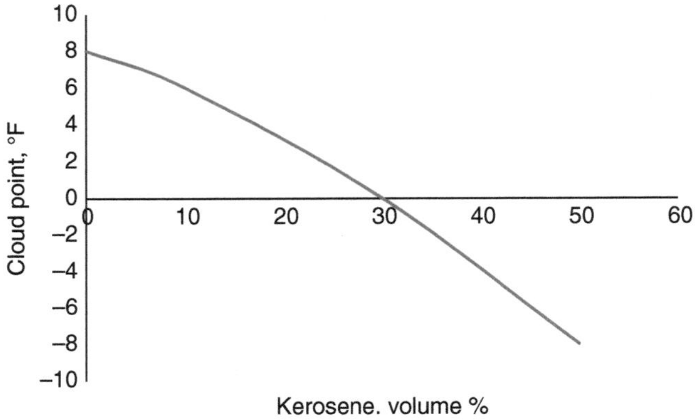

Figure 4.5 Effect of kerosene blending on the cloud point of middle distillate fuel.

(CFPP) is similar to the effect of kerosene blending on the cloud point of the fuel. A reduction in CFPP from 1 to 2°C (2 to 4°F) for each 10 percent of kerosene added is typical.

4.2.2.1.4 Low-Temperature Flow Test Reduction Reduction in the low-temperature flow test (LTFT) temperature of distillate fuel is possible with the addition of kerosene. Typically, kerosene addition will consistently provide a reduction in LTFT temperature. However, prediction of the response rate in lowering the LTFT temperature using kerosene is not readily estimated and testing is required.

In some fuels, the cloud point, pour point, and CFPP reduction temperatures obtained through kerosene blending can be more pronounced than the typical values described in the sections above.

4.2.2.2 Wax Crystal Modifier Addition

The addition of a wax crystal modifier to diesel fuel is a common and well accepted alternative to kerosene blending. Wax crystal modifiers are typically polymeric compounds that have the ability to co-crystalize with wax to alter the size, shape, and structure of the wax crystal lattice that forms in cold fuel. The terms pour point depressant and cold flow improver are often used in conjunction with wax crystal modifier to describe the same polymeric compounds used to improve the low-temperature flow properties of fuels and oils.

The most common type of wax crystal modifier used to reduce the pour point and filtration temperature of distillate fuel is based on ethylenevinylacetate (EVA) copolymer chemistry. The primary differences in EVA copolymers are found in the variation in molecular weight and the acetate ratio of the polymer. The performance of the EVA copolymer can sometimes be enhanced by the addition of a low concentration of a terpolymer or a wax modifier of a different chemical type such as those used in lubricating oils or crude oils. However, performance enhancement is usually specific and not broad based for use in all fuels.

Problems associated with the use of wax crystal modifiers may also involve improper application into fuel. Most EVA copolymers are highly viscous in nature and are diluted in a solvent by the manufacturer to improve product handling and storage properties. Even after dilution, some copolymers must be mildly heated at the point of application to ensure ease of pumping and handling.

The following examples describe actual situations that developed and resulted in the failure of a wax crystal modifier to perform as expected in fuel:

- *Problem 1:* Wax crystal modifier failed to modify wax in fuel, and fuel filter plugging resulted.
- *Possible cause of 1:* Wax modifier was blended into fuel below the fuel's cloud point.

For a wax crystal modifier to function properly, it must be present to co-crystalize with wax as it forms crystals in cold fuel. This requires that the modifier be added to fuel at a temperature higher than the fuel's cloud point and well before wax crystal formation begins. Because of this, wax crystal modifier should be added at a temperature at least 10°C (18°F) above the cloud point temperature of the fuel. Addition at this higher temperature helps to ensure that the wax modifier is present and completely soluble in fuel prior to formation of wax crystals.

Wax crystal modifiers added after wax crystals begin to form in fuel will have only minimal effect at modifying the size, shape, and structure of wax crystals. Consequently, large wax crystals will form and accumulate on the surface of a fuel filter. Eventually, fuel flow through the filter will slow significantly or will halt. Failure of a fueling station dispenser to deliver fuel into a customer tank will occur or plugging of a primary fuel filter on diesel engine with wax can occur and result in engine shutdown are described as follows:

- *Problem 2:* Wax modifier accumulated onto the surface of a fuel filter and halted flow of fuel through the filter.
- *Possible cause of problem 2:* Concentrated wax modifier was added to cold fuel, and the modifier formed a gel. The gelled additive failed to mix into the fuel, accumulated on the fuel filter, and halted flow of fuel through the filter.

When wax crystal modifiers are added to cold fuel, even fuel well above the fuel cloud point, modifiers may not dissolve properly. The polymeric nature of wax crystal modifiers makes them quite viscous at low temperatures. If a wax modifier is added to fuel, which is at a cold temperature from −10 to −5°C (14 to 23°F), the modifier may not dissolve completely in this fuel. The result will be additive accumulation as a viscous layer at the bottom of the fuel storage tank. Ultimately, the undissolved additive will most likely be trapped by the filter as it flows with fuel from the tank.

A second, and even more severe possibility, would be the addition of cold additive to cold fuel. In this circumstance, the additive would fail to mix into fuel and settle to the bottom of the storage tank as a viscous, gelled product. The gelled wax modifier may remain on the bottom of the tank or plug a fuel line or fuel filter. If allowed to reach a pump, the gelled additive could cause sticking of barrel and plunger components or other pump parts.

- *Problem 3:* Wax modifier fails to perform in pour point or CFPP reduction.
- *Possible cause of problem 3:* Fuel was previously treated with a wax crystal modifier.

Occasionally, wax crystal modifiers will not provide the performance anticipated. Either the response to additive treatment was much less than expected, or no response was obtained. When this occurs, it is quite possible that the fuel was previously treated with a wax crystal modifier. Under these circumstances, the expected performance of secondary treatment with wax modifier will be minimal.

Analysis to determine the presence of a wax modifier in fuel can be quite time consuming and expensive. However, it is possible to determine whether a fuel already contains a wax crystal modifier by utilizing the following comparisons:

- *Compare cloud and pour point temperatures:* Fuel that does not contain a wax crystal modifier will have temperature differences between the cloud and pour point values typically from 8 to 11°C (15 to 20°F). If the difference between the cloud and pour point value is greater than 14°C (25°F), it is reasonable to believe that the fuel contains a wax crystal modifier.
- *Compare cloud and CFPP temperatures*: The cloud point and cold filter plugging temperatures for fuel that does not contain a wax crystal modifier can often be the same. Typically, untreated cloud point and CFPP temperatures will be within 1 to 2°C (2 to 4°F) of each other. If the temperature difference between an untreated fuel's cloud point and CFPP differ by 5.5°C (10°F) or more, the fuel probably contains a wax crystal modifier.

4.2.3 Problems with Reproducing the Pour Point of Crude Oil and Heavy Residual Products

Pour point testing of crude oil and certain residual fuel products using manual pour point techniques requires an understanding of how these oils can behave under certain conditions of heating and shearing.

4.2.3.1 Effect of Applied Shear

When pour point testing is performed on crude oil, little to no mechanical shear is applied to the oil in the pour point tube. Under these conditions, the wax crystal lattice matrix that forms in the crude oil normally remains intact, and the oil gels at the pour point.

However, when disturbed by pumping, mixing, or agitation, the loosely formed wax crystal lattice can sometimes be broken with applied shear. If this occurs, some crude oils may again begin to pour and continue to flow at temperatures below the initial reported pour point.

4.2.3.2 Effect of Heating Crude Oil

Shearing due to pumping and pipeline flow can destroy the loosely formed wax lattice network of some gelled crude oil. This action may disturb wax lattice structure so that the oil flows below its natural pour point. Heating can often restore the oil to its original pour point. By heating the crude oil to temperatures 10 to 17°C (20 to 30°F) above the wax appearance point, waxes can be melted, solubilized, redistributed, and crystals reformed in the oil. When the pour point is again measured for the heated oil, the result obtained is often higher than the pour point temperature obtained for the same oil that was not heated prior to pour point testing.

4.2.3.3 Effect of Heating Residual Oil

When determining the pour point of certain heavy residual products such as No. 6 fuel oil, marine residual fuels, atmospheric residual oil, and vacuum gas oil, it is important to pay close attention to the temperature applied to the oil prior to pour point testing and the time the oil is exposed to the preheating temperature. For example, preheating residual oils to temperatures greater than 85°C (185°F) prior to pour point testing may result in a pour point value which is lower than the value obtained for the same oil preheated to a lower temperature such as 50°C (122°F). Also, an extended preheating time verses a shortened preheating time may have a similar impact on residual oil pour point values.

4.3 Fuel Volatility

Volatility is important to consider when discussing the combustibility of automotive spark ignition fuel and the burning quality of kerosene. Fuel volatility is crucial to the combustion process. In order to initiate and ensure smooth combustion of gasoline and petrol, volatile compounds such as low molecular weight, branched paraffins, and aromatics must be present. Factors important to consider concerning gasoline and kerosene volatility and combustion are reviewed in further sections.

4.3.1 Vapor Pressure of Spark Ignition Fuel for Engines

There are various volatility classes of spark ignition fuel to help ensure good starting, warmup, and operation of automotive engines under various seasonal conditions and geographic distributions. Vapor pressure is evaluated by ASTM methods such as ASTM D5191, ASTM D6378 or others and reported as kPa at 37.8°C (psi at 100°F). The purpose of controlling fuel vapor pressure is to help ensure that fuel will vaporize and ignite readily within the combustion chamber of an engine, yet not contribute to vapor lock due to excessive volatility.

Vapor pressure is provided by low molecular weight organic components in spark ignition fuel such as dissolved butane and the presence of pentanes, hexanes, and benzene. Because these compounds have a relatively high vapor pressure, they can escape naturally from fuel. When gasoline or petrol is stored for long periods of time, greater than 6 months for example, the fuel may lose volatility. The cold temperature engine starting properties and warmup time may be affected because of the loss of volatile, higher vapor pressure compounds. The vapor pressure of butane at various temperatures is shown in Fig. 4.6.

4.3.2 Volatility of Kerosene for Burners and Lamps

Kerosene used for illumination and in burners must contain enough volatile, low viscosity compounds to freely travel up the wick in kerosene burners and lamps. If the 10 percent distillation point of kerosene is above 205°C (401°F), the fuel will likely have poorer

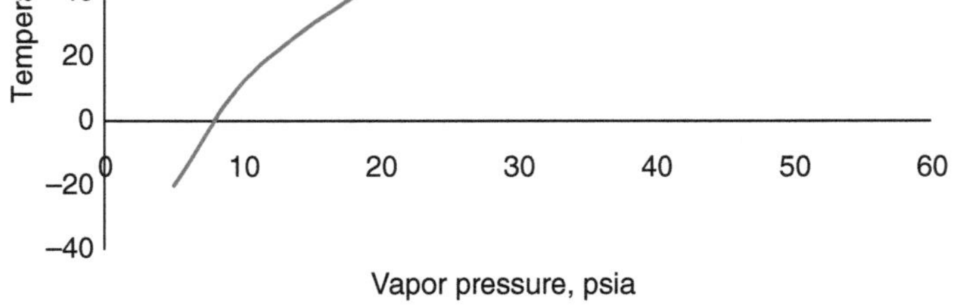

FIGURE 4.6 Vapor pressure of butane.

Percent Distilled at 70°C (158°F)	Minimum Outside Temperature, °C (°F)
3	26.7 (80)
11	15.6 (60)
19	4.4 (40)
28	−6.7 (20)
38	−17.8 (0)
53	−28.9 (−20)

TABLE 4.6 Spark Ignition Fuel Distillation Parameters and Ambient Air Temperature Parameters Affecting Engine Warmup Time

illuminating and burning properties than a fuel with a lower 10 percent distillation temperature. Also, kerosene which contains high boiling point components can carbonize and degrade lamp wicks. A combination of proper volatility and viscosity helps ensure good burning characteristics of kerosene.

4.3.3 Fuel Volatility Relationship to Ambient Temperature

Problems with engine operation can occur if volatile components in spark ignition fuel are lost due to evaporation. Contamination of fuel with compounds that may change the volatility characteristics can also be problematic. For example, engine warmup time will be extended in an engine if the percent distilled at 70°C (158°F) is less than the volumes reported for the temperatures presented in Table 4.6.

Refiners do change the distillation parameters of fuel to meet expected regional changes in temperature. However, these changes are typically based on seasonal averages and are not typically altered to meet a sudden decrease or increase in temperature. Using this information, it is possible to explain why rough running and warmup problems with engines can occur when early season cold weather systems develop to cause a sudden drop in temperature.

4.4 Diesel Fuel Cetane Number and Cetane Index

Diesel cetane engine testing has been used for many years to evaluate diesel fuels for their cetane number. However, some questions related to cetane number testing exist and are usually associated with the following:

- Differences between cetane number and cetane index.
- Performance provided by cetane improver.
- Effect of kerosene blending
- Reproducibility variation due to operator and engine differences.

4.4.1 Cetane Index and Cetane Number

The calculated cetane index methods ASTM D976 and ASTM D4737 were developed to estimate diesel fuel cetane number without extensive engine testing. Since the calculated cetane index is determined from physical property values such as the specific gravity and the 10, 50, and 90 percent distillation points, it is readily measurable from laboratory data.

The cetane number, however, is an engine measurement that requires a skilled operator and a well performing engine to determine accurately. Because of the variability among engines and the differences in operator skill, engine number determinations can differ. Automated cetane number analyzers are available under ASTM method designations ASTM D6890, ASTM D7170, and ASTM D7668. These analyzers produce a derived cetane number (DCN) upon injection of fuel into heated air and measurement of ignition delay.

As a general rule, it is usually safe to assume that the calculated cetane index will be higher than the actual cetane number. On occasion, the numbers will be the same, but rarely. Because of this, if a calculated cetane index value is known, it is reasonable to believe that the cetane engine number is usually one to two numbers lower than the calculated cetane index. For this reason, refiners and marketers of diesel fuel often set cetane index specifications higher than the desired number by one to two numbers or more.

4.4.2 Performance Provided by Cetane Improver

Throughout the world, the most commonly utilized and available cetane improver is based on octylnitrate chemistry, typically 2-ethylhexyl nitrate. Although other compounds have been shown to provide better performance, the cost, availability, and other performance factors make them less common.

Octylnitrate effectively boosts diesel fuel cetane number by aiding in the autoignition of fuel, thus minimizing ignition delay. Some typical guidelines for use in anticipating cetane improver performance are presented below:

- Cetane improver addition will not change the calculated cetane index of diesel fuel; only the cetane number will change.
- An increase of one to two cetane numbers is typically achieved for every 250 ppm of cetane improver added to diesel fuel. In higher base cetane number fuels, an increase of 3 numbers can be obtained for each 250 ppm added, but this is an exception.
- The greatest boost in cetane number improvement is obtained within the first 1000 ppm of cetane improver addition.

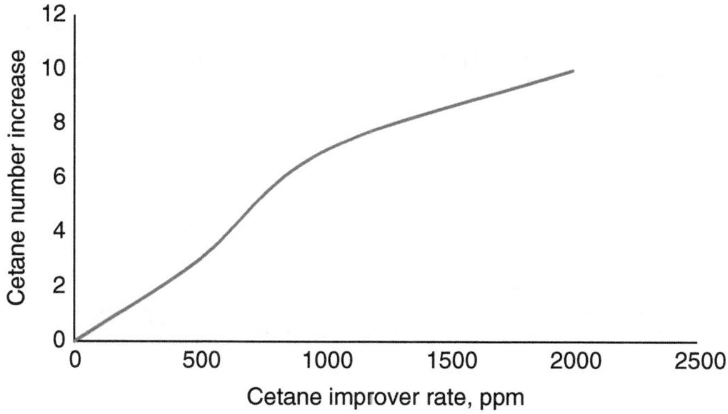

FIGURE 4.7 Cetane improver response in a base 42 cetane number fuel.

- A maximum cetane number increase no more than 12 to 14 numbers from the base number should be expected. This extreme level of performance improvement, however, is not common.
- High base cetane number fuels will typically respond more effectively to cetane improver treatment than lower base cetane fuels. An example of the response of a base 42 diesel fuel to cetane improver treatment is demonstrated in Fig. 4.7.

4.4.3 Effect of Kerosene Blending

The cetane engine number values for kerosene meeting ASTM D3699 specifications are typically lower than for No. 2 diesel fuel. This is primarily due to the greater number of lower molecular weight compounds present in kerosene compared to No. 2 diesel fuel. Information in Table 4.7 demonstrates this fact and provides a comparison of

Chemical Type	Carbon Number	Cetane Number
n-paraffin	10	82
n-olefin	10	68
Alkyl cyclohexane	10	50
Alkyl benzene	10	<20
n-paraffin	16	100
n-olefin	16	95
Alkyl cyclohexane	16	80
Alkyl benzene	16	55

TABLE 4.7 Impact of Fuel Chemistry and Carbon Number on Cetane Number

Average Cetane Number Level	Repeatability Limits, Cetane Number	Reproducibility Limits, Cetane Number
40	0.8	2.8
44	0.9	3.3
48	0.9	3.8
52	0.9	4.3
56	1.0	4.8

Table 4.8 ASTM D613 Cetane Engine Number Repeatability and Reproducibility Limits

the cetane number of compounds of different carbon number and varying chemical composition.

When kerosene of unknown cetane number is blended into diesel fuel to help improve the cold temperature handling properties of the fuel, evaluation of the cetane index or cetane number of the blend is recommended.

4.4.4 Cetane Number Repeatability and Reproducibility

Diesel cetane engine number repeatability and reproducibility are the result of variability in operator technique and engine operation. The typical repeatability and reproducibility variance accepted by ASTM will change with increasing cetane number. The following repeatability and reproducibility limits listed in Table 4.8 have been established by ASTM for method ASTM D613.

Cetane engine number determinations between operators in different laboratories can vary. If the variance is within the limits shown in Table 4.8, the variance is acceptable and should not be regarded as erroneous. If the difference is outside of the reported limits, then the results can be fairly questioned.

Repeatability limits pertain to the difference between two test results obtained by the same operator for the same test sample using the same equipment under constant operating conditions. Reproducibility limits pertain to two test results obtained by different operators on the same test sample in different laboratories under constant operating conditions. If test values fall outside of these limits, the results are considered unreliable.

4.4.5 Ignition Delay and Cetane Number

Since diesel fuel combustion is an autoignition process, the quality and characteristics of the fuel can have a dramatic impact on the efficiency of engine operation. Ignition delay and cetane number go hand in hand to influence the quality of the fuel combustion process. They are primary engine operability parameters directly related to fuel composition.

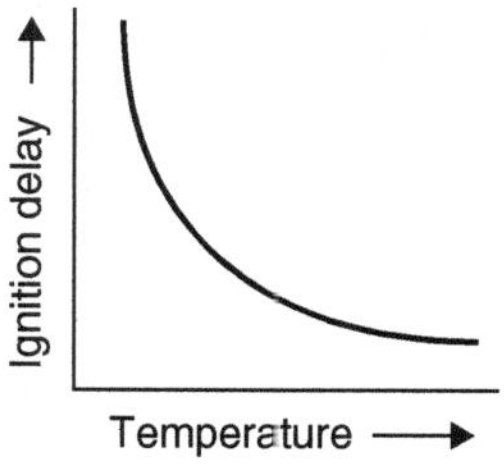

FIGURE 4.8 Effect of temperature on ignition delay.

Ignition delay can be defined as the time period between the injection of fuel into the combustion chamber and ignition of the fuel. A long ignition delay period is characteristic of low cetane fuels; a shorter delay is characteristic of higher cetane number fuels.

Ignition delay between injection of fuel and autoignition can be controlled by engine design, fuel and air temperature, fuel atomization, and fuel composition. The ignition delay period is typically shorter than the fuel injection period. The effect of temperature on ignition delay is shown in Fig. 4.8.

The fuel cetane number is one of the basic characteristics required to ensure proper engine operation and influence ignition delay. Fuel cetane number can impact the following engine combustion properties:

- High cetane number fuels burn with a shorter ignition delay and lower peak pressure. Engine noise is reduced.
- High cetane number fuels typically burn with less smoke and odor.
- Cold temperature starting is improved with high cetane number fuels.
- Engine knock and vibration can be due to rapid pressure rise resulting from the combustion of low cetane number fuels.
- Straight chain paraffins will autoignite more readily than branched chain paraffins and aromatics of the same carbon number.

4.4.6 Smoke and Hydrocarbon Emissions from High Cetane Number Fuel

This atypical problem of smoke and emissions resulting from burning high cetane number diesel fuel results when fuel is injected into the combustion chamber before properly mixing with air. Injection of fuel too early in the combustion cycle is more of a mechanical problem than a fuel performance problem. It can occur when engines timed to run effectively on low cetane number fuel are

Reference Fuel	Characteristics
n-cetane	Cetane number of 100
Heptamethylnonane	Cetane number of 15
T Fuel	Typical cetane number between 73 and 75
U Fuel	Typical cetane number between 20 and 22

TABLE 4.9 Characteristics of Primary and Secondary Cetane Reference Fuels

switched to operate on high cetane number fuel without modification of timing.

If fuel autoignition begins before desirable combustion air pressures and temperatures are reached, less carbon is oxidized during combustion and fuel efficiency is lost. For this reason, the fuel injection timing must always be monitored and adjusted to ensure proper mixing of fuel with fully compressed and heated air. Without retarding the injection timing to account for the shorter delay character of high cetane number fuels, exhaust smoke and hydrocarbon emissions may result.

4.4.7 Cetane Engine Reference Fuels

Either primary or secondary reference fuels can be utilized in the cetane engine when determining the cetane number of distillate fuel. Primary reference fuels are *n*-cetane and heptamethylnonane. Secondary reference fuels are identified as T Fuel and U Fuel. The characteristics of each of these fuels are identified in Table 4.9.

The T Fuel and U Fuel references are commonly used for routine cetane engine determinations. Care should be taken to ensure that both T and U reference fuels are not exposed to long-term storage below temperatures of 0°C (32°F). Both fuels may gel and solidify at cold temperatures.

4.5 Hydroperoxides and Olefins

4.5.1 Hydroperoxides

Certain fuel components can be oxidized by the free radical process. Benzyl, allyl, and tertiary compounds can all be oxidized at room temperature to yield hydroperoxide species.

Through the process of hydrogen abstraction, and consequential formation of peroxy radicals and alkyl radicals, fuel components can be completely consumed through oxidation. The chain reaction which progresses through the formation of organic hydroperoxides can yield insoluble gums and sludge.

4.5.1.1 Hydrogen Abstraction

Generation of an alkyl radical and a hydroperoxide radical as shown in Eq. (4.3) is the first step in the hydrogen abstraction process.

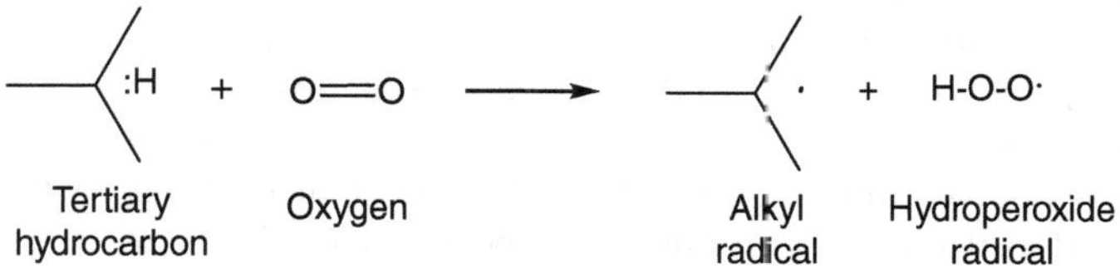

(4.3)

The newly generated alkyl radical can react with oxygen to produce a peroxy radical as presented in Eq. (4.4).

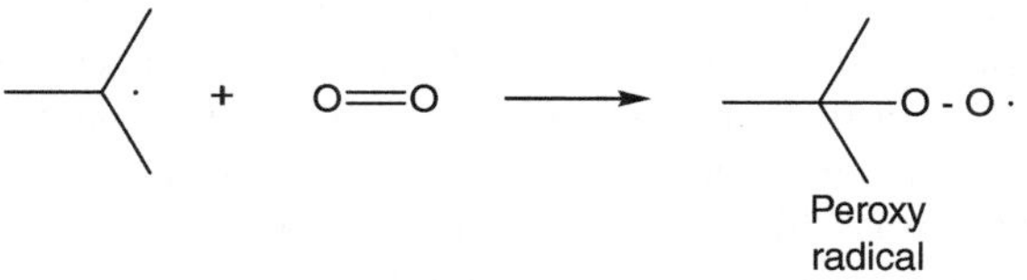

(4.4)

The peroxy radical can further react with a paraffin to form an alkyl hydroperoxide as presented in Eq. (4.5).

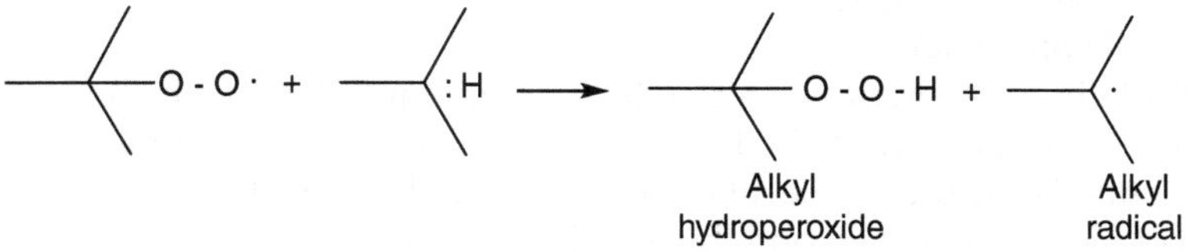

(4.5)

4.5.1.2 Homolytic Cleavage

Organic peroxides are very reactive and unstable. The oxygen-oxygen bond readily cleaves to yield two radical species which further react to form a wide variety of compounds such as ethers, esters, alcohols, and organic acids. The homolytic cleavage of an alkyl peroxide is demonstrated in Eq. (4.6).

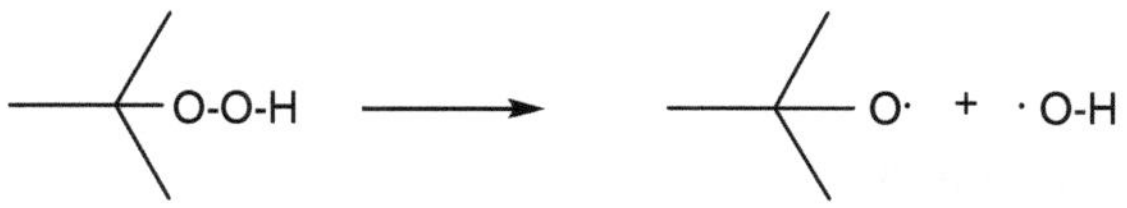

(4.6)

4.5.2 Olefins

Crude oil typically contains little to no olefinic compounds. Through refining and processing, however, olefins are produced and become a part of various crude oil fractions. Olefins can be found in thermally cracked and catalytically cracked naphtha fractions as well as FCC cycle oils and coker gas oils. For this reason, it is not unusual for finished gasoline and distillate blends to contain a high-olefin content stream.

A variety of fuel performance problems can be directly linked to the presence of olefinic compounds in a fuel. Problems such as darkening of fuel color, gum and sludge formation, and combustion system deposits can be linked directly to the presence of olefins.

Double bonds between carbon atoms in fuel components are active sites for oxidative attack and polymerization. During long-term storage and during the combustion process, oxidation and polymerization of olefinic compounds can result in the following.

4.5.2.1 Varnish Formation on Metal Parts

The oxidation of olefins can result in the formation of organic hydroperoxides. These compounds readily decompose to form alcohols, carbonyl compounds, and other oxidized species. These oxidized hydrocarbons can further react to form highly cross-linked, oxygen-rich materials. Some of these species can adhere to metal surfaces to form a hard varnish-like film or coating on metal parts. This varnish can act as a site for further deposition and eventual corrosion of metal. In severe cases, varnish can interfere with the hydrodynamic lubrication of moving metal parts and efficiency of component operation.

4.5.2.2 Dropout of High Molecular Weight, Polymerized Fuel Components

As fuel olefins polymerize, intermediate compounds form, which often further react to form high molecular weight—fuel insoluble components. These high molecular weight materials can settle to the bottom of fuel storage tanks and transfer lines as sludge. Also, sludge deposits may act as initial sites for corrosion and microbial growth.

As presented in Eq. (4.7), the decomposition of a hydroperoxide to an alkoxy radical can result in the reaction of the alkoxy radical with an olefin. A carbon radical intermediate then forms. Olefin chain propagation and polymerization can follow to yield high molecular weight deposits.

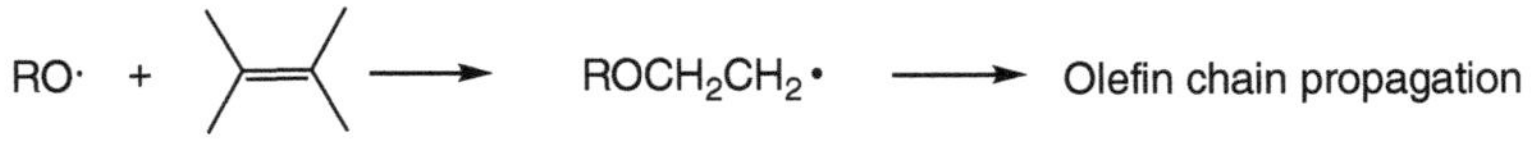

(4.7)

4.5.2.3 Color Bodies

Shifting and rearrangement of bonds within a molecule due to reaction with oxygen can result in the formation of compounds which impart color to fuel. These compounds often have some degree of aromatic functionality. The color imparted to fuel is typically amber to brown. Some distillate fuels turn deep red in shade while others appear pale green.

Chromophore	Auxochrome
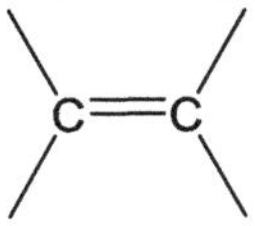	$-OH$
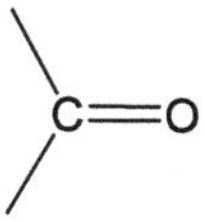	$RO\cdot$
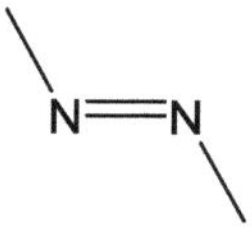	$-NH_2$
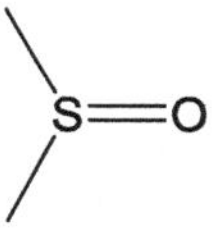	$-SH$
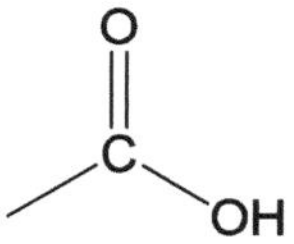	$-NH-R$
	$-NR_1R_2$

Table 4.10 Common Chromophoric and Auxochromic Groups

As a general rule, a fuel component will impart color if it contains four or five conjugated chromophoric or auxochromic groups. Listed in Table 4.10 are examples of common chromophoric and auxochromic groups.

Fuel components possessing conjugated olefins, conjugated carbonyl compounds, or any combination of chromophores and auxochromes can be the source of color bodies in fuel. Examples are provided in Table 4.11.

4.6 Microorganisms, Sediment, and Water

Fuel problems related to microorganisms, sediment, and water in fuel systems can develop for a variety of reasons. Examples include:

- Contamination of fuel due to carryover of water and dissolved solids from crude oil refining and fuel processing.
- Accumulation of fuel degradation products that develop over time as fuel ages in storage.

Compound	Description
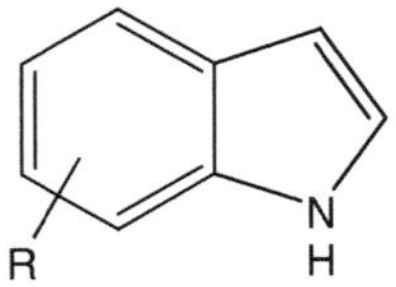 C_1 to C_4 indoles	White to yellow in color turning red on exposure to light
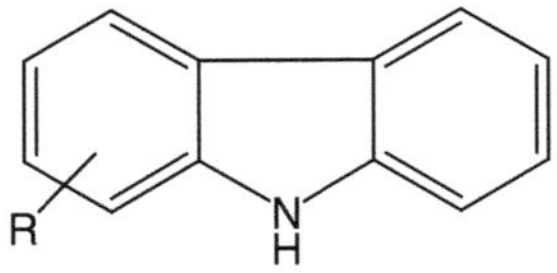 C_1 to C_4 carbazoles	Exhibits fluorescence and long phosphorescence on exposure to UV
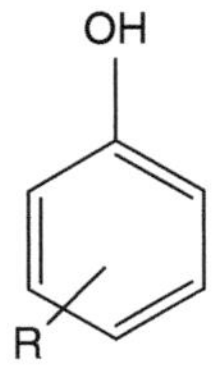 C_1 to C_4 phenols	Phenols turn pink or red if not perfectly pure or when exposed to light. This change is hastened in an alkaline environment.
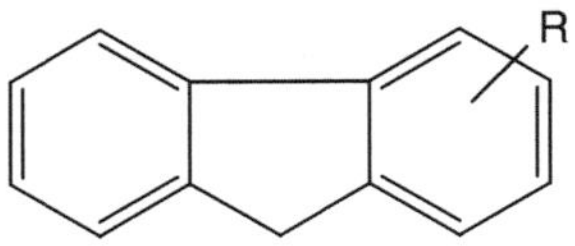 C_1 to C_2 fluorenes	Fluorescent when impure

TABLE 4.11 Examples of Color Bodies Identified in a High Sulfur Distillate Fuel

- Buildup of rust and corrosion by-products in metal piping and fuel storage tanks
- Growth of microorganisms in water that may reside in the bottom of fuel storage vessels and accumulation of products of microbial metabolism

Of these conditions, microbiological contamination seems to have the most diverse impact on fuel quality by contributing to metal corrosion, fuel odor, sediment and deposits, and degradation of fuel components. Once microbes have multiplied and established themselves as a colony, they can adhere to metal parts forming "microbial plaques." Underneath these plaques, severe corrosion of metal is often found.

4.6.1 Products of Microbial Metabolism

In fuel systems containing water, certain microorganisms can survive. Typical species are the bacterial anaerobe *Desulfovibrio* and the fungal species *Hormoconis resinae.* These microorganisms are called Hydrocarbon Utilizing Microbes and often referred to as "HUM" or "HUM-bugs." These species can initiate corrosion in fuel systems.

Fuel sulfur is reduced to H_2S by *Desulfovibrio.* Hydrogen sulfide can attack iron to form an iron sulfide scale and result in the loss of iron from a metal surface. In the presence of water, iron sulfide scale can act as a site for severe pitting corrosion. This type of corrosion is often uncontrollable.

The term *sulfate-reducing bacteria* (SRB) is frequently used to describe organisms that metabolize sulfates in fuel. Upon metabolism, the oxygen bound to sulfate sulfur is consumed by the SRB and utilized in cellular respiration. The sulfur is reduced to H_2S gas. Once liberated, H_2S can react with fuel olefins to form mercaptans, contribute to microbial induced corrosion, or escape into the fuel. The formation of mercaptan through the light catalyzed reaction of hydrogen sulfide with an olefin is shown in Eq. (4.8).

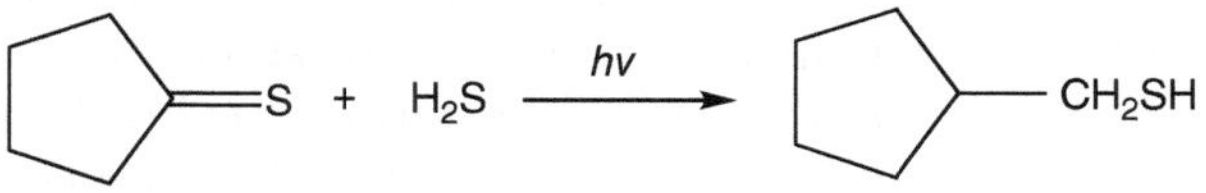

(4.8)

4.6.1.1 Microbiologically Influenced Corrosion

Active microbiologically influenced corrosion (MIC) of metal due to bacteria is quite complex. It can involve several species of microorganisms and is affected by temperature, TOC, pH, and other factors. Examples of bacterial species which are associated with MIC are described below.

4.6.1.1.1 Sulfate Reducing Bacteria Bacterial anaerobes including *Desulfovibrio, Desulfomonas,* and *Desulfotomaculum* are known sulfate-reducing species and are often identified as sulfate reducing bacteria (SRB). They can survive in fresh, brackish, and seawater and are present in most soils and sediments containing sulfate and sulfite compounds. Enzymes that promote the conversion of sulfate and sulfites to metal sulfides through chemical reduction are present in these bacteria. Iron sulfide, FeS, is a product of this process.

4.6.1.1.2 Acid Producing Bacteria Organisms such as *Thiobacillus thiooxidans* and *Clostridium* species have been linked to accelerated corrosion of mild steel. Aerobic *Thiobacillus* oxidizes various sulfur-containing compounds such as sulfides to sulfates. This process promotes a symbiotic relationship between *Thiobacillus* and sulfate-reducing bacteria.

Also, *Thiobacillus* produces sulfuric acid as a metabolic by-product of sulfide oxidation.

Clostridium species are anaerobic, spore-forming microbes. The formic, acetic, propionic, and butyric acids produced as a result of their metabolic activity can enhance the corrosion of steel.

Gluconacetobacter liquefaciens is unique in its ability to convert ethanol into acetic acid. The species is aerobic and is associated with spoiling fruits and with flowers, fruits, and plant sugars used in the fermentation process to produce ethanol. If the ethanol contains *Gluconacetobacter*, conversion to acid can occur resulting in ethanol or E10 storage tank corrosion, especially in the tank overhead region exposed to air.

4.6.1.1.3 Metal-Depositing Bacteria Bacteria that oxidize ferrous iron (Fe^{+2}) to ferric iron (Fe^{+3}) such as *Gallionella* and *Leptothrix* species are termed *metal-depositing bacteria*. The result of this process is the formation of ferric hydroxide.

4.6.1.1.4 Slime-Forming Bacteria Most slime-forming bacteria are aerobic. Species such as *Pseudomonas* produce an extracellular, gel-like polysaccharide capsule, which acts to protect and shield the organism. When in combination with other metabolic by-products, bacteria, and water, *Pseudomonas* species form slime-like films and deposits on surfaces exposed to the air. Sulfate-reducing species and acid formers can frequently be found in high concentrations beneath these slime layers. Also, slime layers can form masses large enough to plug filters and strainers on pumps, lines, and hoses.

4.6.1.1.5 Other Microbiological Factors As bacteria die, ammonia is typically produced as a product of protein decomposition. Aerobic nitrifying bacteria such as *Nitrosomonas* and *Nitrobacter* can oxidize ammonia (NH_3) to nitrate (NO_3^-). As a result of this process, the pH of the system can be reduced. The water analysis provided in Table 4.12 provides a listing of typical microbes that may be found in a corroded system.

4.7 Catalyst Fines

Residual fuel oils and heavy marine fuels are composed of high boiling point petroleum fractions, gas oils, and cracked components. Clarified oil streams (also termed *slurry oil*) from the FCC process can contain degraded alumina/silica catalyst fines. The 20- to 70-micron-diameter fines are known to contribute to a variety of problems in fuel injection and combustion systems. In marine engines, excessive injector pump wear, piston ring wear, and cylinder wall wear can all be due to the abrasive action of catalyst fines on these fuel system parts.

Organisms	Identified in Water	Identified in Deposits and Corrosion Products
Total aerobic bacteria		1,100,000
Enterobacter	490,000	<1,000
Pigmented	30	<1,000
Mucoids	<70	<1,000
Pseudomonas	<10	210,000
Spores	100,000	8,700
Total anaerobic bacteria	35	150,000
Sulfate reducers	30	120,000
Clostridia	5	30,000
Total fungi	2	<100
Yeasts	2	<100
Molds	None	<100
Iron depositing organisms	None	None
Algae		
Filamentous	Very few	None
Nonfilamentous	Few	Few
Other organisms		
Protozoa	Few	None

*All counts expressed as colony-forming units per milliliter for water or gram for deposits.

Table 4.12 Typical Microbiological Analysis in a Service Water System Pipe Experiencing Microbiologically Influenced Corrosion*

4.7.1 Blue Smoke

One effect of valve sleeve, piston ring, and cylinder wall wear is leakage of lubricating oil into the combustion chamber. When lubricating oil accumulates in the combustion chamber and burns with the fuel, the exhaust smoke appears blue in color. Wear due to abrasion or corrosion has the same effect. At times, unburned fuel can also appear as blue smoke if fuel droplets are finely dispersed in the exhaust. Blue smoke is a symptom of a mechanical problem rather than a fuel performance problem.

4.8 Fuel Oxidation and Degradation Products

Certain fuel components, especially olefins, alkylated naphthenes, and various heterocycles can react with oxygen or with one another in fuel. Once reacted, these compounds can further interact to form fuel insoluble sludge and gums.

Component	Concentration, g/kg
Bicarbonate	0.142
Bromide	0.065
Calcium	0.400
Chloride	18.980
Magnesium	1.272
Potassium	0.380
Sodium	10.561
Sulfate	2.649
TOTAL	34.449

Table 4.13 Concentration of Major Inorganic Components Found in Seawater

The term varnish is used to describe the hard, amber colored coating of fuel oxidation products adhering to engine components. The term *sludge* describes higher molecular weight, dark-colored deposits that form and settle out of the fuel. Sludge can accumulate in areas of low turbulence and act as a prime site for initiating corrosion.

If water is present in a fuel system, sludge and gums also lead to plugging of fuel filters. In contact with water, sludge and gums can aggregate into larger gel-like particles, which are trapped by fuel filters. The aggregates can be as large as 100 microns in diameter and will not pass through standard 5- to 15-micron fuel filters but may pass through higher porosity 130-micron tank filters.

4.9 Seawater Contamination

Water from the sea is known to contain high levels of sodium and chlorine. Seawater, however, differs from other forms of saltwater or freshwater in the ratio of magnesium:calcium. In freshwater, the ratio of magnesium:calcium is in the range of 1:2 to 1:3. In seawater, the ratio can range from 3:1 to as high as 4:1. This difference helps to distinguish seawater from freshwater that may be present in a fuel terminal storage tank, transportation vessel, or any other container where seawater contamination is a possibility. Table 4.13 provides information on the concentration of various inorganic components found in seawater.

4.10 Contaminants and Sediment in Lubricating Oil

4.10.1 Metals from Wear and Corrosion

Oils used to lubricate machinery, motors, hydraulic systems, and other mechanical devices can sometimes contaminate fuel systems. These oils often carry with them low levels of the metals, which

Metal	Possible Sources of Origin
Aluminum	Pistons, turbocharger/blower, pump vanes, thrust washers, bearings
Chromium	Compression rings, antifriction bearings, shafts
Copper	Oil cooler tubes, bearings, bushings, thrust washers, valve guides, injector shields
Iron	Cylinders, liners, pistons, rings, valves, valve guides, gears, shafts, clutch plates, antifriction bearings, rust
Lead	Bearings
Nickel	Valves, valve guides, antifriction bearings
Silver	Silver solder, wrist pin bushings, antifriction bearings
Tin	Bearings, plating

Table 4.14 Possible Sources of Metals Found in Used Lubricants

originate from surface wear of the mechanical device. Some of the common wear metals and possible sources of origin are provided in Table 4.14.

Other elements such as boron and silicon can be found in fuel and oil system deposits. They can originate from the following sources:

- *Silicon:* Gasket sealant compounds, antifreeze, oil antifoam, sand and dirt
- *Boron:* Antifreeze, lubricant additive

Element	Engine and Industrial Oil Additive Source
Zinc	Zinc dithiophosphate antiwear/antioxidant additive in most engine oils and hydraulic oils
Calcium	Overbased calcium sulfonate and calcium phenate detergent and acid neutralizing additive used to formulate engine oils
Phosphorus	Zinc dithiophosphate or phosphonate compounds in engine oils and hydraulic oils; phosphorus containing extreme pressure additives found in gear oils
Sulfur	Zinc dithiophosphate; sulfonate compounds in engine oils and hydraulic oils; sulfur containing additives are found in a wide variety of industrial oil products such as metal working fluids, slide way oils, bar and chain lubricants, and gear oils

Table 4.15 Elements Found in Crankcase and Industrial Oil Additives

4.10.2 Elements from Lubricant Additives

If deposits found on vehicle filters, pump parts, and engine systems contain high levels of zinc, calcium, phosphorus, and sulfur, the possible source of the contamination is lubricating oil. These elements are part of the composition of additives used to formulate engine oils, hydraulic oils, gear oils, and other industrial oils. The elements and the products in which they can be found are listed in Table 4.15.

CHAPTER 5

Utilizing Physical and Chemical Property Measurements to Identify Sources of Fuel Problems

Fuel physical and chemical properties such as distillation profile, paraffin content, and viscosity can provide a wealth of information about product quality and performance. The information provided in this chapter can be used to help identify how specific physical and chemical property measurements can be used to identify and solve fuel problems.

5.1 Distillation Profile

One of the easiest ways to determine the character, and sometimes the quality, of a fuel is to examine the distillation profile. Certain characteristics of fuel performance and potential fuel problems can be determined from the distillation profile. Several fuel performance and problem-related examples are provided in this discussion.

5.1.1 Smoke and Odor during Diesel Fuel Combustion

Two factors important to consider when identifying and solving fuel combustion problems are the amount of carbon present and the amount of oxygen available. It is well known that the boiling point of hydrocarbon molecules will generally increase as the number of carbon atoms increases. It is also known that in order for fuel to burn cleanly and completely, enough oxygen must be present to react with all carbon atoms to form either CO_2 or CO. Therefore, if a fuel contains a greater number of carbon atoms, more oxygen will be required to ensure complete combustion.

Diesel fuel with a high T-50, for example greater than 302°C (575°F), will tend burn with more smoke, soot, and hydrocarbon odor than fuel with a lower T-50. This is basically due to incomplete combustion and oxidation of a great number of high boiling, high carbon content fuel components in a limited oxygen content environment. Soot from combustion can result in the formation of carbon deposits in piston ring grooves.

5.1.2 Poor Calorific Value or Btu Content Heating Oil

The heating value or calorific/Btu rating of fuel oil used to generate heat is related to the amount of carbon available in the fuel for burning. If the T-50 of heating oil is less than 232°C (<450°F), the amount of carbon available to serve as an efficient energy source is less than that of a higher T-50 fuel. Problems related to poor energy efficiency of fuel oil could be traced back to the low carbon content evident by the T-50 temperature.

5.1.3 Diesel Engine Warmup

In order to warmup a cold diesel engine effectively, fuel components must be present, which have the proper autoignition characteristics to maintain smooth combustion. If a fuel contains a large portion of components with widely different ignitibility characteristics, smooth warmup will be difficult to achieve.

A T-50 in the range of 232 to 280°C (450 to 535°F) is desirable for warmup in most high-speed diesel fuel applications.

5.1.4 Poor Diesel Engine Startability

Initial starting of a cold diesel engine can be related to the volatility of the fuel. The IBP of diesel fuel and the T-10 provide some information on the startability of fuel in a diesel engine. If the T-10 is too high, for example greater than 177°C (>350°F), cold startability problems may occur.

5.1.5 Dilution of Crankcase Motor Oil/Carbon Deposits in the Combustion Chamber of Vehicle Engines

If fuel contains a large amount of high-boiling, high carbon content compounds, it is possible that some of these compounds will not combust. In both gasoline and diesel engines, fuels with high T-90 temperatures will burn incompletely.

As a result, unburned fuel can remain in the combustion chamber to be drawn into the crankcase with the oil lubricating the piston cylinder walls. The unburned fuel compounds may also remain within the upper part of the cylinder to carbonize and form combustion chamber deposits. Piston ring and land groove deposits may also form.

Fuel enters combustion chamber
↓
Incomplete combustion/oxidation of high boiling point components results
↓
Partially combusted/oxidized fuel remains as a liquid on piston and cylinder walls
↓
Liquid collects in piston grooves, lands, and lubricating oil
↓
Continued oxidation results in deposit formation

The long-term effects of utilizing high T-90 fuel will lead to noticeable deposits, engine operability problems, and possible engine damage. Refiners understand that spark ignition fuel will tend to cause dilution of engine oil if the T-90 point is high, and they will adhere to established ASTM specifications related to maximum T-90 temperatures. For engines operating in cold temperature environments where outside air temperatures fall significantly, the T-90 temperature specifications of refined fuel are also reduced. Lowering of T-90 helps minimize the percentage of compounds in fuel that may not readily vaporize and combust in extreme cold climates.

5.1.6 Inaccurate Cetane Index Ratings

It may not be possible to accurately determine the cetane index of low end point No. 1 diesel and kerosene. This is because the ASTM D976 or ASTM D4737 calculated cetane index scales cannot be utilized effectively for products having an end point less than 260°C (<500°F).

5.1.7 Knock in Spark Ignition Fuel Engines

Problems related to engine knock can be readily diagnosed by measuring the octane number of spark ignition fuel. However, it is possible to look at the distillation profile of fuel and anticipate problems linked to engine knock.

5.1.7.1 High T-90

High molecular weight fuel components have a longer ignition delay than most lower molecular weight components. Therefore, components of the T-90 fraction will naturally have more resistance to knock than other fractions. By reducing the T-90, resistance of spark ignition fuel to knock may also be reduced.

5.1.7.2 Front-End Volatility

Both motor octane number (MON) and research octane number (RON) values of finished fuel blends are typically measured using the whole fuel, not individual fuel fractions. However, the octane number contribution of various fuel fractions can influence the overall knock resistance. Finished fuels with identical RON values may contain fractions with either very similar or widely different octane number values.

For example, the T-10 through T-40 fractions of available gasolines may possess significantly different octane numbers. This difference may be manifested in various levels of knock resistance under low-temperature engine starting and warmup operation.

During cold weather engine warmup and acceleration, volatile fuel components would tend to vaporize and combust more readily than higher boiling point, less volatile compounds. Therefore, in cold weather, fuels with low octane number T-10 through T-40 may knock more readily during initial warmup and acceleration than fuels possessing higher octane number T-10 through T-40 fractions. Although modern engine design has helped to minimize this problem, cold weather engine knock problems still exist.

5.2 High Paraffin Content

Linear, branched, and cyclic paraffins all exist in refined fuel. Fuel performance problems can often be directly related to the type and concentration of paraffins present. For comparison, the typical carbon number range and boiling point temperatures of paraffins found in representative fuels and other petroleum products are provided in Table 5.1.

Several of the important characteristics associated with a high paraffin content in fuel are described in the following sections.

Fuel	Paraffin Carbon Number Range	Paraffin Boiling Temperatures, °C (°F)
Gasoline/petrol	4 to 12	–0.5 to 216 (31 to 421)
Mineral spirits	9 to 12	151 to 216 (303 to 421)
Kerosene/jet fuel	9 to 16	151 to 287 (303 to 548)
Diesel fuel	9 to 17	151 to 303 (303 to 575)
No. 6 Fuel oil	12 to >20	216 to >343 (421 to >650)
Petroleum wax	≥18	>316 (>601)

TABLE 5.1 Typical Carbon Number Range and Boiling Temperatures of Paraffins in Fuels and Other Petroleum Products

Product	KB Value	% Aromatics
Hexane	31	<1
No 2. Kerosene (3000 ppm sulfur)	30	<25
Mineral spirits (Stoddard)	39	15
Xylene	98	99+
Heavy aromatic naphtha	105	86+
140 solvent	35	15
VM&P naphtha	38	10

Table 5.2 Kauri-Butanol Value (KB Value) of Various Hydrocarbon Solvents

5.2.1 Poor Solvent/Low Kauri-Butanol Value

Paraffins function poorly as a solvent for some organic compounds. This fact can have various consequences. For example, gums, deposits, and fuel degradation products will not be held in solution by high paraffin content fuels. As a result, gums and degradation products will fall from solution and settle onto fuel system parts such as storage tank bottoms and fuel system lines. The Kauri-Butanol (KB) value for selected petroleum products is provided in Table 5.2. A high KB value indicates good solvent power.

Also, additives that are used to enhance the performance of fuels and oils are usually highly polar organic compounds. Their solubility in paraffinic systems can be quite low unless formulated with paraffinic side chains or appropriate co-solvents. Additive dropout and decomposition can occur unless solubility is maintained.

5.2.2 Diesel Fuel Cetane Number Increase

As a general rule, paraffinic hydrocarbons have a higher engine cetane number than aromatic compounds with the same number of carbon atoms. Also, linear paraffins and alkenes typically have higher cetane numbers than cycloparaffins of the same carbon number.

For this reason, fuels refined from paraffinic crude oils will have naturally higher cetane numbers than fuels refined from naphthenic or more aromatic crude oil. Some typical cetane number values for selected *n*-paraffins, olefins, and aromatics are shown in Table 5.3.

5.2.3 Diesel Fuel Wax Related Problems

At low temperatures, usually less than –5°C (23°F), waxy paraffins may begin to crystalize in highly paraffinic diesel fuel. These paraffin

Compound	Type	Number of Carbons	Cetane Number
Octane	*n*-Paraffin	8	64
1-Octene	Olefin	8	41
Decane	*n*-Paraffin	10	77
1-Decene	Olefin	10	60
n-cetane	*n*-Paraffin	16	100
1-hexadecene	Olefin	16	84
Decahydronaphthalene	Aromatic	10	42
n-nonylbenzene	Aromatic	15	50

Table 5.3 Approximate Cetane Number of Various Hydrocarbons

wax crystals can create a variety of fuel handling and performance problems. Examples include:

- Plugging of fuel lines and orifices.
- Accumulation on fuel filter media to slow and eventually halt the flow of fuel.
- Pumping difficulty due to an increase in the viscosity of fuel.
- Wax dropout or settling from fuel during long term storage. The result can be accumulation of a wax layer on the bottom of a storage tank. This accumulated wax can cause line and filter plugging when pumped from the tank bottom into vehicles.

In cold-temperature environments and during winter months, fuel refiners and marketers will often blend kerosene or No. 1 diesel fuel into No. 2 diesel fuel to help combat the problems associated with paraffin wax in fuel. This technique is effective in wax control, but it is sometimes limited due to a reduction in the Btu and specific gravity values of the blended fuel.

The paraffins in diesel fuel can vary in molecular weight, concentration, and crystallization temperature. For this reason, fuels with similar physical properties may behave differently at low temperatures. This difference in behavior can be related to the low-temperature properties of fuel paraffins present.

5.2.4 Less Smoke during Burning and Combustion

When compared to highly aromatic fuels, paraffinic fuels combust and burn with less smoke and soot. This is primarily due to the presence of hydrogen rich paraffins compared to lower hydrogen content aromatics. The hydrogen rich nature of paraffins enables the combustion process to progress more completely to form CO_2 and H_2O. This process is illustrated in Eqs. (5.1) and (5.2).

$$-CH_2- + 1.5\,O_2 \xrightarrow[\Delta]{} H_2O + CO_2 \tag{5.1}$$

$$2\,-CH- + 1.5\,O_2 \xrightarrow[\Delta]{} H_2O + CO_2 + C^0 \tag{5.2}$$

As demonstrated, two aromatic carbon atoms must be burned to yield enough hydrogen to produce one water molecule. This process yields two carbon atoms which must be oxidized to either CO or CO_2. Without excess oxygen, free carbon reactions can occur to produce amorphous soot and smoke. By increasing the amount of oxygen delivered into the fuel combustion process, the complete oxidation of aromatic carbon to either CO or CO_2 is possible.

Paraffin combustion, however, can proceeded more completely to H_2O and CO_2 because each carbon atom carries with it at least two hydrogen atoms. By burning only one paraffin carbon atom, enough hydrogen is released to react with oxygen to produce a molecule of water. The paraffin carbon can further react to form CO or CO_2.

5.2.5 Gasoline Engine Knock

Gasoline containing high concentrations of linear paraffinic compounds will not have the ability to prevent knocking as effectively as more aromatic gasoline or gasoline containing branched paraffins. Aromatic compounds and branched paraffins have a longer ignition delay period than most linear paraffins, and therefore, can inhibit engine knock more effectively.

Knock can occur over the entire range of engine speed. During acceleration, a large volume of fuel enters the intake manifold and mixes with air. Vaporization and evaporation of the more volatile, low octane compounds occurs readily. These compounds enter the combustion chamber and detonate quickly to initiate knock. The higher octane, less volatile compounds do not vaporize as rapidly from the fuel volume. Deposit formation within the combustion chamber can also lead to knock.

5.2.5.1 Increase in Cylinder Pressure

The deposit in a cylinder acts to reduce the physical volume of the combustion chamber through air displacement. As a result, during the compression-ignition cycle, the cylinder air volume is compressed to a higher pressure. When ignition occurs, engine knock will be enhanced due to higher cylinder pressures.

5.2.5.2 Increase in Cylinder Temperature

Engine cylinder deposits retain heat and inhibit cooling of the combustion chamber. Under high pressure combustion chamber conditions, volatile fuel components will ignite and begin to combust before spark ignition begins. This premature ignition enhanced by hot deposits can lead to knock. Knock and other combustion chamber abnormalities are presented in Fig. 5.1. The abnormalities are identified as: (a) knock, (b) run-on, (c) preignition, (d) misfire, (e) exhaust seat valve recession.

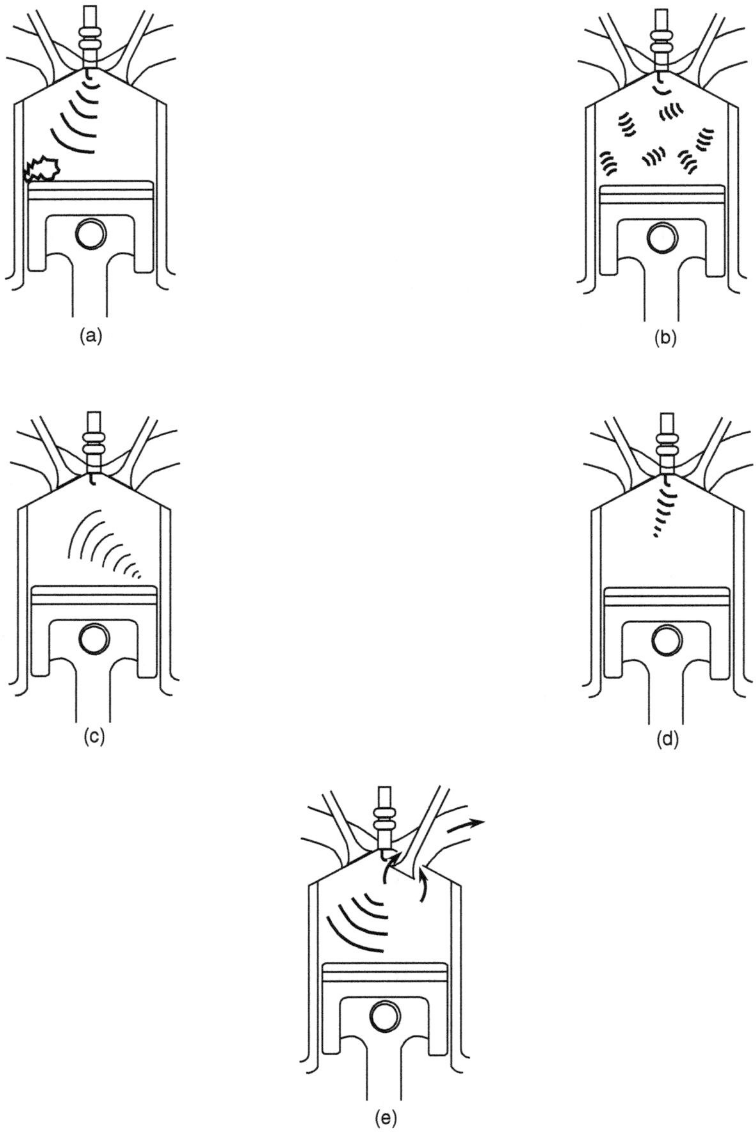

FIGURE 5.1 Fuel combustion chamber abnormalities.

5.3 High Aromatic Content

Monoaromatic, diaromatic, and polyaromatic compounds can all be found in fuels. Although concentrations vary in different fuels, the presence of these aromatics can have both desirable and undesirable consequences.

5.3.1 Low-Temperature Flow Properties

As a general rule, low molecular weight aromatic compounds help to improve the low-temperature flow properties of fuels and oils. Since highly aromatic compounds have little wax, they possess better natural low-temperature handling properties than paraffinic fuels. Also, cloud point, pour point, and cold filtration temperatures of aromatic diesel fuel will typically be much lower than a paraffinic diesel fuel.

5.3.2 Poor Cetane Engine Performance

Aromatic hydrocarbons do not have good performance in the cetane engine. Compared to paraffins of the same carbon number, aromatic compounds have a very low cetane number.

If a diesel fuel is reported to have a lower than expected cetane number, the first place to look for an explanation would be the total aromatic content of the fuel.

5.3.3 Low Cetane Index

A widely accepted relationship exists between the cetane index of a distillate fuel and total aromatic content. Typically, as fuel aromatic content increases, the fuel cetane index decreases. It is generally believed that distillate fuel containing more than 35 percent aromatics will have a cetane index less than 40. Likewise, fuel having less than 35 percent aromatics will have a cetane index greater than 40.

5.3.4 High Calorific Value or Btu Value

The heating value of a fuel, often referred to as the calorific value or Btu value, can be linked to the aromatic content of the fuel. Since aromatic molecules are more highly compacted than linear or branched paraffins, it is possible to "pack" more of them in the same volume or space.

To illustrate this, 1 L of *m*-xylene, an eight-carbon aromatic compound, weighs approximately 868 g (1.91 lb) while 1 L of *n*-octane, an eight-carbon paraffin weighs approximately 665 g (1.47 lb).

- Fuel sold as heating oil typically has a higher aromatic content than fuel sold for use in an internal combustion engine. The higher aromatic content signifies that more carbon atoms are present per volume of fuel than fuel with a lower aromatic content. As a result, more potential heat is available from aromatic heating oils.

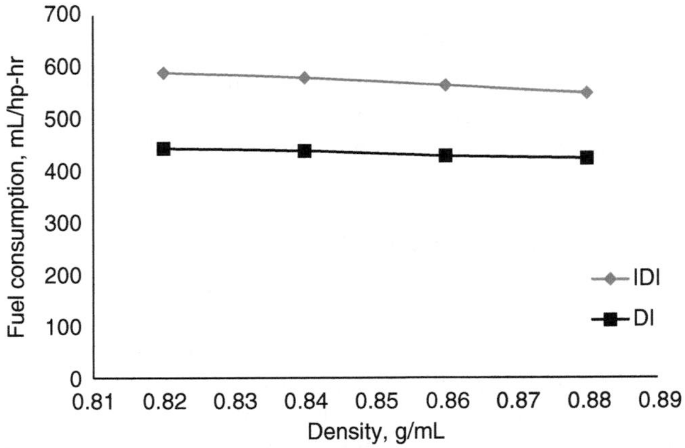

Figure 5.2 Fuel density and volumetric fuel consumption relationship in direct-injected (DI) and indirect-injected (IDI) engines.

5.3.4.1 Volumetric Fuel Consumption

Studies documented by the Council of the Institution of Mechanical Engineers have compared fuel density with volumetric fuel consumption. Test results have shown that there is a general trend in both DI and IDI diesel engines toward lower volumetric fuel consumption whenever higher-density fuel is utilized. The engine results, presented in Fig. 5.2, illustrate the relationship between fuel density and the volume of fuel consumed per horsepower-hour. The results were obtained under full load operating conditions.

5.3.5 Low °API

The American Petroleum Institute (API) established an arbitrary rating system to measure the °API gravity of crude oils and petroleum products at a standard temperature of 15.5°C (60°F). This system assigns water to the °API value of 10. Products heavier than water have °API gravity values less than 10. Most petroleum products are lighter than water and have °API values greater than 10. For example, diesel fuel has a typical °API of 30 to 40 and gasoline has a typical °API of 50 to 60. Some asphaltic oils are heavier than water and have °API values less than 10.

Because fuel and oil weights vary per standard volume, they should always be handled on a weight basis. If sold on a volume basis, the volume should always be corrected to 15.5°C (60°F) by utilizing °API volume correction tables. These tables are used to standardize all volumes of petroleum products to a 15.5°C (60°F) volume. Standard °API volume correction tables for petroleum products are contained in App. 3.

5.3.6 Seal Swelling

Certain elastomeric materials used in the production of gaskets, seals, and hoses can be degraded by aromatic fuel components. Elastomers such as natural rubber, neoprene, Buna-N, and ethylene-propylene will all swell in the presence of aromatic fuel components.

Sometimes, seal swelling is desirable if the purpose of the elastomer is to help provide a tight seal between parts, which have been bolted together or mounted onto an engine. Swelling helps to prevent fuel leakage or air aspiration problems. As an example, swelling of seals used in the assembly of high pressure fuel pumps is desirable. Swelling helps prevent fuel leakage.

5.3.7 High Solvency/High KB Value

There are both advantages and disadvantages of fuels having a high KB value or high solvency. For example, if a fuel with a high KB value were to be introduced into a system containing sludge or gum-like deposits, the fuel may act to dissolve and solubilize some of these deposits. As a result, the fuel may appear dark in color because it now contains dissolved degradation products. The dissolved fuel deposits could possibly lead to tank filter plugging problems due to entrapment of dissolved material by the tank filter.

The fact that naphthenic lubricants have a higher solvency power than paraffinic lubricants is used to an advantage in the formulation of certain oils. Originally, lubricants used in railroad engines were all naphthenic based. Naphthenic oil is effective at solubilizing deposits formed within the engine crankcase, thus preventing the settling and accumulation of deposits. Another example would be the use of naphthenic oils in the formulation of air compressor lubricants, textile processing oils, and other industrial lubricants. In these formulations, solubilizing and dispersing compounds within oil is quite important.

5.3.8 Asphaltene Incompatibility/Dropout in Marine and Residual Fuel Oil Blends

Heavy residual components used in marine fuel applications can contain significantly high levels of asphaltenes. When marine fuels and other heavy fuel oils of widely different viscosities are blended together, the asphaltenes may drop out of the final fuel blend. This phenomenon would be more pronounced if a low viscosity, paraffinic fuel were to be blended with a heavy residual fuel containing vacuum tower bottoms, heavy coker gas oil, or visbreaker bottoms. Consequences of asphaltene dropout could include excessive centrifuge sludge accumulation, injection system deposit formation, and exhaust port fouling. Power loss could result from deposit formation and fouling.

5.3.9 Black Smoke in Diesel Engine Exhaust

Black exhaust smoke is typically produced in diesel engines operating at or near full load. This occurs because fuel volume injected exceeds the volume of air available in the combustion chamber needed for complete combustion of fuel carbon. Particles of carbon and soot form to give the smoke a black or gray color.

Highly aromatic, low °API diesel fuel can also enhance black exhaust smoke production. This is due to the fact that more fuel mass is injected per volume of fuel than would be injected from higher °API fuel. Also, there are a variety of mechanical factors which can influence black smoke production such as poorly functioning fuel injectors and air intake system problems.

5.4 Sulfur Content

The sulfur content of fuel continues to fall as environmental regulations are implemented in various countries worldwide to monitor and reduce atmospheric sulfur dioxide, sulfur trioxide, and the reaction products they can yield. This is being accomplished by:

- Hydroprocessing fuels to remove sulfur by conversion of fuel sulfur to hydrogen sulfide and ultimately to elemental sulfur.
- Burning and combusting lower sulfur content fuel oils in industrial burners and in vehicle engines.

Other fuel issues associated with fuel sulfur are described in the topics that follow.

5.4.1 Combustion Chamber Deposits

Under low temperature, low speed, and low load driving conditions, the fuel combustion system operation is not optimal. These conditions are commonly encountered during the winter months in stop-and-go city driving. Whenever fuel burns under these conditions, the water of combustion does not exhaust completely from the combustion chamber and condenses as fuel system moisture. Also, the SO_x formed upon burning fuel sulfur can be retained within the combustion chamber.

Together, water and SO_x can combine to form sulfur-bearing acids. These acids can accumulate to initiate corrosive wear, degradation of the lubricating oil, and formation of piston lacquer deposits within the combustion chamber. These deposits can result in operability problems such as preignition, knock, dieseling, and component wear.

5.4.2 Copper, Bronze, Brass and Silver Corrosion

Elemental sulfur, mercaptan sulfur, and hydrogen sulfide can all attack silver, copper, and copper alloys present in fuel systems components. Silver is a component of some automotive fuel tank level gauges. Corrosion of silver by sulfur compounds can result in malfunction of the gauge to produce inaccurate readings. Copper, brass, or bronze found in heating coils, cooling coils, valves, and fittings are all susceptible to sulfur initiated corrosion.

5.4.3 Mercaptan Sulfur

Mercaptans or thiols are well known for their unpleasant odors. If they are present in fuel, even at low levels, the role they play can be significant. For example, mercaptans can degrade almost any elastomer, and for this reason, a 30-ppm mercaptan limit is placed on all aviation turbine fuel such as Jet A and Jet A-1. In addition, mercaptans have unpleasant odors, and some mercaptans can corrode copper and its alloys.

As presented in Eq. (5.3), when exposed to air, mercaptans are oxidized to yield disulfides and sulfonic acids.

$$\text{R-CH}_2\text{SH} \xrightarrow{[\text{O}]} \text{R-CH}_2\text{S-SCH}_2\text{-R} \xrightarrow{[\text{O}]} \text{R-CH}_2\text{SO}_3\text{H} \tag{5.3}$$

Under extreme conditions of heat and an acidic environment, aromatic sulfonic acids can form sulfuric acid and an aromatic compound. Problems related to acid buildup include corrosion of metal system parts and degradation of the stability of fuel.

5.4.4 Possible Fuel Stability Problems

Substituted and condensed thiophenes and thiophenols are usually the most abundant sulfur-containing compounds in refined fuels. These compounds are known to react with oxygen to form peroxides and eventually result in color bodies and gum-like fuel deposits. Through hydroprocessing, most of these compounds can be removed from fuel. However, their presence in high-sulfur diesel, high-sulfur fuel oil, and other non-hydroprocessed fuel streams, create fuel color and deposit problems in stored fuel blends. The reaction of thiophenol with oxygen for form color and deposit precursors is presented in Eq. (5.4).

$$\underset{\text{Thiophenol}}{\text{C}_6\text{H}_5\text{SH}} + \text{R}\cdot \xrightarrow{\text{O}_2} \text{C}_6\text{H}_5\text{S}\cdot + \underset{\text{Hydroperoxide}}{\text{ROOH}}$$

$$\text{C}_6\text{H}_5\text{S}\cdot + \underset{\text{Olefin}}{\text{C=C}} \longrightarrow \text{C}_6\text{H}_5\text{S-C-C}\cdot \xrightarrow{\text{O}_2} \text{C}_6\text{H}_5\text{-C-C-O-O}\cdot$$

$$\text{C}_6\text{H}_5\text{SH} + \text{C}_6\text{H}_5\text{S-C-COO}\cdot \longrightarrow \text{C}_6\text{H}_5\text{S}\cdot + \underset{\text{Gum-Color Body Precursor}}{\text{C}_6\text{H}_5\text{S-C-COOH}} \tag{5.4}$$

5.4.5 Elemental or "Active" Sulfur

Elemental sulfur in fuels and oils is sometimes termed "active" sulfur because of its highly aggressive and corrosive nature toward metal, especially copper, bronze, brass, and silver. Although uncommon in finished fuels, elemental sulfur can appear as a reaction product when fuel containing hydrogen sulfide is exposed to oxygen, rust, or acids. Reactions that can liberate sulfur are presented in Eqs. (5.5) to (5.7).

$$2H_2S + O_2 \longrightarrow 2S + 2H_2O \tag{5.5}$$

(Air sparging or reaction of dissolved hydrogen sulfide with atmospheric oxygen could yield elemental sulfur.)

$$3H_2S + Fe_2O_3 \cdot H_2O \longrightarrow S + 4H_2O + 2FeS \tag{5.6}$$

(Rust and corrosion products present in fuel storage and transportation equipment could yield elemental sulfur.)

$$3H_2S + H_2SO_4 \longrightarrow 4S + 4H_2O \tag{5.7}$$

(Although infrequent, low concentrations of sulfuric acid could carry over from acid processing or alkylation)

5.4.6 Exhaust Odor and Catalytic Converter Degradation

The sulfur-containing combustion gases sulfur dioxide and sulfur trioxide both have unpleasant odors. These gases form during the combustion of fuel sulfur. Although less common, other sulfur-containing compounds such as carbon disulfide, carbonyl sulfide, hydrogen sulfide, and various mercaptans could form. All of these compounds have foul odors.

5.4.7 Diminished Effectiveness of Alkyl Lead Antiknock Compounds

Worldwide, alkyl lead compounds are blended into aviation gasoline grades as an antiknock additive. In some regions, alkyl lead is a component of automotive spark ignition fuel. Although being regulated out of use due to its toxicity and exhaust emission contribution, alkyl lead and other organometallics are used to improve engine antiknock performance.

By decomposing to form lead oxide compounds during fuel combustion, lead alkyls interrupt the rapid chain scission reactions involved in combustion knock. Also, lead alkyls help to prevent exhaust valve seat wear and may minimize octane requirement increase. However, unless utilized in conjunction with lead scavengers such as 1,2-dichloromethane, lead deposits can accumulate within the gasoline combustion chamber.

Most Effective	Polysulfides
	Thiols
↓	Alkyl disulfides
	Alkyl sulfides
	Elemental sulfur
↓	Aryl disulfides
	Aryl sulfides
Least Effective	Thiophenes

Table 5.4 Effect of Fuel Sulfur Compounds at Degrading Alkyl Lead Compounds

Fuel sulfur can decompose alkyl lead compounds and can diminish the effectiveness of lead compounds at inhibiting knock. The effect of fuel sulfur at degrading lead compounds is provided in Table 5.4.

5.5 High Viscosity

Problems associated with difficulty in filtering, mixing, and pumping fuel can usually be linked to an increase in fuel viscosity. Some examples of fuel performance problems associated with high fuel viscosity are described in this section.

5.5.1 Inadequate Filling of Diesel Fuel Injection Pump

The barrel and plunger of a distributor-type diesel fuel injection pump rapidly discharges and refills with fuel during normal operation. If fuel viscosity is high, fuel flow can be impaired enough to prevent adequate filling of the barrel. The result is poor fuel distribution to the cylinders and poor engine performance. Contamination of fuel with high-viscosity products or operation at excessively low temperatures can increase the viscosity of diesel fuel and result in fuel pumping problems. Also, if fuel viscosity at −18°C (0°F) is greater than 45 cSt, fuel pumping problems are likely to occur. Listed in Table 5.5 are minimum and maximum viscosity limits for a variety of diesel fuel and fuel oil grades.

5.5.2 Poor Fuel Atomization

For fuel to combust and burn efficiently, it must be atomized into extremely small droplets. Fuel injectors aid tremendously in performing this role. However, if fuel viscosity is high, atomization into small droplets becomes difficult. Highly viscous fuel will not disperse freely after being sprayed from the injector. The fuel instead sprays as a stream or large drops rather than as a fine mist. The result

Fuel Type and Grade	Minimum Viscosity @ 40°C, cSt	Maximum Viscosity @ 40°C, cSt	Other Viscosities
Kerosine No. 1-K and No. 2-K	1.0	1.9	
Aviation turbine fuel			Max. 8.0 mm^2/s @ −20°C
No. 1-D Diesel fuel oils	1.3	2.4	
No. 1 Fuel oils	1.3	2.4	
No. 2-D Diesel fuel oils	1.9	4.1	
No. 2 Fuel oils	1.9	4.1	
No. 4-D Diesel fuel oil	5.5	24.0	
No. 4 Fuel oil (light)	1.9	>5.5	
No. 4 Fuel oil	5.5	24.0	
No. 5 Fuel oil (light)			Min. 5.0 @ 100°C Max. 8.9 @ 100°C
No. 5 Fuel oil (heavy)			Min. 9.0 @ 100° Max. 14.9 @ 100°C
No. 6 Fuel oil			Min. 15.0 @ 100°C Max. 50.0 @ 100°C

Table 5.5 ASTM Viscosity Limits for Kerosine, Aviation Turbine Fuel, Diesel Fuel, and Fuel Oil

is a decrease in fuel efficiency and power due to incomplete burning of larger fuel droplets.

5.5.3 Distortion of Fuel Injection Pump Parts

The tolerances and clearances between the moving parts of a high-pressure fuel injection pump are quite small. Because of this, high fuel viscosity can result in a friction coefficient increase within the high-pressure fuel injection pump. Frictional energy can lead to excessive heat buildup and distortion of pump components. The Streibeck curve presented in Fig. 5.3 demonstrates how an increase in fuel viscosity can have a dramatic effect on the friction coefficient of a moving system, especially under high load conditions.

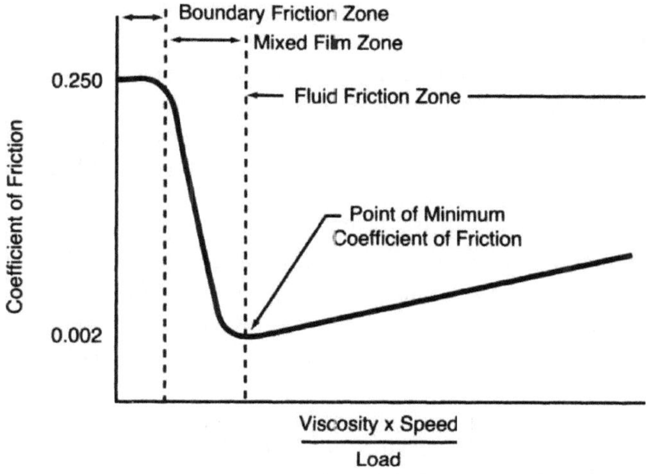

FIGURE 5.3 Streibeck curve: relationship between viscosity, speed and load.

5.5.4 Pumping Problems with Residual Fuel and Marine Fuel Oils

Any time residual fuel or marine fuel pumping problems occur, the first place to begin looking for the source of the problem would be the viscosity of the fuel. Pumps are designed and rated to move fluids within a fixed viscosity range. When fuel viscosity is either above or below the recommended viscosity range, pumping problems can eventually develop.

The presence of high molecular weight compounds such as waxes, polycyclic aromatics, highly branched hydrocarbons, and certain polymeric additives can increase the viscosity of fuel. These compounds have a high resistance to flow. Their presence can be detected indirectly through either ASTM D86 distillation end point temperatures or by high ASTM D445 viscosity measurements.

5.5.5 Low Flame Height in Wick Fed Kerosene Burners

In order to maintain a constant feed rate of fuel to a burning flame, fuel viscosity must be low enough to travel effectively through the wick feeding a burning flame. The rate required to maintain adequate supply of fuel to the flame can be affected by fuel viscosity. If viscosity is high, the rate at which fuel is supplied to the flame will be diminished. Consequently, the height of the flame is lessened due to the lower amount of fuel feeding the flame. A secondary effect of using high-viscosity fuel in wick fed systems is a decrease in fuel consumption.

5.6 Low Viscosity

Low fuel viscosity can be due to the presence of low boiling point, low molecular weight compounds in fuel. Contamination with low boiling point compounds such as solvents, gasoline, and petroleum naphtha can dramatically reduce the viscosity of distillate fuel and residual fuel oil.

A reduction in both the initial boiling point and end point of a fuel can result in a lowering of fuel viscosity. Accompanying a reduction in these distillation parameters is a consequential increase in the concentration of lower molecular weight and often, more volatile compounds. The following fuel performance problems can be due to low fuel viscosity.

5.6.1 Poor Distribution of Fuel Injected into Combustion Chamber

When fuel is injected into the high-pressure environment of a combustion chamber, it must fully penetrate into the pressurized air in order to completely combust. Fuel with a low viscosity does not have enough mass to penetrate effectively through the pressurized air and disperse. The result is a fuel spray that does not spread throughout the chamber. Fuel combustion efficiency and lower engine power may result.

5.6.2 Wear of Fuel Injection System Components

Diesel fuel provides lubrication, or lubricity, and prevents surface wear of the high-pressure fuel pumping and injection equipment parts. If the viscosity of diesel fuel is low, its ability to form a hydrodynamic lubricating film between moving metal parts diminishes. If diesel fuel is low in viscosity due to kerosene dilution or solvent dilution, its lubricity is probably poor, and the possibility of wear of the high-pressure fuel pump parts may increase.

5.6.3 Hot Restart Problems in Engines

When fuel is moving through lines and engine parts which are hot, the fuel can "thin out" or behave as a less viscous fluid. This phenomenon can lead to startability and engine operability problems in worn equipment.

For example, diesel-powered vehicles operating under stop-and-go, high speed, and high load conditions, can experience problems with restarting after the engine is turned off. This is because fuel retained in the hot fuel injection pump can leak past the pump plunger and out of the barrel assembly. Restarting the engine may not be possible until the engine cools down so fuel can again be held and maintained in the barrel and plunger assembly of the fuel pump.

5.7 High Carbon Residue Values

ASTM testing methods such as ASTM D189 (Conradson Carbon), ASTM D524 (Ramsbottom Carbon), and ASTM D4530 (Micro Method) provide useful information on residual carbon remaining after fuel undergoes high-temperature thermal cracking and coking reactions. The results are especially useful in determining the tendency of fuel to form deposits in vaporizing pot-type and sleeve-type burners. The Ramsbottom carbon measurement on the 10 percent distillation residue for diesel fuel grades is an important property for fuels meeting ASTM D975 specifications.

5.7.1 Contamination with Residual Fuel

A high carbon residue value is an indication that fuel may have been contaminated with residual fuel or another heavy fuel fraction. Heavy streams such as VGO, coker gas oil, and No. 6 fuel oil can unintentionally contaminate fuel during transfer through hoses, pipelines, and loading into storage containers. These heavy products tend to form carbon residue when pyrolyzed and can be potentially identified as contaminants in lower boiling fuels through carbon residue testing.

5.7.2 Diesel Fuel Contains Alkyl Nitrate Cetane Improver

Diesel fuel containing an alkyl nitrate cetane improver such as amyl nitrate, hexyl nitrate, or octyl nitrate can produce higher carbon residue values than measured in untreated fuel. If Ramsbottom carbon residue values for diesel fuel exceed the ASTM D975 limits for No. 1-D and No. 2-D fuel, it is possible that the fuel contains an alky nitrate cetane improver.

5.7.3 Fouling in Marine Fuel Exhaust Systems

High carbon residue values for marine diesel fuel, marine gas oil, and marine residual fuels can contribute significantly to exhaust system deposit problems. Deposit formation on exhaust ports and exhaust turbines have been linked directly to high carbon residue in fuel.

CHAPTER 6

Solving Fuel Problems by Using Chemical Additives

A wide range of fuel performance problems can often be solved by the use of chemical additives. Although additives are no substitute for refining techniques such as hydroprocessing, alteration in crude oil distillation parameters, and fuel blend component selection, additives can frequently provide an alternative to further processing.

Some chemical additives such as corrosion inhibitors, wax crystal modifiers, and electrical conductivity improvers provide performance, which is difficult to duplicate through refining without affecting some other fuel property. Fuel additives such as antioxidants, lubricity improvers, metal deactivators, and cetane improver can also be used to solve fuel performance problems. Several of the most frequently used additives effective at solving fuel-related problems are described in this chapter.

6.1 Antioxidants

Over time, oxygen will react with fuel components to degrade the fuel into a dark, viscous liquid and sludge-like mass. Fuel components most susceptible to oxidation are olefins and alicyclic naphthenes. Paraffins and mono-aromatics are less susceptible to attack but may eventually be consumed by the free-radical chain reaction process. The free radical process of fuel oxidation and degradation is outlined in Table 6.1.

Hindered phenol and phenylenediamine (PDA) compounds are commonly used antioxidants and are quite effective at preventing free-radical initiated oxidative degradation of fuel. These antioxidants can be used in gasoline, kerosene, and certain distillates. In some applications, a synergistic effect can be obtained using a combination of hindered phenol and a PDA, or in blending antioxidant with a metal chelating compound.

Process Description	Reaction Steps
Energy initiated radical formation occurs.	R'H → R· + H·
Radical reaction with oxygen to form a peroxy radical.	R'· + O_2 → R'OO·
Peroxy radical abstracts hydrogen from a fuel compound to form an organic hydroperoxide plus a new fuel hydrocarbon radical.	R'OO· + RH → R'OOH + R·
New fuel radical reacts with oxygen to form a new peroxy radical and propagates the process.	R· + O_2 → ROO·
New peroxy radical reacts with additional fuel hydrocarbons to continue the propagation and fuel degradation process.	ROO· +RH → ROOH + R·
An alternate pathway is shown whereby an antioxidant or a radical-trap reacts with a peroxy radical to form a hydroperoxide and a more stable radical compound, thus halting the chain-reaction process.	ROO· + RSH → ROOH + RS·
The more stable radical compound can then react to form less active or inactive compounds.	2RS· → Stable or inactive compounds

TABLE 6.1 Steps in Fuel Oxidation Processes

Although hindered phenol and PDA antioxidants can increase the Rancimat lifetime of B100 FAME biodiesel fuel, alternate antioxidant compounds including more complex phenols and hydroquinones are used instead. These alternate compounds usually require significantly lower dose rates to increase Rancimat lifetime hours to specified values, thus reducing overall treatment cost.

6.1.1 Hindered Phenol Antioxidants

Hindered phenol antioxidants usually possess alkyl functional groups on ortho and para sites. The alkyl groups are typically *t*-butyl or methyl in functionality. The lower cost and good performance of most hindered phenol antioxidants make them attractive for use in fuel applications. Addition rates into fuel vary according to the base fuel oxidation lifetime, but rates are sometimes specified, as in ASTM and various military specifications for aviation gasoline and jet turbine fuel. As an example, Ministry of Defence Standard 91-91 and ASTM D1655 both state that the total active material in phenolic antioxidants used in aviation kerosene or jet turbine fuel, such as Jet A, Jet A-1, and Jet B shall not exceed 24.0 mg/L.

When contaminated or exposed to temperatures of 15.5°C (60°F) or below for an extended period of time, some hindered phenol

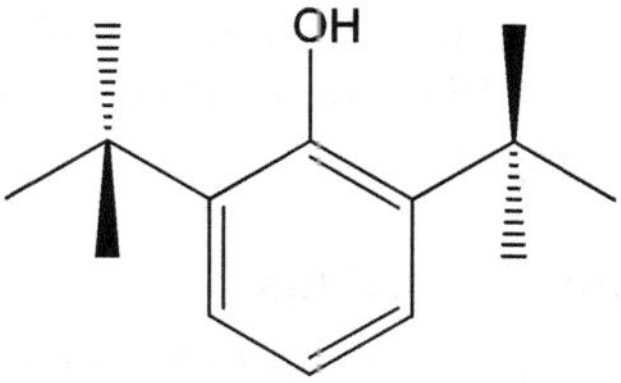

FIGURE 6.1 2,6-Ditertiarybutyl phenol.

antioxidants begin to form solid crystals. If the antioxidant is in a solvent carrier, the crystallization temperature is lowered, but as cooling continues, crystals form and can often fall from solution as a relatively pure compound. Although, high dilution rates of phenolic antioxidant in solvent can help inhibit crystallization, it may not completely eliminate this problem. The structure of a typical hindered phenol antioxidant is presented in Fig. 6.1.

6.1.2 Phenylenediamine Antioxidants

Phenylenediamine antioxidants can be used when hindered phenols do not perform or whenever their cost performance overcomes that of hindered phenol compounds. Although PDA compounds can have a higher cost than a hindered phenol antioxidant, the performance provided can frequently exceed that of a hindered phenol.

Phenylenediamine compounds are relatively weak organic bases and can cause burns if left on the skin for more than a few minutes. Their dark color may make them undesirable for use in fuels or fuel fractions which have a +15 or greater Saybolt color specification. Due to their alkalinity, PDA compounds may react with acidic compounds that may be present in fuel or storage tank bottom water. Reaction with acid may result in a reduction in performance of the PDA antioxidant.

Although both hindered phenol and PDA antioxidants are both used in commercial gasoline and petrol, PDA are not approved for use in aviation kerosene or jet turbine fuel (Fig. 6.2). However, PDA antioxidants are approved for use in aviation gasoline grades such as

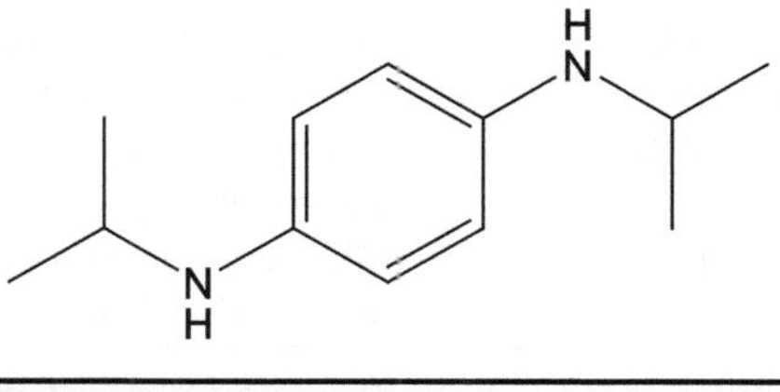

FIGURE 6.2 N,N′-Diisopropyl phenylenediamine.

Grade 80, Grade 91, Grade 100VLL, Grade 100LL, and Grade 100. The maximum rate of active PDA antioxidant component permitted in aviation gasoline is 12 mg/L.

6.1.3 Organosulfur Compounds

Some organosulfur compounds can function as fuel antioxidants by reacting to decompose peroxides. Organosulfides are believed to react with hydroperoxides to form sulfoxides and thiosulfoxides. These compounds decompose to sulfenic acid and further react with hydroperoxides to form other more acidic compounds. These newly formed acids continue the process of hydroperoxide decomposition. Thus, organosulfur compounds function in the process of oxidation inhibition through hydroperoxide decomposition. However, in most fuel applications, sulfur-containing antioxidants are not utilized due to restrictions on total fuel sulfur content.

6.1.4 Characteristics of Hindered Phenol and Phenylenediamine Antioxidants

General characteristics of hindered phenol and PDA antioxidants are compared and listed in Table 6.2.

Antioxidant use in fuel is most effective in retarding the formation of deposits and gums, improving the resistance of fuel to oxidation, and minimizing color change in fuel during transportation and storage.

Antioxidants are less effective or not effective at all in reversing the effects of fuel degradation and color change, reducing the existent gum level of fuel, and significantly improving jet fuel JFTOT ratings.

Hindered Phenol	Phenylenediamine
Effective in most finished fuels and refined fuel fractions at 5 to 50 ppm	Effective in most finished fuels and refined fuel fractions at 1 to 20 ppm
Will not significantly degrade fuel color	May degrade color of fuel
Lower cost per kilogram	Higher cost per kilogram
Can crystallize during storage, especially at temperatures below 15.6°C (60°F); dilutions are available to help lower crystallization temperatures	Can crystallize during storage; crystallization temperature can be controlled by alkyl group functionality; products are available with a crystallization range between 15.5°C (60°F) and −40°C (−40°F)

Table 6.2 Characteristics of Hindered Phenol and Phenylenediamine Antioxidants

6.2 Distillate Fuel Stabilizers

Widespread use of hydroprocessing to reduce fuel sulfur and to saturate reactive carbon-carbon double bonds has yielded fuel that is more stable and less prone to degradation in storage than distillate fuel produced a generation ago. The implementation of hydrocracking has also produced fuels that are more oxidatively stable. Although some high-sulfur heating oils and burner fuels are still produced, the percentage of these high-sulfur fuels in the marketplace will continue to diminish as sulfur limits and environmental emission controls take effect.

These clean fuels no longer require the volumes of stabilizing compounds that earlier fuel blends once needed. However, the production, transportation and storage of various unhydroprocessed refined components such as FCC cycle oils, coker distillates, heavy cracked naphtha, and vacuum gas oils are still ongoing. Present in these unhydroprocessed refined components are unstable olefinic compounds, naphthenic acids, sulfur- and nitrogen-containing heterocycles, and naphthenoaromatic compounds. These components can react by participating in condensation reactions to produce high molecular weight, deposit-forming compounds. Reaction with oxygen to produce organic hydroperoxides can yield insoluble gums and sludge in fuel.

These fractions often require treatment with additives to prevent fuel degradation, deposit formation, and color change during transportation and storage prior to use. A single additive with one function or multiple additives with different functions are sometimes required to help prevent degradation of these refined components. The types of additives used include an oxidation inhibitor, dispersant, acid neutralizing compound, metal chelating agent, or biocide. Some stabilizing additives are formulated to possess two or more of these properties on a single molecule. These component properties are described in the paragraph sections to follow.

6.2.1 Antioxidant and Acid Neutralization Control

The antioxidant and acid neutralization properties of most fuel stabilizer formulations help to prevent the initial formation of most organic acids as fuel components age in an atmospheric environment. Phenolic-, amine-, and hydroquinone-based antioxidant compounds act to prevent the initial formation of organic acids and aldehydes that develop as fuels react with oxygen and begin to degrade. Also, compounds such as tertiary alkyl primary amines, as shown in Fig. 6.3, are quite effective at neutralizing organic acids that form in fuel as a result of oxidation. Together, the processes of oxidation control and acid neutralization help to limit the development of high molecular weight deposits and acid catalyzed polymerization reactions in distillate fuel.

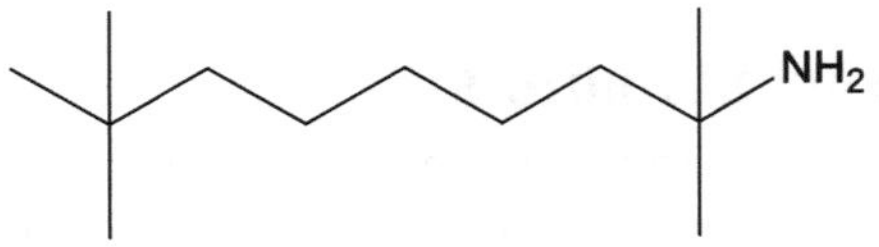

Figure 6.3 1,1,7,7-Tetramethyl octylamine.

6.2.2 Peroxide Inhibition in Diesel Fuel

Some distillate fuel stabilizers minimize the formation of peroxides in hydroprocessed ultralow-sulfur diesel fuel (ULSD). Figure 6.4 presents the results of an aging study whereby 500 mL of hydroprocessed diesel fuel stored in an open-top amber bottle at 65°C was evaluated for peroxide formation. Although significant levels of peroxides formed in the diesel, addition of 40 ppm of stabilizing additive was effective at minimizing the formation of peroxides.

6.2.3 Paraffin Degradation and Metal Chelation

Mechanical components used in fuel systems such as pumps, valves, and bearings may contain copper or copper alloys such as brass and bronze. As a fuel system component, copper is especially undesirable because it acts as a catalyst in promoting the oxidation of fuel paraffins to form oxygen-rich, gum-like deposits. The following reaction

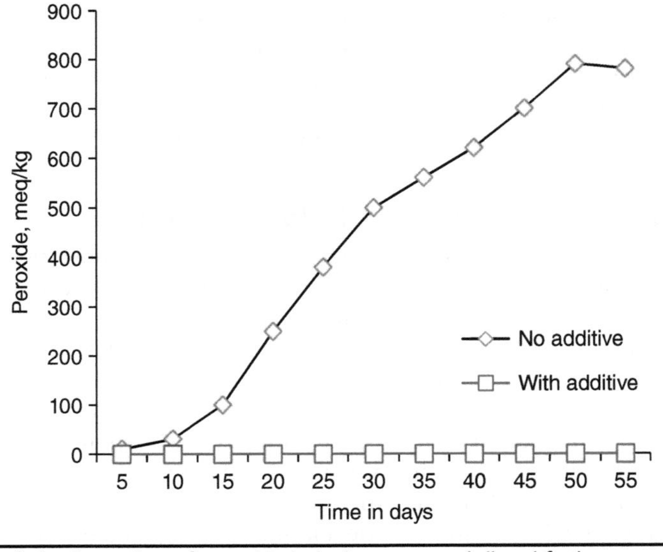

Figure 6.4 Formation of peroxide in hydroprocessed diesel fuel.

sequence in Eq. (6.1) represents how copper ions can catalyze the oxidation and degradation of hydrocarbons.

$$Cu^{+2} + R\text{-}H \longrightarrow Cu^{+} + H^{+} + R\cdot \xrightarrow{O_2} Cu^{+2} + ROOH$$

Step 1. Hydroperoxide Formation

$$Cu^{+2} + ROOH \longrightarrow Cu^{+} + H^{+}\ ROO\cdot$$

$$Cu^{+} + ROOH \longrightarrow Cu^{+2} + OH^{-} + RO\cdot$$

Step 2. Hydroperoxide Degradation (6.1)

The structural design of a fuel stabilizer molecule can enable some compounds to function in metal ion chelation. Most stabilizers have nitrogen atoms within the structure of the compound. The electron pair on nitrogen can function in metal chelation if the nitrogen atoms are oriented in the appropriate conformation.

Also, separate compounds known to perform as effective metal chelators can be blended into stabilizer formulations. Compounds such as N,N′-disalicylidene-1,2-propanediamine, as shown in Fig. 6.5, and N,N′-disalicylidene-1,2-cyclohexanediamine are commercially available and quite effective. Care must be taken when using these compounds in cold environments because they can crystallize and precipitate as relatively pure compounds. At temperatures of 20°C (68°F) and below, some metal deactivator formulations can form crystals in storage. These crystals dissolve and go back into solution upon warming to about 35°C (95°F).

6.2.4 Dispersant Properties of Fuel Stabilizers

The chemical composition and structure of some distillate fuel stabilizers act together to give the compounds dispersant-like properties. By acting as a dispersant, fuel degradation products that may develop will be suspended within the fuel by the stabilizer. As a result, degradation products do not form deposits or accumulate within fuel-storage and fuel-handling system components such as storage tanks, transfer pumps, and pipelines. The degradation products remain dispersed in the fuel. In some fuels, the dispersing action of the stabilizer may darken the fuel color.

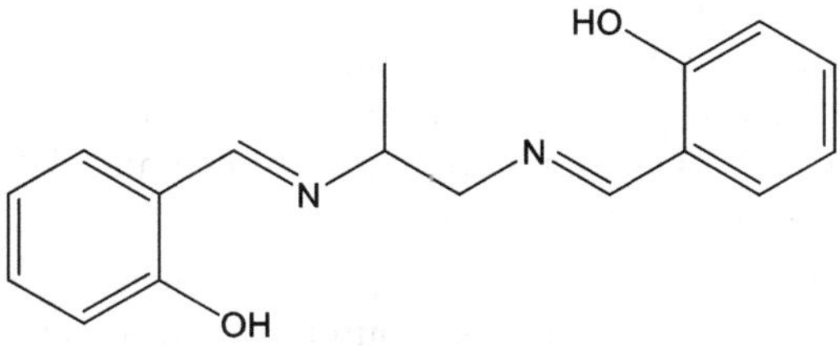

Figure 6.5 N,N′-disalicylidene-1,2-propanediamine.

Corrosion inhibition can be a secondary effect of stabilizers having dispersant properties. Acting as a dispersant, water and water-soluble compounds are held in suspension within the fuel by the stabilizer. This action prevents water from settling onto metal surfaces to initiate corrosion.

6.2.5 Stabilizers and Antioxidants for Specialized Applications

6.2.5.1 Federal Diesel Fuel Oil Stabilizer

A two part diesel fuel oil stabilizer blend containing both stabilizing and biocide components is specified under MIL-S-53021 and listed in QPL-53021-10. The stabilizer blend is identified as SA-15 to be used at a recommended addition rate into fuel of 1:5000. The product is not intended for use in routine applications but for situations where increased stability protection is required. Typical applications include fueled equipment undergoing long-term storage in a warehouse or depot, prepositioned equipment, or equipment maintained in a high-temperature environment.

6.2.5.2 Naval Distillate Fuel

Hydroprocessed synthesized paraffinic diesel (SPD) components, such as those obtained from Fischer-Tropsch synthesis or hydrogenation of fatty acid esters can be used in the production of naval distillate fuel at a maximum rate of 50 volume percent as described under MIL-DTL-16884. These components must be treated with one of the military approved phenolic antioxidants as soon as possible after refining and before storage and blending to prevent peroxide and gum formation in the finished naval distillate fuel blend. The antioxidant rate must not be less than 17.2 mg/L or more than 24.0 mg/L of the active antioxidant component.

6.2.6 Stabilizer Performance Properties and Limits

6.2.6.1 Problems Typically Solved by a Distillate Stabilizer

- Control of color darkening and deposit accumulation during the long-term storage of distillate fuel and distillate fuel blend components.
- Control of color darkening and sludge formation when cycle oil, cracked gas oil, or other unstable fractions are blended into fuel. These fractions typically contain olefins and aromatics in high percentages.
- Neutralization of acidic compounds which may be present in fuel. Acidic organic compounds can react with other fuel components to initiate the formation of higher molecular weight, fuel insoluble deposits.

- Prevent copper catalyzed oxidation of fuel components. Copper ions can act as catalysts to initiate the rapid oxidation and degradation of fuels and oils. The result is the formation of a darkened, viscous, fuel residue.

6.2.6.2 Problems Typically Not Solved by a Distillate Stabilizer

- Reversal of fuel color from a dark ASTM color to a lighter ASTM color.
- Perform as an effective substitute for a hindered phenol or PDA antioxidant in inhibiting free radical reactions in gasoline or jet fuel.

6.3 Demulsifiers and Dehazers

6.3.1 Water and Contaminants

Water is the primary cause of haziness or cloudiness in fuel. Wax is also a cause of cloudiness in fuel, but wax related haze is usually a cold temperature phenomenon rather than a result of contamination. Water can contaminate fuel through several different routes. Examples include:

- Refinery process water and wash water.
- Condensation of water in fuel storage tanks vented to the atmosphere.
- Barge ballast water carryover.
- Storage tank water bottom carryover of various pH ranges.

Under normal circumstances, refined fuels do not form emulsions with water. The fuel and water readily separate into two distinct phases, a lower water phase and an upper fuel or oil phase. However, when emulsifying agents mix with fuel and water, emulsification can occur.

The energy needed to initiate the formation of a fuel-water emulsion can come from sources such as tank mixers or high-speed centrifugal pumps used to blend and transfer fuels from tanks and through pipelines. Turbulence created during loading and unloading fuel into a storage tank or barge hold containing water may also provide the energy to initiate emulsification.

Examples of common emulsifying agents include any of the following:

6.3.1.1 Microorganisms, Rust, and Dirt

Tanks used to store fuel are usually emptied and filled on a continuous basis. Although most fuel refiners and marketers maintain a routine tank maintenance schedule, water and debris still accumulate.

Products of microbial metabolism, rust, the salts of calcium, sodium, magnesium, and iron as well as silicon in the form of dirt can all be found in tank bottom water. These materials help to stabilize a fuel-water emulsion.

6.3.1.2 Naphthenic Acids and Other Polar Compounds

Distillate fuel fractions can contain naphthenic acids, sulfonic acids, and other hydrophilic compounds. If these hydrophilic compounds are present as sodium salts due to caustic washing of fuel, they become powerful emulsifying agents. Also, heavy resinous compounds in fuel derived from crude oil refining can also act to stabilize existing emulsions.

6.3.1.3 Refining and Process Additives

Certain corrosion inhibitors, antifoulants, and other polar organic compounds used in the refining and processing of fuels may carry over into a finished product. If present at significant levels, these compounds can interact with water to form an emulsion. These emulsions are often quite difficult to break. The result is a cloudy, hazy fuel.

6.3.1.4 Fuel Additives

When used at recommended rates in fuel that is clear and bright and free of water, most fuel additives do not enhance emulsion formation. At excessive treatment levels or in contact with water, especially alkaline pH water, highly polar fuel additives may form emulsions with water. These emulsions typically have the appearance of a dense, paste-like layer at the fuel-water interface. Other types of emulsions remain suspended leading to cloudiness in fuel, in the water, or in both fuel and water phases.

6.3.2 Understanding the Terms Demulsifier and Dehazer

The terms *demulsifier* and *dehazer* are often used interchangeably to describe compounds, which break fuel emulsions or remove water-initiated haze from fuel. The terms can be distinguished as follows.

6.3.2.1 Demulsification

The process of demulsification usually involves removal or dropout of relatively large volumes of water that have contaminated fuel. The amount of water in this circumstance is usually greater than 1 percent of the total fuel volume. Also, when an emulsion has been created in a fuel, breaking this emulsion is termed *demulsification*. Sometimes, the term emulsion breaking is also used to describe the process of demulsification.

6.3.2.2 Dehazing

Dehazing refers to removal of water that has been dispersed throughout a fuel matrix. Water volume of fuel is usually less than 1000 ppm.

Haze created by water droplets can be removed by coalescence or by further dispersion through the action of a chemical dehazer.

6.3.3 Demulsification and Dehazing Mechanisms

Two basic mechanisms have been proposed to explain the performance of demulsifiers and dehazers in finished fuels. These two mechanisms are defined as coalescence and adsorption. Both processes rely on the fact that the demulsifier contains both hydrophobic and hydrophilic sites. The mechanisms can be defined as follows.

6.3.3.1 Coalescence

Given time, water that exists in discrete droplets in finished fuel may coalesce into larger drops and settle by gravity from the fuel. Demulsifiers or dehazers can accelerate this process by functioning as a site for attraction of dispersed water.

Water is first attracted to the hydrophilic sites of the demulsifier through the process of hydrogen bonding. Eventually, enough water accumulates and coalesces into larger drops. The larger drops then separate from the hydrophilic sites and fall-out of the fuel to the bottom of the storage tank or blending vessel.

6.3.3.2 Adsorption

If water is emulsified into fuel as a water-in-oil emulsion, coalescence cannot affect the removal of water from the fuel. The outer oil or surfactant layer surrounding water will not permit water to hydrogen bond to the hydrophilic demulsifier sites.

To break through this oil or surfactant layer and free the retained water, it is necessary to chemically disrupt the stability of this layer. Demulsifiers and dehazers can adsorb onto the protective film and subsequently interfere with the electrochemical forces, which hold this outer layer together. Upon adsorption, demulsifier/dehazer compounds weaken and break the oil or surfactant layer thus releasing the contained water. Once free, the water can then coalesce into larger drops and separate from the fuel.

Dispersion is sometimes considered a dehazing mechanism that involves neither coalescence nor adsorption. Dispersion involves the use of compounds that disrupt hydrogen bonding which exists among visible water drops. This disruption produces numerous, microscopic water droplets that are no longer visible to the eye. The dispersed water, however, is suspended and remains in the fuel.

6.3.4 Preventing Emulsification of Fuel Detergents

Many gasoline and diesel fuel formulations contain deposit control additives or detergents to keep fuel injection, intake, and combustion systems clean. If fuel containing detergent contacts water in fuel transportation and storage systems, emulsification of water with the fuel detergent may occur. The addition of demulsifiers to fuel detergent formulations can prevent these emulsions from forming.

The mechanism of preventing detergent-water emulsion formation is controlled by the ability of a demulsifier to interfere with the formation of a stable, oil-like detergent film around water droplets. The demulsifier is believed to associate with the detergent to disrupt the oil film as the detergent attempts to emulsify with water. By preventing an oil film from forming, surrounding, and encapsulating water, detergent-enhanced emulsions can be inhibited.

6.3.5 Chemistry of Demulsifiers and Dehazers

Quaternary ammonium salts and salts of naphthalene sulfonic acid were some of the first compounds to be used effectively as fuel demulsifiers and dehazers. Today, a wide range of monomeric and polymeric demulsifiers and dehazers exist. These compounds are typically produced by reaction of an acceptor molecule with various adducts. Acceptors such as polyols and various copolymers are commonly used. Epoxides including ethylene oxide, propylene oxide, and butylene oxide are widely used adducts.

Sorbitol and glycerine are common acceptor monomers for oxide addition. Various copolymers possessing functional groups that will accept and react with epoxide adducts are also effectively utilized as demulsifiers for fuels and oils. The composition and extent of oxide adduct addition onto the acceptor molecule will have a significant impact on the performance and solubility of the fuel demulsifier or dehazer produced.

As a general rule, the addition of ethylene oxide to a demulsifier acceptor backbone will tend to increase the water solubility of the compound. The addition of propylene oxide or butylene oxide to the acceptor will tend to increase the hydrocarbon solubility of the molecule. Often, a demulsifier or dehazer can be made to perform selectively in oil-water systems by adding both ethylene and propylene oxide to the same molecule. The performance and solubility of the final oxyalkylated compound can then be finely tuned by closely controlling the amount and order of epoxide addition.

Other factors influencing demulsifier selection include the affinity for oil external or water external emulsion, solubility in oil and surfactants, and level of affinity for water. An example of a dehazer compound is presented in Fig. 6.6.

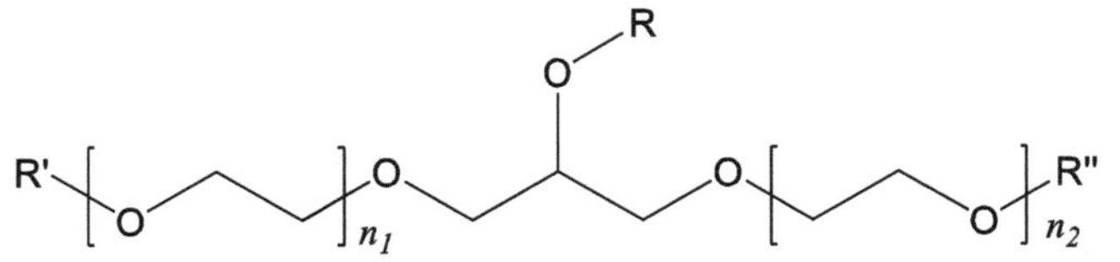

FIGURE 6.6 An ethoxylated polyol.

6.4 Fuel Biocides

The accumulation of water in the bottom of fuel-storage and fuel-handling systems is a common problem within the fuel industry. Condensate water and process water in combination with hydrocarbon fuel can serve as an acceptable medium to support the growth of various microorganisms. Anaerobic species such as *Clostridia, Desulfovibrio,* and *Desulfomonas* can survive in the oxygen-free environment existing at the fuel-water interface of the storage tank. These microbes utilize the oxygen bound up in compounds for their metabolic oxygen needs.

Some bacteria specifically utilize oxygen bound in the sulfate complex of a compound. As a result of this metabolic activity, sulfur is reduced to H_2S. For this reason, these microbes are termed *sulfate reducing bacteria* or SRB. Some SRB species can tolerate temperatures as high as 80°C (176°F) and pH environments ranging from pH 5 to pH 9. *Desulfovibrio* and *Desulfomonas* are examples of SRB.

Many of these anaerobic SRB species possess the ability to utilize hydrocarbon fuel compounds as a nutrition source. These species are termed hydrocarbon utilizing microbes or HUM. Fuel-containing linear paraffins, especially distillate fuel, can be metabolized by these microbes through the β-oxidation process. By-products of anaerobic microbial metabolism can include H_2S, H_2, CH_4 and low molecular weight organic acids. Also, the formation of FeS and the accumulation of microbial decomposition products can complicate the contamination problem.

Colonies established by microbes on the sidewalls and bottom of fuel storage tank systems are called plaques. These plaques can often be the site of microbiologically influenced corrosion, or MIC, of the underlying metal.

A variety of biocides are available for treating water contaminated with microorganisms. However, only those products which are regulated by agencies in various countries such as the Environmental Protection Agency (EPA) in the United States, the Biocide Products Directive in the European Union, and the Canadian Environmental Protection Agency in Canada can be applied in those regions. Regulated biocides are frequently approved for specific applications. For example, a biocide approved for use in treating tank bottom water may not be approved for use in the hydrocarbon fuel contained in the tank. Also, a biocide approved for use in diesel fuel may not be approved for use in jet fuel. Before treating a fuel system with any biocide, the registration limits for the biocide must be confirmed and followed.

6.4.1 Properties of Water Soluble and Fuel Soluble Biocides

Both water soluble and fuel soluble biocides are available for fuel treatment. The advantages and disadvantages of the two treatment methods are provided in Table 6.3.

Water Soluble Biocide		Fuel Soluble Biocide	
Advantages	**Disadvantages**	**Advantages**	**Disadvantages**
Lower treat rates and treat costs due to a low water volume in the tank	Possible concern with disposal of biocide treated water	Long lifetime in fuel; biocide will combust with the fuel	Higher volumes of biocide are needed due to a larger volume of fuel compared to water
Typically has no negative impact on fuel performance properties	Short-term effectiveness due to biodegradation of biocide	Contact with plaques which may be present on sidewalls of fuel tank and tank bottom	Plaques containing killed microorganisms may break free from tank, travel downline and plug fuel filters
Rapid kill at the fuel/water interface	Frequent treatment may be necessary	Usually effective with one single treatment	Higher treat rates may have negative impact on certain fuel performance properties

TABLE 6.3 Advantages and Disadvantages of Water Soluble and Fuel Soluble Biocides

6.4.2 Chemistry and Activity of Fuel Biocides

Some compounds which are approved for various fuel tank biocide applications are provided in Table 6.4.

6.5 Wax Crystal Modifiers

Linear, branched, and cyclic paraffins can all be present in fuel. At low temperatures, fuel filtration, pumpability, and injection problems are often caused by paraffinic wax. Refiners can minimize fuel wax by implementing processing changes and blending with low wax components. Examples include:

- Reducing the T-90 distillation temperature and end point.
- Use of lower percentages of a straight run fraction or waxy distillates in a blend.
- Utilizing a lighter fuel component stream such as kerosene or jet fuel in a blend.
- Lowering the IBP and T-95 temperatures.

All of these refining techniques can be used to some degree of effectiveness at minimizing the problems associated with the low temperature handling of fuel containing waxy paraffins.

Biocide	Bacterial Activity	Typical Dosage*	Stability
Dibrominated propionamide	Typically effective in 15 minutes to 2 hours against all bacteria	100 ppm of a 20% solution used weekly into water	Degrades readily in high temperature, high pH conditions
Glutaraldehyde	Typically effective in 3 to 8 hours against all bacteria	400 ppm of a 50% solution used weekly	Stable under temperature and pH fluctuations; degraded by H_2S; incompatible with polyacrylamide
Hexahydro-1, 3, 5-triethyl-S-triazine	Typically effective in 3 to 8 hours against all bacteria	100 ppm of a 95% solution used weekly into water	At pH values <6.5, activity increases but lifetime diminishes
Isothiazolone	Typically effective in 6 to 38 hours against all bacteria	300 ppm of a 1% to 2% solution for a single application into fuel	Degrades slowly under temperature and pH fluctuations
Methylene bis-methyloxazolidine	Typically effective in 15 minutes to 2 hours; effective against bacteria (including SRB), yeast, and mold	200 to 500 ppm for a single application into fuel volume	Protect from heat and sunlight; slow degradation in fuel
Sodium dimethyl dithiocarbamate/ disodium ethylene bis-dithiocarbamate	Moderate effectiveness in 3 to 8 hours against most bacteria	200 ppm of a 15% solution used weekly into water	Degrades under temperature and pH fluctuations; degraded by H_2S; incompatible polyacrylamide

**Note:* Typical dosage is based on the water volume in the fuel storage tank. Although these recommended dosages are based on a weekly or one time treat rate, lower dosages can be added at a more frequent rate.

Table 6.4 Characteristics of Common Biocides

In addition to refining techniques, wax crystal modifier compounds are available for use in fuels. Wax crystal modifiers, also called pour point depressants or cold flow improvers, are typically polymeric compounds which have the ability to crystalize with fuel wax as it forms. By co-crystallizing with wax, the modifiers typically affect a change in the size, shape, and conformation of wax crystals. Wax crystal modifiers function by dispersing wax or inhibiting the nucleation or growth of wax crystals within a fuel or oil.

6.5.1 Chemistry and Performance of Wax Modifiers

Copolymers of ethylene vinylacetate are the most commonly utilized fuel wax crystal modifiers. Other compounds such as vinyl acetate-fumarate copolymers, styrene-ester copolymers, diester-alphaolefin copolymers are also effective wax crystal modifiers. These compounds differ in both chemical structure, molecular weight, and in the extent of performance provided.

The different chemical compounds used as wax crystal modifiers do not all provide ideal performance under every circumstance. Various tests have been designed to help differentiate the performance of one wax modifier over another. For example, a modifier may be quite effective at controlling wax crystal formation to enable a fuel to flow by gravity from a storage tank to a pump. However, once past the pump, the modifier may not effectively reduce the wax crystal size and shape to allow cold fuel to flow through a fuel filter. The result is wax accumulation on the filter media, plugging of the fuel filter, and halting of fuel flow. A different wax crystal modifier or product with wax dispersant properties may be required to permit effective fuel filtration. See Figs. 6.7 to 6.9 for examples of wax modifier performance in different refined fractions.

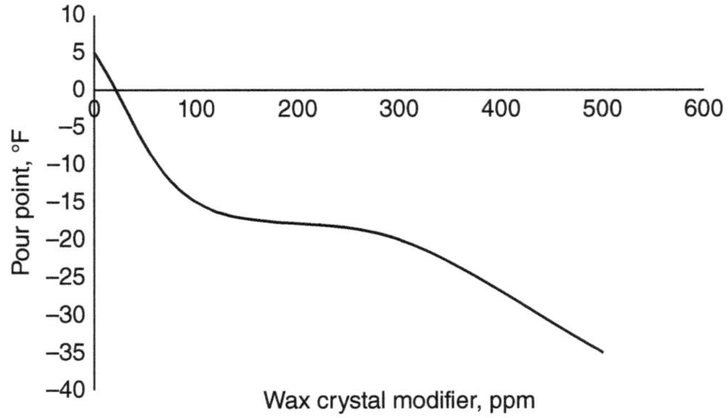

FIGURE 6.7 Pour point response of an EVA-based wax modifier in a hydroprocessed distillate fuel.

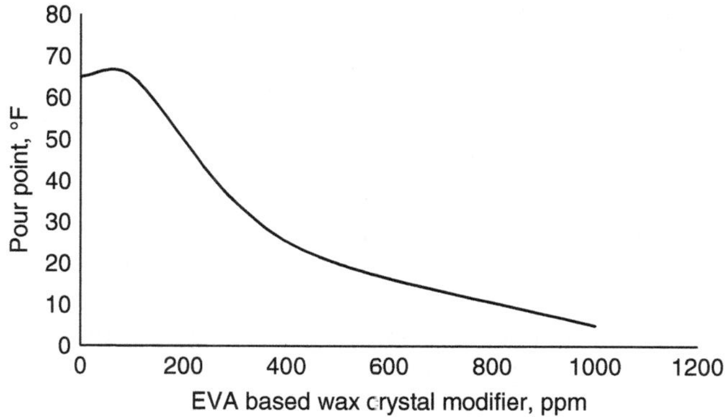

FIGURE 6.8 Pour point response of an EVA-based wax modifier in a heavy atmospheric gas oil (HAGO).

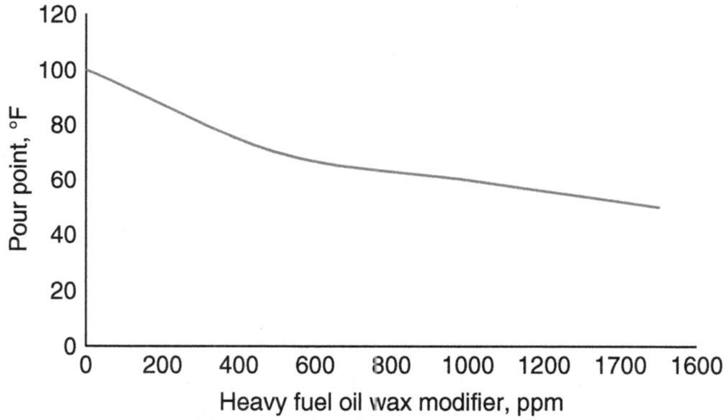

FIGURE 6.9 Pour point response of a heavy copolymer-based wax modifier in a No. 6 fuel oil.

Fuel problems that can typically be solved by a wax crystal modifier are described below:

- Distillate fuel pour point temperatures can often be reduced by 25 to 33°C (45 to 60°F) from the base pour point of the fuel using most wax crystal modifiers. Pour point response is dependent on the additive treat rate.
- Distillate fuel cold filter plugging point (CFPP) temperatures can be reduced by 5 to 10°C (10 to 18°F) from the base fuel CFPP filtration temperature. Greater CFPP response is possible, but the additive treating rate can be quite high. Also, fuels with low initial filtration temperatures often respond more effectively to wax crystal modifier treatment than high filtration temperature fuels.

- The low temperature pumpability of a fuel containing wax crystal modifier additive can be reduced below the pumping temperature of a fuel that contains no additive. Wax crystal modifiers help minimize pumpability problems related to high-yield stress and viscosity.
- The pour point and low temperature viscosity of a residual fuel or heavy fuel oil can typically be reduced by using a wax crystal modifier formulated to interact with higher molecular weight waxes. A phenomenon termed pour point reversion in heavy fuel oils can often be prevented by using the correct type of wax crystal modifier.

Fuel performance problems that typically cannot be solved by a wax crystal modifier include:

- If wax has already formed in fuel, addition of a wax modifier will typically have little to no effect in further reducing the low temperature flow of fuel. Wax in the form of a gel-like or semi-solid mass cannot be effectively modified with additive unless the fuel is warmed to temperatures above the fuel cloud point. Once wax has melted or liquefied, a wax modifier can be added, and the additive will function to inhibit wax-related problems such as deposition onto interior surfaces of cold fuel lines and plugging of fuel filters.
- Wax modifier addition has little effect in reducing the high-temperature viscosity of fuels. At temperatures 25°C (45°F) above the cloud point, wax in fuel is no longer organized into a lattice-like network or crystalline form. Above the cloud point, fuel viscosity is influenced primarily by the chemical composition and concentration of all fuel components.

6.6 Corrosion Inhibitors

In fuel systems, iron and steel corrode when both water and air enter the system. Water can contaminate fuel through refinery processes such as water washing, caustic treating, salt drying, or condensation of water vapor in fuel storage tanks. Entrainment of air from the atmosphere or during pumping, mixing, or blending operations introduces oxygen needed for corrosion to progress.

An extensive amount of information has been written about ferrous and nonferrous metal corrosion and corrosion mechanisms. Topics applicable to corrosion in fuel systems are reviewed in this section.

6.6.1 Corrosion Chemistry

Ferrous metal corrosion can occur quite readily whenever a difference in electrical potential exists between iron and a corroding agent.

Corrosion begins whenever electrons leave iron at the anode, travel to the cathode, and ferrous ions form as shown in Eq. (6.2).

$$Fe \rightarrow Fe^{+2} + 2e^- \tag{6.2}$$

Reaction of water with ferrous ions results in the formation of a gel-like ferrous hydroxide compound and hydrogen ions. Under acidic conditions, the hydrogen ions react with two electrons liberated from the reaction to produce hydrogen gas. This reaction process is presented in Eqs. (6.3) and (6.4).

$$Fe^{+2} + 2H_2O \rightarrow Fe(OH)_2 + 2H^+ \tag{6.3}$$

$$2H^+ + 2e^- \rightarrow H_2 \tag{6.4}$$

Ferrous hydroxide can continue to react with water and oxygen to form colloidal ferric hydroxide as presented in Eq. (6.5).

$$4Fe(OH)_2 + O_2 + 2H_2O \rightarrow 4Fe(OH)_3 \tag{6.5}$$

Also, as shown in Eq. (6.6), ferrous hydroxide can further react with oxygen from the air to form ferric oxide (rust) and water.

$$4Fe(OH)_2 + O_2 \rightarrow 2Fe_2O_3 + 4H_2O \tag{6.6}$$

In the formation of ferric oxide, only iron and oxygen are consumed. Water is conserved. The overall process for this reaction is presented in Eq. (6.7).

$$4Fe + 3O_2 + 8H_2O \rightarrow 2Fe_2O_3 + 8H_2O \tag{6.7}$$

A graphical representation of a ferrous metal corrosion cell is presented in Fig. 6.10.

A rusted anode appears as a raised orange to brown upper layer occupying more volume than the original iron surface. This expansion loosens the rusted layer to expose even more surface to corrosion.

6.6.2 Basic Forms of Corrosion

6.6.2.1 Uniform Corrosion

Metals develop a natural corrosion resistant film when exposed to the environment. Examples include the rusting of iron, tarnishing of silver, and the formation of the patina on copper. These passive films help prevent further corrosion. However, films to not provide complete resistance to chemical attack and are destroyed by various corrosive agents.

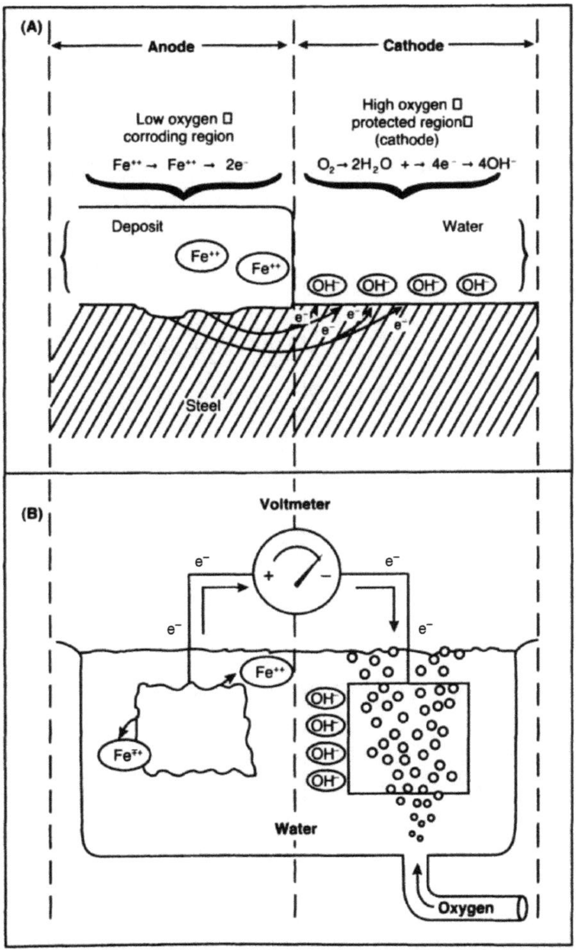

Figure 6.10 Ferrous metal corrosion. (A) Oxygen concentration cell corrosion beneath a deposit; (B) Oxygen concentration cell corrosion in a beaker containing an aerated piece of steel and an unaerated piece of steel. A potential develops between the aerated and unaerated steel pieces. Steel exposed to the lower dissolved oxygen concentration corrodes. Beneath a deposit (A) the oxygen poor environment causes metal loss.

Source: Herro H.M. and R.D. Port, 1993, The Nalco Guide to Cooling Water System Failure Analysis, McGraw-Hill, Inc.

In a corrosive medium, metals and metal alloys form reaction products that are stable and similar to those found in nature. Loss of metal in this manner is uniform and can be measured. Standard corrosion tables contain this type of information. In general, three broad classifications of uniform corrosion based on metal loss are utilized and classified in Table 6.5.

Classification	Metal Loss Rating, Mils Per Year
Good corrosion resistance	<5
Satisfactory if a high corrosion rate can be tolerated	5 to 50
Usually unsatisfactory	>50

Table 6.5 Corrosion Resistance Ratings

6.6.2.2 Galvanic Corrosion

Galvanic corrosion occurs whenever two different metals are contained in a corrosive or electrically conductive medium. It can occur when two similar metals are interconnected but are in contact with two different electrically conductive mediums.

Under either of these circumstances, the more active metal serves as the anode and the less active metal as the cathode. Metal loss from a small anode can be extremely high if connected to a cathode with a larger area. The closer together metals are in the galvanic series, the less the corrosion rate. The further apart, the greater the corrosion rate. A listing of the galvanic series of metals and metal alloys is provided in App. 4.

6.6.2.3 Crevice Corrosion

When metal parts make contact or when metals contact a nonmetallic part such as a gasket, the crevice formed at the point of contact can become a corrosion site. Corrosive attack at this site is usually more severe that the surrounding area due to factors such as oxygen deficiency, acidic changes, or inhibitor depletion. Using nonabsorbent gasket material and avoiding threaded joints can help limit crevice corrosion.

6.6.2.4 Stress Cracking Corrosion

When sensitive metals are exposed to certain environmental factors or tensile stress, corrosion can occur. Exposure to only a few parts per million of a corrosive agent can initiate stress in some metals. High concentrations are often not necessary. Also, temperature and pH can be influencing factors.

Metal components under compression do not usually undergo stress corrosion cracking. Welding, bending, denting, and installation initiate cracking. An applied tensile stress approaching the yield strength of the metal is needed to initiate failure. The addition of nickel as an alloying element helps to prevent stress corrosion cracking of many metals. See Fig. 6.11 for a graphical depiction of stress corrosion cracking.

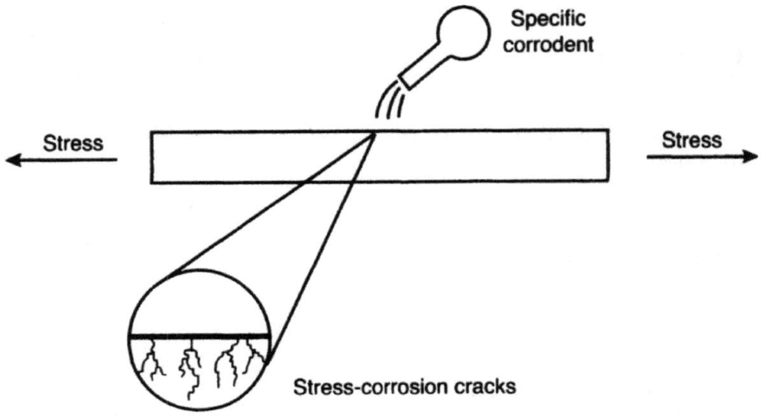

Figure 6.11 Stress corrosion cracking.

6.6.2.5 Pitting

Corrosion described as pitting is usually localized and can be in the form of small, deep pits as well as large, shallow pits. This type of corrosion can occur in areas of stagnation where an anode site can develop.

6.6.3 Hydrogen Sulfide Corrosion

Although less common than water initiated corrosion in fuel systems, ferrous metal corrosion initiated by hydrogen sulfide (H_2S) can occur. In the presence of water, H_2S can react with iron to form iron sulfide as shown in Eq. (6.8):

$$Fe + H_2S + H_2O \rightarrow FeS + 2H^+ + H_2O \tag{6.8}$$

6.6.4 Corrosion Inhibitor Function

Water is the most common cause of internal corrosion of ferrous metal pipelines and storage tanks used in the transportation and storage of fuel. Sulfur and sulfur-containing compounds are the most common causes of corrosion of copper, copper alloys, and silver components found in fuel systems.

Most fuel system corrosion inhibitors function by strongly adsorbing onto the exposed surfaces of metal to form a protective film-like layer. This inhibitor layer acts as a barrier to prevent water from contacting the surface of metal components and initiating corrosion.

Filming corrosion inhibitors are characterized by possessing a polar, electron dense "head" that has an affinity for a metal surface, and a nonpolar, oil soluble "tail" that enables the corrosion inhibitor

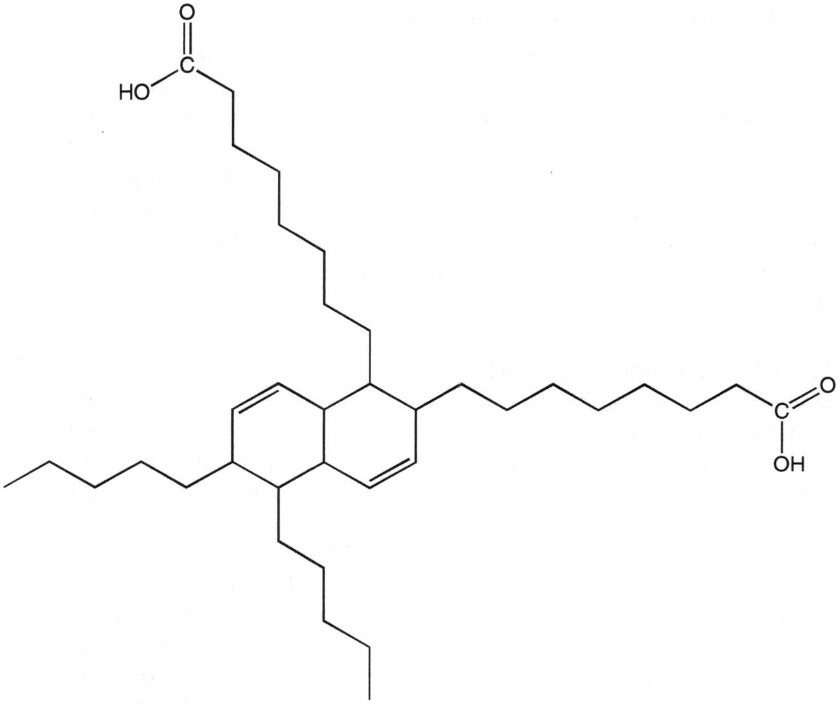

Figure 6.12 Dimer fatty acid corrosion inhibitor with polar head and non-polar tail.

to become soluble in the fuel. In Fig. 6.12, the structure of a dimer fatty acid–based corrosion inhibitor is displayed. The two carboxylic acid functional groups for the dimer fatty acid represent the metal filming "head," and the remaining alkyl groups represent the fuel soluble, nonpolar "tail." In addition to dimer fatty acids, examples of other filming corrosion inhibitors include synthetic diacids, alkyl sulfonates, and alkyl imides. These materials are fuel soluble, nonvolatile, and quite effective at preventing water initiated corrosion of ferrous metal surfaces.

At elevated temperatures of 230 to 260°C (446 to 500°F), for example, film forming inhibitors can lose their effectiveness. However, during transportation and storage, finished fuels are not exposed to these high temperatures, and the corrosion inhibitor film should remain until the fuel is combusted.

Fuel system corrosion inhibitors must have a low tendency to emulsify in the presence of water and toward foaming in a turbulent system. These properties are especially critical whenever inhibitors are used in jet fuel. The sensitivity of jet fuel pumping and injection systems requires that fuel be free of emulsions and foam.

6.6.5 Applying Corrosion Inhibitors

Corrosion inhibitors are used to protect fuel system equipment such as storage tanks and pipelines from internal corrosion initiated by water, acids, and other corrosive compounds. These inhibitors are typically applied to fuel using a continuous application method of addition as fuel passes through the system. An atomizer, injection quill, or a similar type of assembly continuously injects corrosion inhibitor into the fuel. As the fuel moves through the distribution system, the inhibitor is delivered to the exposed metal surfaces of the storage, distribution, and handling system.

6.6.5.1 New System Application

When corrosion inhibitor is first used in new storage or pipeline systems that have not been treated with a film-forming corrosion inhibitor, the rate of inhibitor addition into fuel should be maintained at a level higher than the recommended rate. The recommended rate is usually established by first completing corrosion testing using methods such as NACE TM0172 to determine the level of corrosion inhibitor needed to provide a passing rating such as "B+" or higher. If the corrosion inhibitor is applied using a continuous injection method, then a rate of addition in the range of two times the recommended rate should be used for new systems. An application period of 2 weeks to 1 month at this higher rate is typically adequate in providing a protective film of corrosion inhibitor on new metal surfaces. NACE corrosion testing can be completed during this time by sampling fuel from various points along the pipeline. When all collected samples provide the same passing NACE rating, then this is a good indication that a surface film of corrosion inhibitor has formed within the pipeline. At this point, the inhibitor rate can be reduced to the original recommended rate.

6.6.5.2 Existing System Application

Caution should be used when applying a corrosion inhibitor to an existing storage tank or pipeline that contains rust or to a storage tank or pipeline that has not received corrosion inhibitor treatment in the past. If applied to either of these systems, corrosion inhibitor will loosen and remove rust and rouge present. Rust will travel with the fuel and accumulate on filters or in areas of low turbulence. New metal surfaces will be exposed after rust removal.

Under these circumstances, the corrosion inhibitor should be added initially at low treating levels and gradually increased to the recommended level over a period of several days. This incremental addition will help minimize the negative effect of loosened rust or rouge and will gradually build an inhibitor film onto the newly exposed metal surfaces.

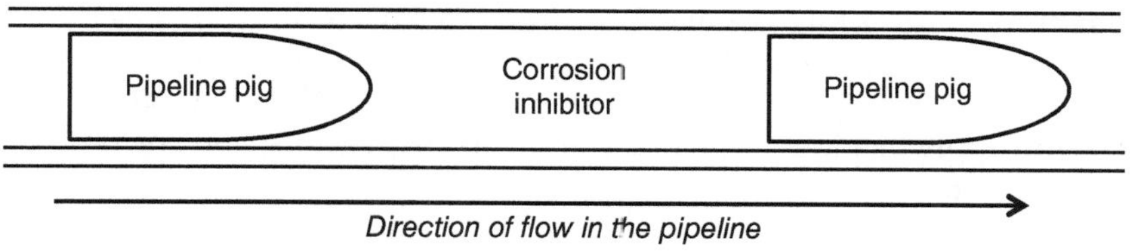

Figure 6.13 Batch treatment of corrosion inhibitor using pipeline pigs.

6.6.5.3 Batch Treatment of Pipelines

Instead of using the continuous application method of applying corrosion inhibitor to fuel, a batch treatment method can be completed. This method utilizes a large volume of corrosion inhibitor that is preceded and followed by fuel and a set of pipeline pigs. The recommended configuration for this type of treatment application is shown in Fig. 6.13.

A cup type of pipeline pig precedes the corrosion inhibitor volume to seal the volume of inhibitor behind the pig, and a squeegee or poly/foam type of chaser pig follows behind. The poly/foam chaser pig will absorb some of the corrosion inhibitor and will act as a swab in applying the corrosion inhibitor to the total internal surface area of the pipeline wall. The use of two pipeline pigs helps ensure that the corrosion inhibitor will contact all internal surfaces of the line.

One formula for use in calculating the appropriate volume of corrosion inhibitor required to protect a fuel pipeline is provided in Eq. (6.9). This formula is based on leaving a 1-mil-thick film of the inhibitor on the internal pipe surface with 20 percent of the inhibitor volume remaining at the end of the line. This remaining volume of inhibitor collects, holds, and carries away the solids dislodged during the pigging operation.

$$V = 1.033\, dl \tag{6.9}$$

where V = volume (US gallons) of corrosion inhibitor
d = inside diameter (inches)
l = length of pipeline (miles)

Therefore, a line that is 10 miles long and 10 in. in diameter will require 103.3 gal of corrosion inhibitor for treatment. A maximum pipeline pig velocity of 10 mph should not be exceeded to help ensure that the volume of corrosion inhibitor in the line is in contact with the pipeline wall for approximately 10 seconds.

Slug treatment is another method of additive application into pipelines when flow velocities or fuel production volumes will not support continuous injection of additive and pipeline pigs cannot be used. The additive is typically diluted with a solvent, such as diesel fuel, increasing the overall volume to help meet the contact time requirement. The additive slug travels through the pipeline between a leading and a trailing volume of fuel. This method is used to apply biocides, asphaltene and wax dispersants, and corrosion inhibitors.

6.6.6 Fluid Friction Effect in Pipelines

The protective film formed by corrosion inhibitors can act as a lubricating film to help in reducing fluid friction within a pipeline. This property can be measured using the Hazen-Williams coefficient and is often termed the *C Factor* expressed as in Eq. (6.10):

$$\left[C = \frac{4.52 \times L \times Q^{1.85}}{d^{4.87} \times p_d}\right]^{\frac{1}{1.85}} \tag{6.10}$$

where L = pipe length in feet
Q = flow in gallons per minute
d = internal diameter in inches
p = pressure drop over the pipe length in psig

Carbon steel piping C Factor ratings can typically range from 90 to 110. For pipelines that have been newly constructed and effectively treated with a film forming corrosion inhibitor, ratings can be significantly higher. A low C Factor indicates there is higher internal friction within the pipeline system and a resulting reduction in efficiency.

6.6.7 Nonferrous Metal Corrosion

Copper and copper alloys such as brass and bronze can be present in fuel systems in the form of heating coils, cooling coils, pump components, transfer lines, valves, fittings, and various connectors. Silver is used as a component in some fuel tank level sensors.

6.6.7.1 Corrosive Agents

Copper and silver are resistant to corrosion by water. However, sulfur compounds such as H_2S and elemental sulfur create copper and silver corrosion problems in fuel systems. Low molecular weight mercaptans can also be present in fuel, but they are less frequently identified as the primary causes of copper and silver corrosion. Report No. 653 issued by the Coordinating Research Council in 2009 found that together, H_2S and elemental sulfur, undergo interactive reactions to produced sulfane and alkyl polysulfide compounds that were corrosive and reactive toward silver, whereby S_8 elemental sulfur alone was not corrosive toward silver in laboratory testing.

The anodic corrosion of copper by H_2S to yield copper sulfide and hydrogen is shown in Eq. (6.11).

$$Cu \rightarrow Cu^{+2} + 2e^-$$

$$Cu^{+2} + H_2S \rightarrow CuS + 2H^+ \tag{6.11}$$

Under acidic conditions, hydrogen gas evolves

$$2H^+ + 2e^- \rightarrow H_2$$

6.6.7.2 Sources of Corrosive Compounds

In finished fuel hydroprocessing to remove sulfur and nitrogen heteroatoms, catalysts such as cobalt-molybdenum (CoMo) and nickel-molybdenum NiMo are often used. The CoMo catalyst yields primarily H_2S in sulfur reduction, while the NiMo catalyst yields primarily ammonia as well as H_2S. Both H_2S and aqueous ammonia are corrosive toward copper. Hydrogen sulfide is also involved in contributing to silver corrosion.

Sulfuric acid is used as a catalyst in the alkylation process to produce a high octane alkylate gasoline blend component. Although not common, trace levels of sulfuric acid can carry into a finished fuel blend and lead to corrosion of copper.

6.6.7.3 Fuel Level Sensor Corrosion

Silver and silver alloys have been used in the manufacture of some fuel level sensor gauges used within automotive gasoline fuel tanks. Corrosion and deactivation of these gauges have been linked to elemental sulfur in fuel, along with lower levels of H_2S and mercaptans, as the primary causes of corrosion of silver. The silver sulfide corrosion products that form on the fuel tank level sensor contact points interfere with the ability of the sensor to accurately measure the fuel volume.

Hydrogen sulfide scavengers, copper corrosion inhibitors, and elemental sulfur scavengers are all used to help prevent corrosion of copper and silver.

6.7 Hydrogen Sulfide and Sulfur Scavengers

During processing of high sulfur crude oil, H_2S and mercaptans or thiols (R-SH) dissolved in the crude can be liberated and released. Cracking reactions and thermal processing of distilled crude oil fractions can result in the formation of a variety of sulfur-bearing compounds such as H_2S, carbonyl sulfide, sulfur oxides, mercaptans, and other species through degradation of organosulfur compounds. These sulfur-containing compounds can accumulate in heavy residual

H_2S Concentration	H_2S Effects
10 ppm	Can detect by rotten egg type odor
100 ppm	Eliminates sense of smell in 3 to 15 minutes May burn eyes and throat
200 ppm	Eliminates smell quickly; stings eyes and throat
500 ppm	Loss of balance; respiratory difficulties in 30 to 45 minutes
700 ppm	Unconsciousness in less than 15 minutes Death will occur if not rescued
1000 ppm	Permanent brain damage or death if not rescued immediately

Table 6.6 Human Toxicity of H_2S in Vapor

products and in the vapor space above oils in storage. Since H_2S is soluble in both oil and water, the possibility exists for high levels of toxic H_2S to reside in heavy oil fractions used in preparing finished fuels such as No. 6 fuel oil and marine residual fuels. Hydrogen sulfide in the vapor space above residual fuel contained in storage tanks and in transport vessels can exist in lethal concentrations. Table 6.6 provides information on the toxicity effects of H_2S.

6.7.1 Hydrogen Sulfide Scavenger Application

Regulations requiring lower sulfur levels in finished fuels have resulted in extensive utilization of hydroprocessing worldwide to convert sulfur-containing compounds in fuel to H_2S and yielding nearly sulfur-free fuel. In most cases, H_2S is recovered as an off-gas from processing or removed by other treating units.

At times, H_2S, mercaptans, and elemental sulfur remain in fuel even after processing has taken place. Levels can sometimes be high enough to contribute to fuel odor or cause corrosion and failure of fuel copper or silver corrosion specifications. In the case of marine distillate fuels and marine residual fuels, ISO 8217 limits H_2S to a maximum limit of 2 mg/kg.

Scavengers for H_2S, mercaptans, and elemental sulfur can be used to help resolve odor, corrosion, or high contaminant levels in fuels. Scavenger compounds can possess reactive nitrogen, oxygen, or even sulfur-containing functional groups. Scavengers with reactive metals, such as iron and zinc, are also available.

It is necessary to know the H_2S level present in the vapor phase, the liquid phase, or both phases of a fuel in order to establish H_2S

scavenger addition rates. Vapor phase measurements such as ASTM D5705 using H_2S gas detection tubes or liquid phase measurements using titration or chromatography methods are the most common. As a general rule, 0.7 to 1.2 ppm of H_2S scavenger is required to react with each 1 ppm of H_2S measured in the fuel vapor. For fuel liquid phase measurements of H_2S, the addition rates of H_2S scavenger can range from 10 to 20 times the rate of H_2S detected in the liquid. These rates, however, may vary from fuel to fuel.

Proper application of a H_2S scavenger should enable direct contact of the scavenger with H_2S in the fuel. Contact time and reaction temperature together are equally important to ensuring that the scavenger removes H_2S. The ability to monitor the overall reaction for completeness is necessary to confirm that the desired specifications for H_2S, mercaptan, or elemental sulfur have been met.

6.7.2 Mercaptan Treatment

Mercaptan removal from fuel using a scavenger provides variable results. The outcome often depends upon the testing methods used to detect the presence of mercaptans. For example, upon applying a mercaptan scavenger to a fuel to remove mercaptan odor, the treatment is often effective. When evaluated by the ASTM D4952 "Doctor Test" method, fuel is odor free, and mercaptans are not detected. However, when evaluating the scavenger-treated fuel by ASTM D3227 potentiometric titration, the chemical bond between the scavenger and mercaptan is broken by the titrant releasing a mercaptide reaction product.

The type of sulfur-bearing products formed upon reaction of scavengers with either H_2S, mercaptans, or elemental sulfur can vary. Examples of reaction product types include organic salts, heterocyclic compounds, metal sulfides, high molecular weight thiols, disulfides, dithiazines, and polysulfides. Reaction products may be hydrocarbon soluble, water soluble, or volatile.

6.8 Cetane Improver

The most commonly used compound to increase the cetane number of distillate fuel is 2-ethylhexyl nitrate, an isooctyl nitrate compound. Dibutyl peroxide, polyunsaturated compounds, certain glycols, and other nitroparaffins are also effective at improving the cetane number of diesel fuel. Although 2-ethylhexyl nitrate is the most commonly used cetane improver, compounds including amyl nitrate, isopropyl nitrate, hexyl nitrate, and cyclohexyl nitrate are also effective. Table 6.7 contains important physical properties of 2-ethylhexyl nitrate cetane improver.

Property	Typical Result
Flash point	>65°C (>149°F)
Pour point	<–45°C (<–49°F)
Specific gravity @ 15.5°C (60°F)	0.0964
Viscosity, cSt @ 15.5°C (60°F)	1.9
Viscosity, cSt @ 0°C (32°F)	2.8
IBP (decomposes on boiling)	137.2°C (279°F)
Autoignition temperature	190°C (374°F)

Table 6.7 Typical Physical Properties of 2-Ethylhexyl Nitrate Cetane Improver

A thorough discussion of cetane improver performance and applications can be found in Chap. 4.

6.9 Fuel Lubricity Improver

Lubricity describes the ability of a fluid, such as a fuel or oil, to minimize the friction and wear between surfaces moving in relative motion under load. Fuel helps to lubricate and prevent wear of high-pressure fuel injection pump components, especially under boundary lubrication conditions. Boundary lubrication is defined as a condition whereby friction and wear between two surfaces in relative motion are determined by the properties of the surfaces and the properties of the contacting fluid, other than bulk viscosity.

For many years, the lubricity performance of various jet fuel grades has been improved by treatment with organic, fatty acid–based compounds that have the ability to provide both lubricity and corrosion inhibiting properties to the fuel.

Hydroprocessing of various fuels to remove sulfur containing organic compounds produces a fuel that has cleaner combustion and exhaust emission properties, but poorer lubricity performance. To restore lubricity, compounds such as fatty acids, esters, imides, and other organics are blended into the fuel at low concentrations.

Testing methods used to evaluate the lubricity of fuel and fuel lubricity improver performance rely on a final measurement of the wear scar diameter obtained on a steel test ball under load and in contact with fuel. Once wear testing has been completed, an average wear scar diameter is measured on the surface of the test ball, and a pass/fail rating is assigned for the fuel. Accepted standardized lubricity testing methods include:

- ASTM D5001 Standard Test Method for Measurement of Lubricity of Aviation Fuel by the Ball-on-Cylinder Lubricity Evaluator (BOCLE).

- ASTM D6079 Standard Test Method for Evaluating Lubricity of Diesel Fuels by the High Frequency Reciprocating Rig (HFRR).
- ASTM D7688 Standard Test Method for Evaluating Lubricity of Diesel Fuels by the High Frequency Reciprocating Rig (HFRR) by Visual Observation.
- ISO 12156 Diesel fuel—assessment of lubricity using the high-frequency reciprocating rig (HFRR).

A comparison of the testing parameters for the ASTM D5001 BOCLE method used in evaluating jet fuel lubricity and the ASTM D6079 HFRR method used evaluating diesel fuel lubricity are provided in Table 6.8. Schematics of the test cells used in rating jet fuel and diesel fuel lubricity are shown in Fig. 6.14. The ball on three seats (BOTS) assembly is also shown for comparison.

Operating Parameter	ASTM D5001-BOCLE	ASTM D6079-HFRR
Fuel sample size	50 ± 1.0 mL	2 ± 0.20 mL
Test cell	A rotating test ring is partially submersed into fuel; a stationary test ball with applied load is lowered onto a rotating test ring	A test disk is immersed into fuel; a vibrating test ball with an applied load is lowered onto the test disk
Frequency	—	50 ± 1 Hz
Vibrational path length	—	1 ± 0.02 mm
Test ring rpm	240 ± 1 rpm	
Applied load	500 ± 1 g	200 ± 1 g
Test time	30 minutes	75 ± 0.1 minute
Test fluid temperature	25 ± 1°C	60 ± 2°C
Additional comments	Fluid flow rate over the test ring must be maintained at 3.8 L/min; conditioned air relative humidity of 10 ± 0.2 percent at 25 ± 1°C must be kept	Relative humidity of air in the system must be between 30 and 85 percent

TABLE 6.8 Comparison of the Jet Fuel BOCLE and Diesel Fuel HFRR Lubricity Methods

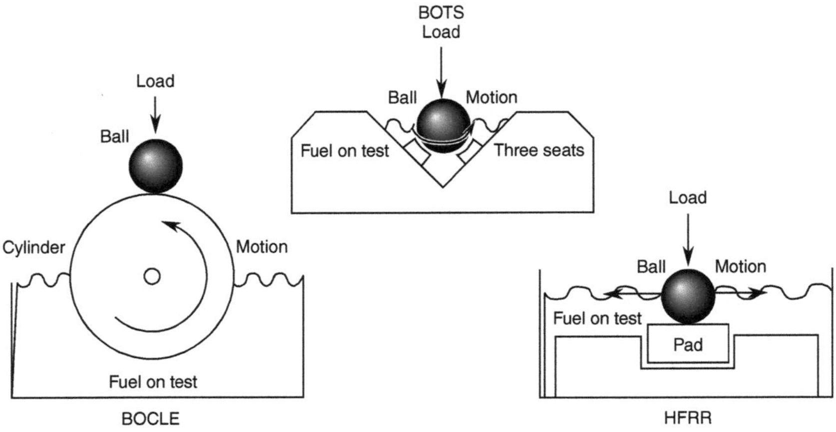

FIGURE 6.14 Schematic arrangements of BOCLE and HFRR test cells.

The BOCLE result is very sensitive to contamination of testing balls, test rings, reference fluid, and hardware. Both semi-automatic and fully automatic BOCLE testing units are available.

6.10 Fuel Deposit Control Additives

The use of deposit control additives to help prevent the formation and accumulation of deposits in spark-ignition and compression-ignition engines today is widespread. Compounds such as succinimides, polyalkyl amines, or other polyamine-type compounds are examples of deposit control additives. The primary purpose of deposit control additives is to help ensure that fuel ignites and combusts properly within the fuel combustion system by limiting the formation of deposits that may form on fuel injection, fuel intake, and combustion chamber components. Deposits can interfere with fuel atomization, fuel and air flow into the combustion chamber, and fuel ignition timing.

The high-pressure diesel fuel pump and injector nozzles must be kept free of deposits and corrosion to help ensure proper injection pressure and atomization of fuel into the combustion chamber. Deposit control additives aid in this process by providing the following performance:

- Maintain cleanliness of the fuel pump and component parts
- Prevent fouling and deposit formation on fuel injector nozzles
- Control acid initiated corrosion and erosion of fuel injector nozzle orifices

Other fuels such as kerosene, heating oil, and marine fuels do not typically contain deposit control additives. However, in some specialty applications, combustion catalysts and burner nozzle antifoulants are utilized.

6.10.1 Types of Deposits in Fuel Systems

6.10.1.1 Carburetor and Fuel Injector Deposits

Fuel olefins in gasoline have been identified as the primary cause of deposit formation on fuel injector nozzles and carburetors. High boiling, high molecular weight aromatic compounds in fuel can further contribute to deposits. Once deposits have formed, additional compounds can adhere to these deposits to form a larger, amorphous deposit mass.

6.10.1.2 Intake Valve Deposits

Olefins, especially diolefins, and higher molecular weight organics in gasoline lead to the formation of intake valve deposits (IVD). The higher-temperature environment of the intake valve region promotes the formation of more carbonaceous deposits than those formed on fuel injector nozzles. Also, components of the lubricating oil additive package are often found within the matrix of intake valve deposits.

The appearance and character of IVD can vary from hard, dry crusty deposits to relatively soft, thick deposits to deposits that are thin and varnish-like. Deposits form on the valve tulip, valve stem, and valve seats. Poor starting, warmup, and initial acceleration are common symptoms of an IVD problem.

6.10.1.3 Combustion Chamber Deposits

Components such as high molecular weight aromatics in gasoline have been identified as sources of combustion chamber deposits (CCD). These compounds can often be found in the unwashed gum fraction of gasoline. Also, high molecular weight aromatics from the lubricating oil, especially bright stock, and some fuel detergent additives are believed to contribute to combustion chamber deposit formation.

Investigation has shown that fuel-induced deposits first form as oxidized compounds which condense onto the surface of the combustion chamber. Heating and polymerization of these condensed compounds result in the formation of a higher molecular weight, liquid-like material. Continued heat transforms this material into a deposit composed of a crust-like outer layer and a liquid underlayer. In addition to fuel and additive chemistry, the impact of engine design on combustion chamber deposit formation has been investigated. Table 6.9 contains a description of the location, typical cause, and description of deposits that may form in the fuel intake and combustion system.

Location of Deposit	Typical Causes	General Description
Fuel Injectors		
• Deposits mostly on interior and exterior of injector pintle or tip	• Olefins • High boiling point and high molecular weight aromatic components	• Gum-like and amorphous solid deposits
Intake Valves		
• Deposits mostly on valve tulip, valve stem, and valve seat	• Olefins, especially diolefins • High molecular weight aromatic components • Components of lubricating oil additive	• Variable in appearance such as: • Dry-hard-crusty deposit • Thick-soft deposit • Thin-varnish like deposit
Combustion Chamber		
• Deposits on piston head, underside of intake valve seat, and combustion chamber surfaces	• Unwashed gum fraction of gasoline • Higher molecular weight lubricating oil aromatics such as those from bright stock fraction • Additive induced deposits	• Variable in appearance • Crust-like, carbonized outer layer with liquid under layer • Crust layer cracks—flakes away exposing liquid—new deposit forms • Deposits near intake valve have waxy appearance

Table 6.9 Fuel Intake and Combustion System Deposit Characteristics

6.10.1.4 Engine Testing of Deposit Control Additives

Deposit control additives can sometimes be a complex mixture of compounds such as a detergent, demulsifier, antioxidant, metal deactivator, and cosolvent. The effective rate for deposit control additives in gasoline and diesel can vary widely.

A number of different engine tests have been used to rate the ability of fuel deposit control additives to keep fuel injection, intake valves, and combustions system components of engines clean and operational. Although carburetor detergents have been available for many decades, development of more sophisticated fuel detergents escalated significantly during the 1980s when fuel injectors replaced

Test Designation—Engine Type	Test Purpose
ASTM D5598 1985–1987 Chrysler 2.2L, 4-cylinder, turbocharged engine	Port fuel injector keep clean test for spark ignition fuels
ASTM D5500 1985 BMW 318i automatic transmission engine	Intake valve keep clean test for spark ignition fuels
ASTM D6201 1994 Ford Ranger 2.3L engine	TOP TIER Gasoline Detergent Test Intake valve keep clean test; combustion chamber deposit test for spark ignition fuels
Peugeot XUD-9 engine Cummins L-10 engine	Deposit prevention in diesel fuel injection systems for compression ignition fuels
1.9L Volkswagen Wasserboxer engine 1990–1995 General Motors V8 engine	Cold start valve sticking test for spark ignition fuels

TABLE 6.10 Engine Tests to Rate Deposit Control Additive Performance

carburetors in spark ignition engines. Problems with fuel injector nozzle fouling and deposit formation on intake valves began to arise, and factors such as fuel olefin content and lubricant interaction were found to play a role in contributing to fuel intake and combustion system deposits.

Engine tests were developed to simulate the formation of deposits on fuel injectors and intake valves. Both stationary and on-road engine tests were utilized, and testing was carried out on a worldwide basis. Combustion chamber deposit control also became part of the effort in fuel detergent additive development. A brief list of various engine tests that became part of the process in development of deposit control additives is provided in Table 6.10.

6.11 Problems Related to Fuel Additive Use

Although uncommon, fuel additives can be a source of various problems in fuel systems. The additives may be the direct cause or a contributing factor in the cause of a problem. Fuel additive related problems are typically due to the following:

- The additive has inherent physical property limitations.
- Improper application of the additive into fuel at the wrong injection point.

- Adding too much additive into the fuel.
- Additive interaction with a fuel contaminant such as water or rust.
- Additive interaction with another chemical component in the fuel or with another fuel performance additive.

Some of the common problems encountered are described in the sections that follow.

6.11.1 Antioxidants

6.11.1.1 Crystallization

When cooled or contaminated, hindered phenol antioxidants and some PDA antioxidant compounds can crystalize and precipitate from solution as concentrated compounds. At temperatures below 15°C (59°F), some antioxidants will begin to crystalize and drop from their solvent carrier as relatively pure materials.

Usually upon warming to temperatures of 25 to 35°C (77 to 95°F), antioxidant crystals will dissolve and return back into the solvent carrier. The performance of the antioxidant is typically not adversely affected by this process.

These antioxidants are also sensitive to contamination by impurities such as rust, dirt, or other particulate-like material. The presence of these impurities can initiate the formation of an antioxidant seed crystal. Once a seed crystal forms within the concentrated antioxidant, it can cause the volume of stored product to crystallize. Keeping antioxidants free of contamination helps ensure product integrity and performance.

6.11.1.2 Reaction with Inorganic Acid or Base

If fuel-containing antioxidant becomes contaminated with the sodium hydroxide solution (caustic) used in mercaptan removal from fuel or with sulfuric acid used as an alkylation catalyst, fuel antioxidant may be deactivated. In reaction with caustic, hindered phenol antioxidants may form a sodium phenate salt. In reaction with sulfuric acid, the PDA antioxidant may degrade to form an organic sulfonate or more complex reaction product. In either case, both the phenolic and PDA antioxidants can lose performance because of contamination and may become water soluble.

6.11.1.3 Saybolt Color Degradation

Most PDA type fuel antioxidants are naturally very dark in color. If used at high concentrations in water white products such as hydroprocessed kerosene, No. 1 diesel, or various hydrocarbon solvents, the Saybolt color values may degrade to failing levels.

6.11.2 Corrosion Inhibitors

6.11.2.1 Rouge Removal

Rouge is a term used to describe the fine, almost powder-like corrosion material that develops on the inside walls of fuel system equipment. Steel pipelines, storage tanks, and transfer lines that have rust buildup and accumulation on the internal surfaces are continually releasing low levels of rouge and rust into fuel.

If fuel-containing corrosion inhibitor is utilized in systems containing rouge, the inhibitor "films out" onto rouge in addition to rust-free steel. As a result of the filming effect of the corrosion inhibitor buildup, rouge can break free from the internal walls of a pipeline or storage tank and be released into the fuel. As the fuel passes through various tank and line filters, rouge can be trapped by the filter media. Filter plugging and halting of the flow of fuel through the filter can result.

For this reason, it is recommended that addition of corrosion inhibitor to previously uninhibited systems be closely monitored. Initial low level addition followed by gradual increase in the corrosion inhibitor treat rate will help to minimize the rapid removal of pipeline rouge and filter plugging problems.

6.11.2.2 Filter Plugging due to Caustic Contamination

When fuel processing involves contact with a caustic NaOH solution, some of the caustic may carry over into the finished fuel blend. Carboxylic acid-based corrosion inhibitors are very sensitive to the presence of caustic. The acid functionality of the inhibitor can react with caustic to form a gel-like organic salt. This sodium salt of the corrosion inhibitor can initiate the formation of an emulsion at the fuel-water interface in a fuel storage tank, or it may travel downstream with fuel to accumulate on fuel filters. Fuel filter plugging and halting of fuel flow may result. Fuel haziness and the presence of particulate matter in the fuel can result.

Finishing processes involving the use of caustic to remove mercaptans and acids from fuel are typical sources of contamination.

6.11.2.3 Fuel Injection Pump Sticking at High Rates of Addition

Substantial overtreatment of the corrosion inhibitor rate in diesel fuel has been shown to lead to sticking of fuel pump plungers. Fatty acid based corrosion inhibitors can form a tenacious film on metal surfaces. The tight tolerances in high-pressure injection pump barrel and plunger assemblies require very little fluid film on the surface to provide proper lubrication. Corrosion inhibitor is part of this film. For this reason, maintaining corrosion inhibitor at the minimum rate needed to ensure proper corrosion protection of metal components is recommended.

6.11.2.4. MSEP or WSIM Degradation of Jet Fuel

The water separation properties of jet fuel evaluated by method ASTM D3948 Determining the Water Separation Characteristics of Aviation Turbine Fuels by Portable Separometer (WSIM or MSEP) are degraded by high addition rates of corrosion inhibitor. Maximum rates of corrosion inhibitor addition allowed in jet fuel are provided in specifications such as ASTM D1655, MIL-PRF-25017, and DEF STAN 91-91.

6.11.3 Demulsifiers and Dehazers

6.11.3.1 Precipitate Formation

Demulsifiers are quite effective at clearing fuel of water-initiated haze. Occasionally, water will interact with a demulsifier to form an emulsion. This emulsion will precipitate from fuel and settle to the bottom of a fuel storage tank. The emulsion can plug filters and clog small lines if not removed from the tank with the water bottoms. Evaluating a range of demulsifiers for their performance in fuel and rating the tendency to form an emulsion is recommended.

6.11.3.2 Creation of Haze in Pipeline Fuel

Fuel refiners will occasionally use dehazers to remove water from cloudy or hazy fuel. Some dehazers disperse water present in fuel into fine droplets so that the water haze cannot be seen. The fuel, thus, appears bright and clear. However, if fuel treated with a high concentration of dehazer is transported through a pipeline system, the dispersed water can sometimes cling to the pipeline wall and remain as water droplets.

A problem may then occur with the batch of fuel that follows in the sequence of pipeline shipments. The water on the pipeline wall can be picked up by the subsequent fuel batch. The result could be the creation of haze in this subsequent shipment of fuel.

6.11.3.3 Dispersion of Line or Tank Deposits

If rates greater than 500 ppm of dehazing compounds are added to fuel to remove haze caused by water, a possibility exists for the dehazer to interact with deposits held within a fuel pipeline, transfer line, or fuel tank. The surfactant like character of the dehazer may enable it to function in removal and dispersion of existing deposits. The dispersed deposits may then be carried with the fuel through the distribution system, and the deposits may darken the fuel color. Because of this, extreme care should be taken when using high concentrations of dehazer to clear fuel of water haze. If deposits removed by the use of high concentrations of dehazer are water soluble, the deposits may remain in the tank bottom water layer. Unless the water is removed, subsequent volumes of fuel entering the tank may become contaminated by the water soluble compounds.

6.11.3.4 Insolubility in Additive Packages

Gasoline and diesel detergent additive packages contain compounds to prevent deposit formation and fouling within internal combustion engines. Demulsifiers are added to prevent water from emulsifying with the detergent component. Because of their high molecular weight and polarity, some demulsifiers are not freely soluble in detergent formulations. Problems in maintaining demulsifier solubility within the detergent package can develop. To resolve this situation, extensive demulsifier reformulation is required to enhance solubility in the detergent along with the use of various cosolvents.

6.11.4 Electrical Conductivity Improver or Static Dissipater

6.11.4.1 Diminished ULSD Conductivity with Lubricity Improver Addition

When lubricity improver is blended into ultralow-sulfur diesel fuel (ULSD) containing electrical conductivity improver additives, the fuel conductivity performance can diminish. Conductivity loss will plateau and can be restored by further conductivity improver addition. Loss of fuel conductivity can occur in the presence of both monoacid and ester based lubricity improvers. The loss of fuel electrical conductivity is typically less pronounced in the presence of an ester-based lubricity improver than a monoacid lubricity improver. The impact of electrical conductivity improver on restoring conductivity loss in the presence of a monoacid-based lubricity improver is presented in Table 6.11.

Conductivity Improver, ppm	Lubricity Improver, ppm	Conductivity pS/m @ 25°C
—	—	5
0.5	—	73
1.0	—	137
1.5	—	210
1.5	100	197
1.5	150	118
1.5	200	104
2.0	200	134
2.5	200	164
5.0	200	368

Table 6.11 Impact of Monoacid Lubricity Improver on Conductivity of ULSD

6.11.4.2 Diminished Conductivity in Cold Diesel Fuel

The electrical conductivity of diesel fuel will fall as the temperature of the fuel cools. The rate of conductivity can be lower by 20 to 40 percent as diesel fuel temperature cools from 25°C (77°F) to −10°C (14°F).

6.11.5 Wax Crystal Modifiers: Cloud Point—Pour Point—CFPP Additives

6.11.5.1 Antagonism between Cloud Point Depressant and CFPP Additive Performance

Cloud point reduction of fuels treated with cloud point depressant may be halted or reversed when additives that lower the CFPP temperature are blended into the same fuel. To compensate for this loss in performance, additional cloud point improver must be added back into the fuel. This additional cloud point depressant may then begin to interfere with CFPP reduction in some fuels.

In most applications where both cloud point depressant and additives for CFPP reduction are used, there is an optimum rate of addition for each additive, where there is minimal to no interference or antagonism in additive performance.

6.11.5.2 Reversion in Pour Point and Filtration Temperatures

To achieve very low pour point and filtration temperatures in some fuels and oils, high rates of wax crystal modifiers are sometimes used. On occasion, high concentrations of wax modifier addition can cause pour point or filtration temperatures to revert. This reversion in flow or filtration properties is primarily due to the influence of the higher viscosity of the inhibitor polymer at the low test temperatures.

6.11.5.3 Poor Solubility in Some Paraffinic Solvents

Many of the concentrated wax crystal modifier polymers used to lower the pour point of distillate and residual fuels are highly viscous. Because of this fact, it is often necessary to heat or dilute the wax modifier with a solvent to enable the product to flow at cold temperatures.

Diesel fuel and kerosene alone are usually not effective solvents for solubilizing most concentrated wax crystal modifiers for cold temperature storage and use. The low KB value of diesel fuel and kerosene makes these products ineffective at holding the wax crystal modifier polymers in solution for extended periods of time at low temperatures. Stratification and separation of the wax modifier from solution can occur and may lead to application problems such as filter plugging by concentrated polymer and incorrect additive addition rates into fuels and oils.

6.11.6 Cetane Improver

6.11.6.1 HFRR Wear Scar Diameter Increase in Fuels Containing Lubricity Improver

The use of alkyl nitrate cetane improver in diesel fuel can result in an increase in the HFRR wear scar diameter in fuels containing lubricity improver. The rate of wear scar diameter increase will vary and can be in a range of 50 to 150 μm. Increasing the addition rate of lubricity improver in fuel frequently overcomes the loss in lubricity due to cetane improver addition.

6.11.6.2 Darkening of Diesel Fuel Color after Thermal Stability Testing

A commonly used method for evaluating the stability of diesel fuel is test method ASTM D6468 High Temperature Stability of Distillate Fuel. This method involves heating fuel to a temperature of 149°C (300°F) for either 90 or 180 minutes and afterward, measuring the change in fuel color and the formation of deposits.

Cetane improvers based on alkyl nitrate compounds can interact with fuel at the 149°C (300°F) test temperature. The products of this interaction can initiate the formation of dark, insoluble compounds in fuel. This color change does not occur in all fuel-containing cetane improver. The addition of an antioxidant to diesel fuel can often prevent color change from occurring in fuel-containing cetane improver.

6.11.6.3 Increase in Conradson Carbon Number

Under testing conditions described in ASTM D189 Conradson Carbon Residue of Petroleum Products, the presence of cetane improver may give higher than normal carbon residue values. If there is question about the presence of cetane improver in fuel, method ASTM D4046 Alkyl Nitrate in Diesel Fuel by Spectrophotometry can be used for detection.

6.11.7 Distillate Fuel Stabilizers

6.11.7.1 Fuel Color Darkening and/or Filter Plugging

Some distillate fuel stabilizer formulations can function as dispersants to prevent settling and accumulation of deposits in fuel pipelines and storage tanks. If fuel containing a stabilizer with dispersant properties is stored or transported in a system having existing deposits, the dispersant may act to break free and suspend the deposits within the fuel. The resulting fuel would appear darker in color.

6.11.7.2 Short-Term Performance in Reactive Distillate Fuel Components

Highly unstable distillate fractions such as FCC light cycle oil, heavy cycle oil, coker distillate, and some vacuum gas oils can be used as fuel blend components. If these fractions are not hydroprocessed, their presence in fuel may cause rapid darkening in fuel color and formation of sludge-like deposits affecting the appearance and quality of the overall blend. Distillate stabilizers can be added to these unstable fractions to help slow degradation and color change. However, the length of performance of stabilizers and antioxidants in these unstable fractions is limited and may provide only a few weeks of color and deposit control.

CHAPTER 7

Test Methods Used to Identify and Solve Fuel Problems

To begin the process of identifying the source of a fuel problem and recommending a solution, it is usually necessary to collect information about the physical, chemical, and performance properties of the fuel. A variety of test methods are available for this purpose. Described in this chapter are some of the primary testing methods that are commonly used to evaluate the composition and performance of fuels.

7.1 Testing Storage Stability and Deposit Forming Tendency of Fuels

Test methods in this section are useful in determining the following:

- Possible cause of deposit formation on fuel system components (i.e., line varnish, fuel injector deposits, valve deposits, filter deposits, etc.).
- Identify fuel blending components that rapidly darken in color over time.
- Predicting the long-term storage stability of distillate fuel.
- Screening of antioxidant and stabilizer compounds to enable fuels to meet various required specifications.
- Determine the potential for distillate fuels such diesel fuel, heating oil, biodiesel blends, and B100 FAME to degrade in long-term storage.
- Evaluate the tendency of fuel to form sludge in fuel tanks and pipelines leading to possible filter plugging.

Fuel degradation due to oxidative reactions, condensation reactions, and contamination is a common cause of fuel stability problems. Gasoline, jet fuel, diesel fuel, heating oil, and biodiesel fuel are all susceptible to degradation and deposit formation. The test methods presented in this section can be used to help anticipate or determine the cause of fuel stability problems.

7.1.1 ASTM D525/IP 40 Standard Test Method for Oxidation Stability of Gasoline (Induction Period Method)

7.1.1.1 ASTM Summary

The sample is oxidized in a pressure vessel initially filled at 15 to 25°C with oxygen pressure at 690 to 705 kPa and heated at a temperature between 98 and 102°C. The pressure is recorded continuously or read at stated intervals until the breakpoint is reached. The time required for the sample to reach this point is the observed induction period at the temperature of the test, from which the induction period at 100°C can be calculated.

7.1.1.2 Break Point

The break point is defined as the point in the pressure-time curve that is preceded by a pressure drop of exactly 2 psi within 15 minutes and succeeded by a drop of not less than 2 psi in 15 minutes.

The result of this testing provides information on whether fuel components are susceptible to attack by oxygen. If fuel components do react with oxygen, then further degradation of the fuel can be expected. The final result could be gum formation, fuel discoloration, and fuel system deposits. Experience has shown that fuel with induction period times reported to be less than 240 minutes are unstable and can degrade to form deposits. These fuels tend to perform poorly in the field under long-term storage conditions.

The ASTM D4814 gasoline specification requires an induction time greater than 240 minutes, and most worldwide specifications for spark-ignition fuel have similar or higher induction time ratings. If refined gasoline does not meet this specification, antioxidants can be added to the fuel to provide an increase in induction time. Antioxidant rates ranging from 5 to 50 ppm are typical. Also, the addition of a metal deactivator at rates of 1 to 2 ppm may further increase the induction time.

7.1.2 ASTM D381/IP 540 Standard Test Method for Gum Content in Fuels by Jet Evaporation

The ASTM D4814 gasoline specification states that an existent gum weight be measured. Existent gum in fuel is evaluated by test method ASTM D381, IP 131, or similar methods. ASTM D381 is summarized as follows:

7.1.2.1 ASTM Summary

When testing either aviation or motor gasoline, a 50 ± 0.5 mL quantity of fuel is evaporated under controlled conditions of temperature and flow of air. When testing aviation turbine fuel, a 50 ± 0.5 mL quantity of fuel is evaporated under controlled conditions of temperature and flow of steam. For aviation gasoline and aviation turbine fuel, the resulting residue is weighed and reported as milligrams per 100 mL. For motor gasoline, the residue is weighed before and after extracting with heptane and the results reported as milligrams per 100 mL.

This method is helpful in determining whether or not aviation gasoline, motor gasoline, or aviation turbine fuel contains a significant amount of nonvolatile compounds. The compounds can be contaminants that are organic or inorganic in nature and are classified as unwashed and washed gums and described below:

7.1.2.2 Unwashed Gums

- High boiling point fuel components
- Oxidized, high molecular weight organics
- Fuel additives such as detergents and corrosion inhibitors
- Inorganic materials such as iron and calcium salts

7.1.2.3 Washed Gums

- Oxidized high-molecular weight organics
- Inorganic materials such as iron and calcium salts

The ASTM D4814 motor gasoline specification limits existent gums to 5 mg/100 mL in fuel. A similar existent gum measurement rating is in effect for aviation gasoline. The ASTM D1655 specification for aviation turbine fuel limits existent gums to a maximum rating of 7 mg/100 mL of fuel. If fuel does not meet the existent gum specification, addition of antioxidant to fuel typically does not further reduce the existent gum level.

7.1.3 Dupont F21 Stability Test and Nalco Pad Test

The Dupont F21 Stability Test and the Nalco Pad Test are two of the earliest of the rapid test methods developed to rate fuel stability. The tests are similar. The Nalco Pad Test was developed initially to rate the stability of railroad diesel fuel, and the Dupont F21 Stability Test was an expansion of the method to incorporate all distillate fuels.

7.1.3.1 Test Summary

A 50-mL sample of distillate fuel is immersed in a 149°C (300°F) oil bath for 90 minutes. After removal from the bath and cooling to room temperature, the color of the fuel is measured and compared to the initial color. The fuel is then vacuum filtered through two No. 1

Whatman filter pads and rinsed with heptane. The color of the pad is compared to a standard color chart and rated. The color of the pad may also be measured by a light reflectance meter.

Although no direct correlation has been established between rapid, short-term stability testing and other longer-term testing results, these short-term test methods can be used to identify potentially unstable distillate fuels that may degrade in color and form deposits upon storage and use.

7.1.4 ASTM D2274/IP 388 Standard Test Method for Oxidation Stability of Distillate Fuel Oil (Accelerated Method)

7.1.4.1 ASTM Summary

A 350-mL volume of filtered middle distillate fuel is aged at 95°C (203°F) for 16 hours while oxygen is bubbled through the sample at a rate of 3 L/h. After aging, the sample is cooled to approximately room temperature before filtering to obtain the filterable insoluble quantity. Adherent insolubles are then removed from the oxidation cell glassware with trisolvent. The trisolvent is evaporated to obtain the quantity of adherent insolubles. The sum of the filterable and adherent insolubles, expressed as milligrams per 100 mL, is reported as total insolubles.

ASTM D2274 is commonly utilized throughout the world to determine the oxidative stability of distillate fuel. Although this method is not as effective at predicting the long-term stability of distillate fuel as ASTM D4625, it is useful for measuring the resistance of fuel to rapid degradation by contact with air. Metal catalysts such as copper and iron are sometimes added to the fuel during ASTM D2274 testing to further accelerate degradation and to also test the performance of corrosion inhibitors and metal deactivators at preventing metal catalyzed fuel degradation.

7.1.5 ASTM D4625/IP 378 Standard Test Method for Middle Distillate Fuel Storage Stability at 43°C (110°F)

7.1.5.1 ASTM Summary

Four-hundred milliliter volumes of filtered fuel are aged by storage in borosilicate glass containers at 43°C (110°F) for periods of 0, 4, 8, 12, and 24 weeks. After aging for a selected time period, a sample is removed from storage, cooled to room temperature, and analyzed for filterable insolubles and for adherent insolubles.

This test method is effective for use in predicting the long-term storage stability of distillate fuel. The results obtained closely approximate those obtained after fuel storage at ambient temperature. For example, fuel storage at 43°C (110°F) for 12 to 13 weeks has been shown to correlate well with one year of storage at ambient temperature.

Early work on the development of this method was carried out by the U.S. Navy. It was found that approximately the same amount of new total insolubles formed in 1 week of beaker storage at 43°C (110°F) as formed in 1 month of ambient bottle storage at 19°C (67°F). There is no standardized rating scale to interpret total insoluble values obtained from ASTM D4625A testing. However, the following general guidelines can be used:

- Fuel stability is considered to be good if the total insoluble level is less than 2 mg/100 mL.
- Fuel stability is considered to be moderate or poor if the total insoluble measurement is in the range of 2 to 7 mg/100 mL.
- Fuel is considered unstable if the total insoluble measurement is greater than 7 mg/100 mL.

7.1.6 ASTM D6468 Standard Test Method for High-Temperature Stability of Middle Distillate Fuels

7.1.6.1 ASTM Summary

Two 50 mL volumes of filtered middle distillate fuel are aged for 90 or 180 minutes at 150°C in open tubes with air exposure. After aging and cooling, the fuel samples are filtered and the average amount of filterable insolubles is estimated by measuring the light reflectance of the filter pads. The 100 and 0 percent extremes of the reflectance rating range are defined by an unused filter pad and a commercial black standard, respectively.

7.1.7 Williams Brothers Pipeline Stability Test for Distillate Fuel

This accelerated stability test method is preferred by some fuel refiners and pipeline companies over other test methods used to determine the stability of distillate fuel. The more severe conditions, relatively short test time period, and similarity of this method to ASTM D873 Oxidation Stability of Aviation Fuels (Potential Residue Method) contribute to its acceptance.

7.1.7.1 Test Summary

A 50-mL sample of distillate fuel is placed into a stainless steel test bomb and sealed. The bomb is pressurized with oxygen at 100 psi and then placed in a heating block maintained at a temperature between 98 and 102°C (209 and 215°F). The test bomb is held at this temperature for 16 hours. After this test period, the bomb is removed from the heating block and cooled to room temperature. The color of the fuel is measured and the insoluble residue is weighed. If the final fuel color is greater than ASTM 6.0, the fuel has failed to pass the test. If the heptane washed insoluble residue is greater than 50 mg / 100 mL, the fuel has failed to pass the test.

7.1.8 ASTM D5304 Standard Test Method for Assessing Middle Distillate Fuel Storage Stability by Oxygen Overpressure

7.1.8.1 ASTM Summary

A 100-mL aliquot of filtered fuel is placed in a borosilicate glass container. The container is placed in a pressure vessel, which has been preheated to 90°C. The pressure vessel is pressurized with oxygen at 800 kPa (absolute) (100 psig) for the duration of the test. The pressure vessel is placed in a forced air oven at 90°C for 16 hours. (*Warning:* Observe all normal precautions while using oxygen under pressure and at high temperatures in the presence of combustible liquids. Appropriate shielding should be used for any container under pressure. Pressurize and depressurize the containers slowly using appropriate personnel shielding. Never attempt to open the pressure vessel while it is pressurized. All fuel and solvent handling should be done in an appropriate fume hood only.) After aging and cooling, the total amount of fuel insoluble products is determined gravimetrically and corrected according to blank determinations.

This short-term distillate fuel stability test method is designed to accelerate the formation of deposits in fuel. Depending upon the user, total deposit ratings less than 25 mg/100 mL are desired while other fuel users accept total deposit ratings less than 50 mg/100 mL.

7.1.9 ASTM D7545 Standard Test Method for Oxidation Stability of Middle Distillate Fuels—Rapid Small Scale Oxidation Test

The rapid small-scale oxidation text (RSSOT) method was developed to determine the stability of fuels such as diesel fuel, heating oil, biodiesel blends, and B100 FAME. The method incorporates less sample volume and typically less testing time than other distillate oxidation stability testing methods.

7.1.9.1 Test Description

At ambient temperature, a 5-mL volume of fuel is added to a pressure testing cell, the cell is sealed using a screw cap, and covered by a locking, insulating hood. Oxygen is introduced into the cell until reaching a pressure of 700 kPa. The cell is then heated to 140°C and the pressure and time are continuously recorded until the pressure drop reaches the breakpoint or the fuel meets a required specification.

7.1.10 EN 14112 and EN 15751 Automotive Fuels—Fat and Oil Derivatives—Fatty Acid Methyl Esters (FAME)—Determination of Oxidation Stability by Accelerated Oxidation Test

Test methods EN14112 and EN 15751 can be used to rate the oxidation lifetime of pure FAME. However, EN 15751 can also be used to rate the oxidation lifetime of fuel blends containing both FAME and petroleum diesel fuel. These methods are based on a similar method used by the producers of animal and vegetable fats and oils. By measuring the time between initial contact of the fat or oil with oxygen in the air and the formation of "rancid" by-products, the storage stability of the natural oils can be determined.

7.1.10.1 Test Summary

Both methods involve passing purified air at a rate of 10 L/h through pure FAME or FAME blend heated to a temperature of 110°C. Reaction of oxygen with unsaturated bonds in the biofuel results in product degradation and the formation of volatile, low molecular weight acidic compounds. These compounds and air pass through water contained in a separate vessel. An electrode immersed into the vessel measures the change in water conductivity. When the measured conductivity begins to rapidly increase signaling the end of the induction period, the time is recorded. The time span between the point of initial introduction of purified air and the end of the induction period is reported as the oxidation stability or Rancimat lifetime.

In the biodiesel industry, the oxidation stability or Rancimat lifetime relates to the hours of resistance to oxidative degradation. The hours measured are used to establish an oxidation stability specification for FAME or FAME blends. Currently, region to region specifications vary for the minimum number of oxidation stability hours required for FAME and FAME blends.

7.2 Testing Copper and Silver Corrosion Properties of Fuel

The test methods in this section are useful in determining the following:

- Confirm whether fuel has been contaminated with agents that corrode copper and silver such as hydrogen sulfide, mercaptans, elemental sulfur, or inorganic acids.
- Rate the ability of copper and silver corrosion inhibitors to prevent corrosion.
- Identify fuel blend components that lead to corrosion of copper and copper alloys such as brass or bronze, and silver components found in fuel systems.

Nonferrous metal hardware found in fuel systems can include copper bearing transfer lines, cooling coils, nonferrous metal fittings, pump components, as well as automotive fuel tank level gauges containing silver. Because of this, copper and silver corrosion tests have been put in place within fuel specifications as a means to rate the impact of these corrosive agents.

Nearly all fuels have a copper corrosion specification that must be met while automotive spark ignition fuel has an additional silver corrosion specification. A representative listing of fuels and their copper and silver corrosion test specifications are presented in Table 7.1.

7.2.1 ASTM D130/IP 154 Standard Test Method for Corrosiveness to Copper from Petroleum Products by Copper Strip Test

7.2.1.1 ASTM Summary

A polished copper strip is immersed in a specific volume of the sample being tested under conditions of temperature and time that are specific to the class of material being tested. At the end of the heating period, the copper strip is removed, washed, and the color and tarnish level assessed against the ASTM Copper Strip Corrosion Standard.

7.2.2 ASTM D1838 Standard Test Method for Copper Strip Corrosion by Liquefied Petroleum Gases

7.2.2.1 Test Description

A copper test coupon is polished with steel wool and three different grades of silicon carbide cloth/paper and suspended within a stainless steel test cylinder. A 100-mL volume of the LP gas is introduced into the cylinder, and the cylinder is sealed and immersed for 1 hour into a 37.8°C (100°F) water bath. After the test period, the coupon is rated.

7.2.3 ASTM D7667 Standard Test Method for Corrosiveness to Silver by Automotive Spark-Ignition Engine Fuel—Thin Silver Strip Method

7.2.3.1 Test Description

A thin silver test strip is polished using a 150-grit aluminum oxide cloth, a 400-grit pad, and centered into a test tube containing 30 mL of spark-ignition test fuel. After heating in a pressure vessel at 50°C for 2 h, the silver strip is rated for color and the type of surface tarnish.

Fuel Specification	Copper/Silver Corrosion Method	Specification Requirement, Maximum Rating	Temperature, °C (°F)	Time, h
ASTM D396 fuel oils	ASTM D130 IP 154	a. No. 3 copper corrosion rating for S500 and S5000 No. 1 and No. 2 fuels only b. No requirement for No. 4, No. 5, and No. 6 fuels	50 (122)	3
EN 590 diesel fuels	ISO 2160	Class 1 copper corrosion rating for all grades	50 (122)	3
ASTM D910 aviation gasolines	ASTM D130	No. 1 for all grades	100 (212)	2
ASTM D975 diesel fuel oils	ASTM D130 IP 154	a. No. 3 copper corrosion rating for S15, S500, and S5000 No. 1-D and No. 2-D fuels only b. No requirement for No. 4-D fuel	50 (122)	3
ASTM D1655 aviation turbine fuels	ASTM D130 IP 154	No. 1 rating for Jet A and Jet A-1	100 (212)	2
ASTM D1835 liquefied petroleum (LP) gases	ASTM D1838	No. 1 rating for all grades	37.8 (100)	1
ASTM D4814 automotive spark ignition fuel	ASTM D130 IP 154	No. 1 rating for copper	50 (122)	3
ASTM D4814 automotive spark ignition fuel	ASTM D7667	No. 1 rating for silver	50 (122)	2
ASTM D4814 automotive spark ignition fuel	ASTM D7671	No. 1 rating for silver	50 (122)	3

TABLE 7.1 Copper and Silver Corrosion Specification for Fuels

7.2.4 ASTM D7671 Standard Test Method for Corrosiveness to Silver by Automotive Spark-Ignition Fuel—Silver Strip Method

7.2.4.1 Test Description

Silicon carbide powder and a 240-grit silicon carbide cloth/paper are used to polish the silver coupon. The coupon is set into position on a suspension assembly and placed into a pressure vessel, as described in Procedure A, or a vented test tube, as described in Procedure B. Using either procedure, the coupon is suspended in 30 mL of spark ignition fuel and heated to a temperature of 50°C for 3 hours. After testing, the silver strip is removed and rated for color and the level of tarnish.

7.3 Testing the Ferrous Metal Corrosion Properties of Fuel

Test methods in this section are useful in determining the following:

- Identify the possible source of a fuel corrosion problem.
- Rate the impact of certain fuel blend components in altering fuel corrosion characteristics.
- Rate the performance of fuel corrosion inhibitor additives.

Although no ferrous metal corrosion requirement exists now in the ASTM specifications for gasoline, diesel fuel, and other fuels, many fuel pipeline companies, refiners, and marketers have adopted the National Association of Corrosion Engineers (NACE) Corrosion Standard Method TM-01-72-2001 as a specification. The NACE corrosion method is summarized as follows.

7.3.1 Standard Test Method Determining the Corrosive Properties of Cargoes in Petroleum Product Pipelines

7.3.1.1 Test Summary

A polished steel spindle is immersed into 300 mL of fuel maintained at 37.8°C (100°F). The fuel is continually stirred during the entire test period. Upon immersion for 30 minutes, a 30-mL volume of distilled water is added to the fuel. After remaining in the fuel for an additional 3½ hours, the spindle is removed and the surface is rated for percent corrosion.

The NACE corrosion scale is a visual rating of surface rusting from 0 to 100 percent. Pipelines, refiners, and marketers of fuel require a NACE surface rust rating of less than 5 percent. The NACE corrosion rating scale is outlined in Table 7.2.

% Surface Rust	NACE Rating
0	A
1 to 3 spots <1 mm in diameter	B++
<5	B+
5 to 25	B
25 to 50	C
50 to 75	D
>75	E

TABLE 7.2 NACE Corrosion Test Spindle Rating Scale

Internationally, corrosion test method ASTM D665/IP 135 is often used. This method allows for the application of either distilled water or synthetic seawater as needed to initiate corrosion. The ferrous metal corrosion rating systems for the NACE corrosion and ASTM D665/IP 135 methods also differ. A summary of the ASTM D665/IP 135 test method appears below.

7.3.2 ASTM D665/IP 135 Standard Test Method for Rust-Preventing Characteristics of Inhibited Mineral Oil in the Presence of Water

7.3.2.1 ASTM Summary

A mixture of 300 mL of the oil under test is stirred with 30 mL of distilled water or synthetic seawater, as required, at a temperature of 60 ± 1°C (140°F) with a cylindrical steel test rod completely immersed therein. It is recommended to run the test for 4 hours; however, the test period may, at the discretion of the contracting parties, be shorter or longer. The test rod is observed for signs of rusting and, if desired, degree of rusting.

Often, this method is further defined as ASTM D665-A, ASTM D665-B, or ASTM 665-C. Procedure A requires the use of distilled water. Procedure B requires the use of a synthetic seawater solution. Procedure C was developed for evaluating heavier than water fluids using either distilled water or synthetic seawater. The inorganic salt composition of synthetic seawater is found in App. 5.

Interpretation of the results of ASTM D665 testing are often defined as either "Pass" with no rusting of the steel test rod, or "Fail," with rusting observed on the rod. However, a degree of rusting scale has been developed by ASTM to better define the rusting severity observed. The scale is provided in Table 7.3.

Rust Rating	Description
Light rusting	Rusting confined to not more than six spots, each of which is 1 mm or less in diameter
Moderate rusting	Rust in excess of the above but confined to <5% of the surface of the test rod
Severe rusting	Rusting covering >5% of the surface of the test rod

Table 7.3 ASTM D665 Rusting Severity Scale

7.4 Testing the Emulsion Tendencies of Fuel

Test methods in this section are useful in determining the following properties:

- Effect of water pH on fuel-water separation.
- Impact of fuel additives on fuel-water separation.
- Effect of high shear on fuel-water separation.
- Impact of free water and particulate such as rust and corrosion products on fuel clarity.

During transportation and storage of fuel, it is important to prevent contact of the fuel with water. Although under most circumstances, if water contact does occur, the fuel should naturally and quickly separate from water. However, there are situations that develop when water and fuel do not readily separate after contact, and the fuel begins to appear cloudy or hazy. Emulsification may also result leaving an emulsion layer at the fuel-water interface. The emulsion layer may eventually break apart on its own, or it may remain as a stable phase between the fuel and water layers.

To describe the ability of fuel and water to separate, the term *demulsibility* is used. Fuels that have good demulsibility properties do not form emulsions when in contact with water, and the two phases quickly separate.

Gasoline containing oxygenated organic compounds such as ethanol or salts of organic acids that may form during fuel processing may solubilize or emulsify water into fuel. Water is especially soluble in ethanol, and care must be taken to ensure that water does not come in contact with fuel-ethanol blends. If gasoline-ethanol blends are exposed to large volumes of water, ethanol may be extracted from the fuel. This can result in removal of most of the ethanol from the fuel. Other oxygenates, such as ETBE and TAME, will not separate from gasoline on exposure to high concentrations of water.

Unless required by ASTM, SAE, IP, ISO or other organizations, pipeline companies, refiners, and fuel marketers may establish their own criteria for rating fuel demulsibility. Also, fuel low temperature water tolerance limits are established in some regions of the world for each month of the year.

7.4.1 ASTM D1094/IP 289 Standard Test Method for Water Reaction of Aviation Fuels

This method provides information on the water separation properties of jet fuel and can be used to rate the tendency of the fuel to form emulsions in the presence of water. A newer test method, ASTM D7451, has been developed to rate the water separation properties of middle distillate and spark ignition fuels.

7.4.1.1 ASTM Summary

An 80-mL volume of fuel and a 20-mL volume of a pH 7 phosphate buffer solution are poured into a clean glass—100 mL graduated cylinder fitted with a glass stopper. The mixture is shaken for 2 minutes using a technique specified in the method. The cylinder is placed on a bench top and the fuel, water, and interface layers are rated using a standardized rating scale.

7.4.2 ASTM D1401/IP 79 Standard Test Method for Water Separability of Petroleum Oils and Synthetic Fluids

This ASTM test method is primarily used in rating the water separation properties of lubricants, but it has been adopted by the United States' military for rating *naval distillate* demulsibility.

7.4.2.1 Test Summary

To evaluate naval distillate fuel demulsibility, a 40-mL volume of distillate plus a 40-mL volume of synthetic seawater are added to a 100-mL glass cylinder. The cylinder and contents are placed in a heated testing unit and equilibrated at 25°C (77°F). A specialized stirring paddle is lowered into the cylinder, and the contents stirred at 1500 ± 15 rpm for 5 minutes. After stirring, the fuel-water-emulsion volumes are recorded at 1-minute intervals and the demulsification time reported to the nearest minute. A 10-minute time limit has been specified for complete separation into two phases with no visible cuff at the interface. A lacy emulsion or cuff which does not form a band is disregarded.

7.4.3 ASTM D3948 Standard Test Method for Determining Water Separation Characteristics of Aviation Turbine Fuels by Portable Separometer

This method is frequently described as the micro-separometer or MSEP test. It is also termed the *water separation index—modified* (WSIM), named after the original water separation index procedure.

7.4.3.1 ASTM Summary

A water-fuel sample emulsion is created in a syringe using a high-speed mixer. The emulsion is then expelled from the syringe at a programmed rate through a standard fiberglass coalescer and the effluent is analyzed for uncoalesced water by light transmission measurement. The results are reported on a 0 to 100 scale to the nearest whole number. High ratings indicate that water is easily coalesced, implying that the fuel is relatively free of surfactant materials. A test can be performed in 5 to 10 minutes.

Although water usually does not emulsify readily with jet fuel, trace amounts of surfactants and fuel additives picked up by the fuel during handling, transport, and storage may impact the ability of filters and coalescers to separate free water from fuel. Under various MSEP testing modes, fuels such as Jet A, Jet A-1, Jet B, JP-4, JP-5, JP-7, and JP-8 are evaluated. Under a DESP mode, diesel fuel can also be evaluated for water separation characteristics. Often jet fuel refiners place a minimum passing MSEP specification of 80 percent or greater light transmission measurement on finished jet fuel. The minimum passing value accepted by the U.S. military and the airline industry is a light transmittance rating of 70 percent.

Hardware used in testing includes a portable, self-contained separometer and a one-time use package of test supplies termed a *micro-separometer six pack* that includes a disposable plastic syringe with syringe plug, container of distilled water, premarked sample collection vials, pipet with plastic tip, catch pan for waste fuel, and an Alumicel coalescer.

7.4.4 ASTM D4176 Standard Test Method for Determining Free Water and Particulate Contamination in Distillate Fuel (Visual Inspection Procedures)

This ASTM method can be used to quickly determine the visual clarity of distillate fuel. A similar method, the Colonial Pipeline haze rating, has been used for many years by the finished product pipeline and fuel distribution industry to rate fuel clarity. A fuel clarity rating of "2" or better is considered a passing value throughout the fuel industry when compared against a standardized rating photograph of six fuels with increasing levels of clarity.

7.4.4.1 ASTM Summary

In Procedure 1, approximately 900 mL of fuel is placed into a 1-L clear glass jar and examined visually for clarity. The sample is then swirled and examined for visual sediment or water drops below the vortex.

In Procedure 2, approximately 900 mL of fuel is placed into a 1-L clear glass jar and examined visually for clarity. The fuel clarity is rated by placing a standard bar chart behind the sample comparing its visual appearance with the standardized haze rating photographs.

The sample is then swirled and examined for visual sediment or water drops below the vortex.

Both procedures 1 and 2 are performed immediately after sampling and at storage temperature conditions.

7.4.5 ASTM D7451 Standard Test Method for Water Separation Properties of Light and Middle Distillates, and Compression and Spark Ignition Fuels

Test method ASTM D7451 was developed as a complement to the long-standing ASTM D1094 method for determining the water reaction of aviation fuel. Method ASTM D7451 incorporates specialized rating scales for determining the water separation properties of other fuels including gasoline, kerosene, diesel fuel and distillates such as gas turbine fuel, marine fuel, home heating oil and furnace fuels. ASTM D7451 also accounts for the use of buffer solutions of various pH ranges to simulate fuel storage conditions where acidic water or alkaline water may be present in storage tank bottom layers.

The testing apparatus and operating procedures used in ASTM D7451 are the same as those utilized in ASTM D1094. However, the rating scales are wider in scope and include the following ratings:

- Table 1: Fuel clarity ratings
- Table 2: Fuel-water separation rating
- Table 3: Interface condition rating

7.4.6 Waring Blender Test

The Waring blender test is not an ASTM standard, but it is an optional method sometimes used to rate fuel demulsibility. The test method is described below:

7.4.6.1 Test Summary

A 475-mL volume of fuel and 25 mL of distilled water are placed into a stainless steel, explosion-proof blending vessel. The fuel and water are stirred at low speed (10,000 rpm) for 4 minutes using an explosion-proof Waring blender. The mixture is then poured into 1-quart square bottles and the fuel and water layers are rated for clarity after 1, 3, 5, 8, 12, and 24 hours. A rating scale ranging from 0 to 11 has been developed to describe fuel clarity and appearance. This rating scale is provided in App. 10.

7.4.7 Multicontact Emulsion Test or 10-Day Multiple Contact Test

This procedure is used to determine the emulsion tendencies of finished fuels in frequently filled storage tanks contaminated with small volumes of water. This would include storage tanks at fueling stations, convenience stores, or other fuel stops. The method is most often used to rate the emulsion tendencies of various additives in fuel.

7.4.7.1 Test Summary

Place 200 mL of fuel and 20 mL of test water into a 500-mL or 1-pint clear glass bottle. Test water can be distilled, deionized, saltwater, or buffer solutions of various pH values. Bottles are placed horizontally onto a mechanical shaker having a 1½ in stroke capable of 280 oscillations per minute such as Eberback No. 6010. Run the shaker at its highest setting for 5 minutes. Remove the bottle and stand it upright in the dark for 24 hours.

After the 24-hour period, rate the fuel layer, the interface, and the water layer according to the descriptions in App. 10. With a pipette, decant as much fuel as possible without disturbing the fuel-water interface. Carefully decant within 6.0 mm or ¼ in of the fuel-water interface. Discard the fuel. Leave the water to remain in the bottom of the test container. Add 200 mL of new fuel to the remaining water. Repeat this cycle of fuel addition, agitation, storage, rating, and fuel removal for 10 additional cycles. After the 10th contact cycle has been rated, do not remove the fuel, but store the bottle in an upright position in the dark for 7 days. Rate the fuel again after the 7-day storage period.

7.5 Testing the Low Temperature Performance Properties of Fuels

Test methods in this section are useful in determining the following:

- Impact of various fuel blend components on the cold temperature flow properties of fuel.
- Effect of temperature cycling on the pour point of crude oils and residual fuels.
- Tendency of components in fuel to block or impede the flow of fuel through filters under various cooling rate conditions.
- Performance comparison of different wax modifier chemistries in altering the low temperature flow and filterability properties of fuels.

Distillate fuel: When distillate fuel is cooled, wax crystals begin to form in the fuel. The temperature at which these crystals are first observed is identified as the cloud point of the fuel. Upon further cooling, the crystals begin to coalesce into a lattice-like network to form a semi-solid gel. The temperature at which this gelation occurs is termed the *pour point*.

Lowering the cloud point and pour point values of a distillate fuel can be accomplished by blending the fuel with a low wax content distillate stream such as kerosene or jet fuel. Also, additives are frequently used in conjunction with kerosene blending, or as a substitute for kerosene blending, to reduce the cloud point and pour point of distillate fuel.

Residual fuel: Crude oil and high boiling, high-viscosity petroleum fractions such as No. 6 fuel oil, atmospheric tower bottoms, and vacuum gas oil can contain wax that crystallizes at temperatures often above room temperature. It is not unusual for these oils to have base pour points of 40°C (104°F) or greater. In order to handle, pump, and utilize these heavy oils, the pour point and viscosity of these oils must be reduced. One method that is used to accomplish this is to dilute the residual oil with lower viscosity components such as light cycle oil, light cracked gas oils, or other distillate products.

Another common method that is used to improve the handling characteristics of heavy oils is to treat the oils with a wax crystal modifier. The process is similar to that used in diesel fuel treatment. Wax crystal modifiers used in heavy fuels are typically higher in molecular weight than those used in diesel fuel applications.

Behavior of wax in residual oils can sometimes be influenced by temperature cycling and long-term storage. As wax melts and recrystallizes though repeated cycles of heating and cooling during storage or transport, the pour point may be affected. If the pour point increases after exposure to continued heating and cooling cycles, a phenomenon called *pour point reversion* has occurred. Pour point reversion can sometimes be prevented by the addition of a wax crystal modifier to the residual oil.

Numerous test methods are available to evaluate the low-temperature properties of distillate fuel and residual fuel. Many of these methods are reviewed in this section.

7.5.1 ASTM D97/IP 15 Standard Test Method for Pour Point of Petroleum Products

7.5.1.1 ASTM Summary

After preliminary heating, the sample is cooled at a specific rate and examined at intervals of 3°C for flow characteristics. The lowest temperature at which movement of oil is observed is recorded as the pour point.

The pour point evaluation is used to determine the lowest temperature at which a fuel can be pumped. This manual test method is effective for most oils, but the ASTM D97 pour point can sometimes be misleading, especially when it is used to determine the low temperature handling characteristics of residual fuel oil and crude oil. For residual fuels and crude oil, test method ASTM D5853 is recommended.

7.5.2 Automated ASTM Pour Point Test Methods

A listing and general description of automated ASTM pour point methods are outlined in Table 7.4.

7.5.3 ASTM D2500/IP 219 Standard Test Method for Cloud Point of Petroleum Products

7.5.3.1 ASTM Summary

The sample is cooled at a specific rate and examined periodically. The temperature at which a cloud is first observed at the bottom of the test jar is recorded as the cloud point.

The cloud point test is one of the most commonly used methods to evaluate the low-temperature characteristics of distillate fuel. The cloud point temperature identifies the point when wax begins to form into crystals large enough to become visible in the fuel. At this temperature, wax can settle from fuel, deposit onto fuel filters, and interfere with the flow of fuel through small tubes and pipes. During cold weather months, distillate fuels with lower cloud point values are refined and blended to minimize the cold temperature problems associated with wax.

Additional low-temperature fuel filterability information is often required to confirm whether fuel will pass through fuel filters in cold environments. Tests commonly used to determine this performance include

- ASTM D4539 Low Temperature Flow Test (LTFT)
- ASTM D6371 Cold Filter Plugging Point of Diesel and Heating Fuels
- ASTM D7501 Determination of Filter Blocking Potential of Biodiesel Blend Stock by Cold Soak Filtration Test (CSFT)

7.5.4 Automated ASTM/IP Cloud Point Test Methods

A listing of and general description of ASTM/IP cloud point testing methods are provided in Table 7.5.

7.5.5 ASTM D5853 Standard Test Method for Pour Point of Crude Oils

This test method was developed because upon cooling, the equilibrium between crystallized wax and dissolved wax in crude oils and residual fuel oils develops more slowly than in distillate fuels. In most crude oils and residual fuel oils, though, this state of equilibrium between crystallized and dissolved is reached within 24 hours. Because of this, measuring the pour point of oil too soon after removal from a heating source may provide an inaccurate result.

Test Designation	Name	General Description
ASTM D5949	Automatic Pressure Pulsing Method	Nitrogen gas under pressure is pulsed across the surface of fuel cooled at 1 to 3°C intervals. The lowest temperature whereby movement occurs is considered the pour point.
ASTM D5950	Automatic Tilt Method	Fuel in a pour point tube is cooled in a refrigerated jacket at 1 to 3°C intervals. The jacket containing the tube is tilted at the prescribed temperature intervals and the fuel surface is monitored for movement. The lowest temperature at which movement is detected is considered to be the pour point.
ASTM D5985	Rotational Method	A specimen cup containing cooled fuel that is constantly being monitored for temperature is rotated at approximately 0.1 rpm. A balanced, stationary, spherical pendulum is suspended in the fuel. When wax crystallization occurs in the cooled fuel or the fuel viscosity increases to cause movement of the pendulum, the recorded temperature is reported as the pour point.
ASTM D6749	Automatic Air Pressure Method	One end of a communicating tube is inserted into a test specimen contained in an airtight test container while the other end of the tube is exposed to atmospheric pressure. On cooling at 1 to 3°C intervals, air pressure is applied to the fuel surface resulting in upward movement of fuel within the communicating tube and detection by a pressure sensor. The pour point is the lowest temperature whereby surface movement and pressure change is detected.

Table 7.4 Automated ASTM Pour Point Test Methods

Test Designation	Name	General Description
ASTM D5771 IP 444	Optical Detection Stepped Cooling Method	Fuel is placed into a test jar with a mirrored base and incrementally cooled through a stepwise mode. An optical system emitting light onto the mirrored base detects a change when wax crystals form in the fuel. The system temperature is then recorded within 0.1°C.
ASTM D5772 IP 445	Linear Cooling Rate Method	Fuel in a test cell is cooled through a linear mode while a reflective optical system emits light horizontally through the fuel volume. A change in light transmittance is detected when wax crystals form in the fuel. The system temperature is then recorded with 0.1°C.
ASTM D5773 IP 446	Constant Cooling Rate Method	Fuel in a test receptacle is cooled at a constant rate of 1.5 ± 0.1°C/minute while optically monitored for the first appearance of a cloud of wax. This temperature is recorded to 0.1°C and rounded to the next lowest temperature and reported as equivalent to the ASTM D2500/IP 219 cloud point.
ASTM D7397	Miniaturized Optical Method	Fuel in a low volume test cell is cooled at a rate of 30 ± 5°C/minute until the first appearance of a cloud of wax. Fuel is then warmed at a rate of 15 ± 5°C/minute until wax crystals are no longer detected. The fuel is cooled a second time at a slower rate of 6 ± 3°C/minute until a cloud of wax first appears. This temperature is recorded to 0.1°C and considered as the cloud point.

Table 7.5 Automated ASTM / IP Cloud Point Test Methods

7.5.5.1 Test Description

A preheated crude oil sample is cooled under the specified ASTM D97 conditions and evaluated at 3°C cooling increments for flow. The lowest temperature observed where flow or movement of the oil surface occurs is considered the pour point.

7.5.6 IP 309 Cold Filter Plugging Point

Test method IP 309 is used to determine the low-temperature filterability of distillate fuels including those treated with cold flow improver. The cold filter plugging point (CFPP) is defined as "the highest temperature at which the fuel, when cooled under the prescribed conditions, will not flow through the filter or requires more than 60 seconds for 20 mL to pass through or fails to return completely to the test jar." This method is part of a series of standardized tests developed by the Institute of Petroleum.

7.5.6.1 Test Summary

The fuel sample is cooled at intervals of 1°C, and a vacuum of 200 mm H_2O is applied to draw the fuel through a fine wire mesh filter. As the fuel cools below its cloud point, increasing amounts of wax crystals will be formed. These crystals will cause the flow rate to decrease, and eventually complete plugging of the filter will occur.

The CFPP was developed in 1965 to 1966 to predict the low-temperature filterability performance of diesel fuels. This method is commonly used throughout the world to quickly determine fuel filterability characteristics as the fuel cools. In 2005 ASTM D6371 Standard Test Method for Cold Filter Plugging Point of Diesel and Heating Fuels was issued by ASTM and modeled after IP 309.

7.5.7 ASTM D6371 Standard Test Method for Cold Filter Plugging Point of Diesel and Heating Fuels

7.5.7.1 ASTM Summary

A specimen of the sample is cooled under specified conditions. At intervals of 1°C, fuel is drawn into a pipet under a controlled vacuum through a standardized wire mesh filter. The procedure is repeated as the specimen continues to cool for each 1°C below the first temperature. Testing is continued until the amount of wax crystals that have separated out of solution is sufficient to stop or slow down the flow so that the time taken to fill the pipet jar exceeds 60 seconds, or the fuel fails to return completely to the test jar before the fuel has cooled by a further 1°C. The indicated temperature at which the last filtration was commenced is recorded as the CFPP.

7.5.8 ASTM D4539 Standard Test Method for Low-Temperature Flow Test

7.5.8.1 ASTM Summary

The temperature of a series of samples of test fuel is lowered at a controlled cooling rate. Commencing at a desired test temperature and at each 1°C interval thereafter, a separate sample from the series is filtered through a 17-micron screen until a minimum LTFT pass temperature is obtained. The minimum LTFT pass temperature is the lowest temperature expressed as a multiple of 1°C at which a minimum of 180 mL of sample, when cooled under the prescribed conditions, can be filtered in 60 seconds or less. Alternatively, a single sample may be cooled as described above and tested at a specific temperature to determine whether it passes or fails at that temperature.

Provided in Table 7.6 is a listing of various parameters comparing the ASTM D6371 CFPP and ASTM D4539 LTFT distillate fuel testing methods.

7.5.9 ASTM D7501 Standard Test Method for Fuel Filter Blocking Potential of Biodiesel (B100) Blend Stock by Cold Soak Filtration Test

7.5.9.1 Test Description

A 300-mL volume of B100 biodiesel fuel is held at 4.5°C (40°F) for 16 hours. The biodiesel is then removed from storage and maintained at room temperature until a fuel temperature of 25°C (77°F) is reached. The biodiesel is vacuum filtered through a 0.7-μm glass fiber filter, and the filtration time in seconds is recorded.

Test Criteria	CFPP	LTFT
Cooling rate	35 to 40°C/h drop in temperature	1.0°C/h drop in temperature
Sample size	45 mL	200 mL
Filter diameter	12 mm	16 mm
Filter pore size	45 microns	17 microns
Filtration vacuum	2 ± 0.05 kPa (200 ± 1 mm H_2O)	20 ± 0.2 kPa
Passing value	20 mL of fuel filtered in 60 seconds	180 mL of fuel filtered in 60 seconds

TABLE 7.6 Comparison of ASTM D6371 CFPP and ASTM D4539 LTFT Test Methods

7.5.10 British Admiralty Pour Point Test

This procedure can be utilized to determine whether heavy fuel wax crystal modifiers will lose their performance properties after long-term storage at fluctuating temperatures. Daily heating and overnight cooling may interfere with the ability of wax crystal modifiers to maintain their performance properties in some residual oils and crude oils. This loss of performance is frequently termed pour point reversion. The British Admiralty Pour Point Test can be utilized to help predict these reversion tendencies.

7.5.10.1 Test Summary

1. An oil sample is treated with the proper amount of wax crystal modifier to obtain the pour point desired.
2. A sample of the treated oil is heated to 93°C (200°F) and poured immediately into pour point jars fitted with a thermometer capable of reading temperatures from −37°C (−35°F) to 99°C (210°F).
3. Allow the sample in the pour point test jars to cool to 49°C (120°F), and then place the tubes into a −34°C (−30°F) cold well. Allow the test jars to remain in the bath until the temperature of the oil is −18°C (0°F).
4. Remove the test jars and allow them to warm slightly on the benchtop. Next, place the jars in an oven or bath maintained at 96°C (205°F). When the test samples have reached 93°C (200°F), hold for an additional 30 minutes.
5. Remove the test jars and allow them to cool on the benchtop to 32°C (95°F).
6. Once cooled, place the oil samples into the pour point test well maintained at −1°C (30°F). Conduct the remainder of the test as described in ASTM D97 checking the pour point every 3°C (5°F).
7. Compare the pour point obtained for the oils tested by the reversion test method against the pour point of the original sample containing wax modifier evaluated in Step 1. If the pour point temperature obtained is higher than the Step 1 sample temperature, a reversion in pour point has occurred.

Note: In addition to the 96°C (205°F) oven or bath used in Step 4, additional ovens can be heated to 85, 68, 57, 49, and 41°C or 185, 155, 135, 120, and 105°F. A total of six separate test samples could be evaluated for reheating temperature effect.

7.5.11 Shell Amsterdam Pour Point Test

This method describes a procedure for determining the critical pour point of residual fuel oils.

7.5.11.1 Test Summary

Apply heat to a stirred sample of oil to be tested. Once heated to 104°C (219°F), pour the oil into a pour point test jar to a height of approximately 57 mm. Place a cork carrying a thermometer onto the test jar and place set the sample into a –40°C (–40°F) test well.

When the oil reaches a temperature of –20°C (–4°F), remove the test jar from the well. If the oil is not solid at –20°C (–4°F), continue cooling the test sample until it becomes solid, and then remove the sample. Place the test jar in a water bath held at a temperature between 47°C and 47.5°C (116.6°F and 117.5°F). When the oil sample has reached a temperature of 46°C (115°F), remove the sample and perform the ASTM D97 pour point to obtain the critical pour point.

7.6 Cetane Number Testing

In a diesel engine, air is compressed to pressures of 1000 psi or greater. Upon compression, the air temperature increases to the range of 315 to 425°C (599 to 797°F). At close to maximum compression, diesel fuel is injected into this hot, compressed air. Ignition of fuel within this environment is initiated whenever fuel components with the lowest autoignition temperature begin to combust.

Occasionally, cetane improvers are added to diesel fuel and are capable of increasing the cetane number of diesel fuel from 1 to 12 numbers or higher. The reason for this range of response is due to the differences in the chemical nature of the fuel. If fuel contains a higher level of aromatic compounds, it will typically respond less effectively to cetane improver treatment than fuels with fewer aromatic compounds. Fuel cetane number is determined by standard test methods including cetane engine testing and automated constant volume combustion chamber testing methods.

7.6.1 ASTM D613/IP 41 Standard Test Method for Ignition Quality of Diesel Fuel by the Cetane Engine Method

7.6.1.1 ASTM Summary

The cetane number of diesel fuel is determined by comparing its ignition quality with those for blends of reference fuels of known cetane number under standard operating conditions. This is done by varying the compression ratio for the sample and each reference fuel to obtain a fixed delay period, that is, the time interval between the start of ignition and ignition. When the compression ratio for the sample is bracketed between those for two reference fuel blends differing by more than five cetane numbers, the rating of the sample is calculated by interpolation.

Test Designation	ASTM Title
ASTM D6890	Standard Test Method for Determination of Ignition Delay and Derived Cetane Number (DCN) of Diesel Fuel Oils by Combustion in a Constant Volume Chamber
ASTM D7170	Standard Test Method for Determination of Derived Cetane Number (DCN) of Diesel Fuel Oils—Fixed Range Injection Period, Constant Volume Combustion Chamber
ASTM D7668	Standard Test Method for Determination of Derived Cetane Number (DCN) of Diesel Fuel Oils—Ignition Delay Combustion Using a Constant Volume Combustion Chamber

TABLE 7.7 Automated Cetane Number Test Methods

7.6.2 ASTM Automated Cetane Number Testing Methods

ASTM automated cetane number testing methods are listed in Table 7.7. These methods produce a derived cetane number (DCN) upon injection of fuel into heated air and measurement of combustion ignition delay. There is variation in the DCN range produced by each method.

7.7 Hydrogen Sulfide and Mercaptan Determination

Test methods in this section are useful in determining the following:

- Possible cause of failure of fuel to meet a copper, brass, bronze, or silver corrosion specification.
- Confirm the likely cause of a fuel odor problem.
- Identify and quantify the type and amount of sulfur containing compounds present in fuel.
- Measuring the rate of hydrogen sulfide scavenger required to reduce the H_2S level in both the fuel vapor and liquid phases.

Hydrogen sulfide and mercaptans can accumulate in the vapor and liquid phases of fuels contained in storage tanks, rail cars, ground transportation vehicles, and storage compartments of seafaring transport tankers. Vapor phase head space measurements are available to evaluate the amount of free H_2S or mercaptans that may accumulate in these areas. Liquid phase titration methods and chromatographic methods are used to detect low levels of hydrogen sulfide, mercaptans, and elemental sulfur in crude oils, fuels, and various refinery process fractions. Some of the more frequently utilized test methods for detecting these sulfur compounds in petroleum products are described in this section.

7.7.1 ASTM D2420 Standard Test Method for Hydrogen Sulfide in Liquefied Petroleum Gases (Lead Acetate Method)

7.7.1.1 Test Description

Specially produced lead acetate paper that changes color when exposed to H_2S in LP gas vapor is moistened with water and suspended in a glass test cylinder. Gas vapors are passed through the cylinder at a rate of 2.3 L/min for 2 minutes. Based on paper color, the LPG is then rated as either pass or fail.

7.7.2 ASTM D3227/IP 342 Standard Test Method for (Thiol Mercaptan) Sulfur in Gasoline, Kerosene, Aviation Turbine, and Distillate Fuels (Potentiometric Method)

7.7.2.1 Test Description

The test sample must not contain H_2S. Mercaptan level is determined by dissolving the fuel in an alcohol solution of sodium acetate and using a solution of silver nitrate in potentiometric titration. The potential difference between a glass reference electrode and a silver/silver-sulfide indicating electrode serves as an indicator. Mercaptan sulfur is precipitated as silver sulfide. A large change in cell potential indicates an end point.

7.7.3 ASTM D4952 Standard Test Method for Qualitative Analysis for Active Sulfur Species in Fuels and Solvents (Doctor Test)

7.7.3.1 Test Description

Test fuel is added into a glass cylinder followed by the addition of an aqueous solution of sodium plumbite. The cylinder is shaken for 15 seconds, and pure flakes of sulfur are sprinkled onto the solution surface. The cylinder is shaken again for 15 seconds. A brown to black discoloration of the floating sulfur flakes or of the test solution indicates the presence of mercaptans or hydrogen sulfide. Peroxides may produce a false positive. Avoid UV light.

7.7.4 ASTM D5623 Standard Test Method for Sulfur Compounds in Light Petroleum Liquids by Gas Chromatography and Sulfur Selective Detection

7.7.4.1 Test Description

A gas chromatography analyzer fitted with a proper sulfur selective detector, such as a sulfur chemiluminescence detector or an atomic emission detector operated in the sulfur mode, is used to evaluate fuel for the presence of sulfur-containing compounds. To ensure that all sulfur compounds detected produce a comparable response as sulfur, both internal and external calibration standards are utilized. The various sulfur compounds eluting from the column of the gas chromatograph are quantified by the detector yielding an equimolar and linear response to the presence of sulfur compounds.

7.7.5 ASTM D5705 Standard Test Method for Measurement of Hydrogen Sulfide in the Vapor Phase above Residual Fuel Oils

7.7.5.1 ASTM Summary

A 1-L H_2S insert test container (glass test bottle) is filled to 50 volume percent with fuel from a filled H_2S insert container (glass sample bottle) just prior to testing. In the test container, the vapor space above the fuel oil sample is purged with nitrogen to displace air. The test container with sample is heated in an oven to 60°C (140°F) and agitated on an orbital shaker at 220 rpm for 3 minutes. A length-of-stain detector tube and hand operated pump are used to measure the H_2S concentration in the vapor phase of the test container. The length-of-stain detector tube should be close to but not in contact with the liquid surface.

7.8 Additional Tests Used to Solve Fuel Problems

There are a number of analytical tests that can be used to help identify the cause of a fuel performance problem. Some of these tests are described below:

7.8.1 ASTM D86/IP 123 Standard Test Method for Distillation of Petroleum Products

Use:

- Determine whether fuels are within specification (IBP, T-90, residue, etc.)
- Identify possible contamination with low boiling or high boiling components
- Address questions about poor fuel performance such as startability or power
- Identify a possible source of high deposit level in engines
- Determine whether diesel fuel distills within a narrow distillation profile, indicating a possible wax related, cold flow problem

7.8.2 ASTM D482 Standard Test Method for Ash from Petroleum Products

Use:

- Determine the level of catalyst fine contamination in residual fuel
- Identify whether fuel has been contaminated with lubricating oil
- Address concerns about organometallics in fuel

7.8.3 ASTM D664 Standard Test Method for Acid Number of Petroleum Products by Potentiometric Titration

Use:

- Useful in determining whether a fuel has aged significantly
- Identify whether fuel has been contaminated with acidic or alkaline components

7.8.4 ASTM D873/IP 138 Standard Test Method for Oxidation Stability of Aviation Fuels (Potential Residue Method)

Use:

- Predict the gum or deposit forming tendencies of kerosene, jet, or diesel. Due to high pressures generated, use with caution for predicting deposit forming tendency of gasoline or naphtha fractions
- Confirm complaints of fuel instability
- Evaluate the performance of fuel antioxidants and stabilizers

7.8.5 ASTM D1319/IP 95 Standard Test Method for Hydrocarbon Types in Liquid Petroleum Products by Fluorescent Indicator Adsorption

Use:

- Identify the paraffin, olefin, naphthene, and aromatic content of fuel; good for gasoline; not as accurate for jet, diesel, and other middle distillates
- Identify the source of a fuel gum, sludge, or deposit
- Identify the source of a fuel color degradation problem

7.8.6 ASTM D2425 Standard Test Method for Hydrocarbon Types in Middle Distillates by Mass Spectrometry

Use:

- Identify compounds that affect cold temperature flow properties
- Identify paraffins, naphthenes, and aromatics that contribute to fuel cetane number
- Identify compounds that impact fuel stability, deposits, and color
- Identify the reason for low cetane number; octane number
- Addressing fuel performance problems such as smoke, soot, and power loss

7.8.7 ASTM D2624 Standard Test Method for Electrical Conductivity of Aviation and Distillate Fuels

Use:

- Determine if fuel meets basic conductivity specifications, especially hydroprocessed fuel
- Measure the rate of conductivity increase provided by conductivity improver
- Monitor the possible interference of corrosion inhibitor or lubricity improver on conductivity

7.8.8 ASTM D3120 Standard Test Method for Trace Quantities of Sulfur in Light Liquid Petroleum Hydrocarbons by Oxidative Microcoulometry

Use:

- Identifying the sulfur content of most fuels and oils

7.8.9 ASTM D3227/IP 342 Standard Test Method for Mercaptan Sulfur in Gasoline, Kerosene, Aviation Turbine Fuel, and Distillate Fuels (Potentiometric Method)

Use:

- Identifying the cause of copper, bronze, or brass corrosion
- Identify the cause of elastomer degradation
- Identifying the source of a fuel odor problem

7.8.10 ASTM D3703 Standard Test Method for Peroxide Number of Aviation Turbine Fuel

Use:

- Predict fuel stability or instability versus time
- Identify the cause of elastomer degradation

7.8.11 ASTM D6423 Standard Test Method for Determination of pHe of Ethanol, Denatured Ethanol, and Fuel Ethanol

Use:

- Determine whether ethanol has aged to become acidic and possibly corrosive
- Measuring the impact of ethanol corrosion inhibitor on maintaining pHe between 6.5 and 9.0.

CHAPTER 8

Identifying and Solving Specific Fuel Problems

Specific problems related to fuel handling, performance and use are presented in this chapter. Problems reviewed are taken from actual field applications. Some of the problems described have more than one possible cause. By reviewing all possible causes given for each problem, it may then be possible to develop a cause and effect scenario using known information. If more information is needed, the steps suggested in the "What to Do" section can be taken.

8.1 Cold Temperature Problems

8.1.1 Difficulty in Pumping or Filtering Diesel Fuel at Low Temperatures

8.1.1.1 Possible Cause: Wax

Measuring the cloud point, pour point, CFPP, or LTFT will help to confirm whether pumping or filterability problems are due to wax. Fuels at their cloud point, CFPP, or LTFT temperatures will eventually plug filters when pumped. When a fuel is very near its pour point, pumping will be quite difficult:

What to Do:

- Determine the ASTM D97, D1500, D4593, D6371, or IP 309 temperatures.
- Look for higher than expected filtration or pumpability temperatures.
- Look for wax on fuel filters, within tanks or lines, etc.

Potential Solutions:

- Blend fuel with kerosene or No. 1 diesel to dilute the effect of wax.
- Treat fuel with wax crystal modifier to control the growth of wax crystals.
- Use fuel with a lower initial boiling point (IBP), end point (EP).
- Heat trace system components.
- Add solvents such as propyl alcohol or diethylene glycol to disperse wax.
- As a last resort, add gasoline in low concentrations to dissolve wax.

8.1.1.2 Possible Cause: Ice

Water contamination in diesel fuel is common. When diesel fuel cools, ice crystals may form in the fuel well before the fuel reaches its pour point. These ice crystals will settle to the bottom of fuel tanks and may result in fuel filter plugging and pumping problems.

What to Do:

- Check water in diesel fuel by ASTM D1744 Karl Fischer method.
- Hold fuel at –12 to –6°C (10 to 20°F) overnight in a centrifuge tube; look for ice crystals in the bottom of the tube.

Potential Solutions:

- Drain or filter water from the fuel system routinely.
- Use an anti-icing agent or solvent in fuel such as propyl alcohol or diethylene glycol compounds.

8.1.1.3 Possible Cause: Narrow Distribution of Fuel Paraffins

Fuels with a narrow paraffin distribution often respond poorly to treatment with a wax crystal modifier or may require high rates of wax crystal modifier to improve filtration and pumpability performance. This problem seems to occur whenever two situations exist, as listed below:

1. The *n*-paraffin carbon number distribution for a fuel is narrow.
2. Few low molecular weight *n*-paraffins are present in the fuel.

An example is provided in Table 8.1.

The response of fuels to wax modifier treatment may also be manifested in the fuel distillation profile. More responsive fuels may have a lower IBP and EP than poorly responsive fuels. Also, in poorly responsive fuels, a larger volume of fuel may distill within mid-boiling

Responsive Fuel	Poorly Responsive Fuel
>1% *n*-paraffins at C_7	No *n*-paraffins at C_7
4% paraffins < C_8	<1% paraffins < C_8
Highest % *n*-paraffins at C_{11}	Highest % *n*-paraffins at C_{15}
5% *n*-paraffins at C_{20}	7% *n*-paraffins at C_{20}

TABLE 8.1 Response of Fuels to Wax Crystal Modifier Treatment Based on *n*-Paraffin Content

point temperature fractions of the distillation profile. The wax crystals that may form from these narrow boiling fractions may tend to crystallize rapidly at about the same temperature. Diesel fuel pumpability problems can be due to these narrow boiling range fractions.

What to Do:

- Characterize the carbon number, distribution, and concentration of fuel paraffins.
- ASTM D-86 distillation and look for narrow boiling point fractions.

Potential Solutions:

- Blend with low-viscosity kerosene or No. 1 fuel.
- Treat with a wax crystal modifier effective at modifying high carbon number or narrow boiling point temperature range wax.

8.1.1.4 Possible Cause: Viscosity Limited, High IBP/EP Fuel

The distillation profile will provide some indication of the low temperature performance of diesel fuel. IBP values greater than 177°C (350°F) and EP temperatures greater than 371°C (700°F) indicate that the fuel viscosity will probably be high at cold temperatures. Even fuel treated with a wax crystal modifier may be difficult to pump due to the high viscosity of the fuel components.

What to Do:

- ASTM D86 distillation; look for high IBP and EP values.
- If possible, measure the viscosity versus temperature increase between the cloud and pour point values; look for a dramatic increase in viscosity rather than a gradual increase.

8.1.1.5 Possible Cause: Biodiesel Blended into Fuel

Biodiesel produced from palm oil or animal fats may form a fuel insoluble gel or wax-like precipitate when stored in cold temperature environments. If blended into diesel fuel during cold weather months, methyl palmate and methyl stearate components present in biodiesel may form a gel and accumulate in cold lines and on fuel filters to halt the flow of fuel to a fuel pump or engine.

What to Do:

- Know the handling properties and basic composition of various oil sources used to produce B100 FAME such as soy, canola, tallow, RME, palm, jatropa, and others.
- If a detailed problem solving analysis is needed, complete a carbon number distribution of the fuel or precipitate gel formed in cold fuel and look for an increase in C_{16} and C_{18} carbon content, ester content, or unsaturation common to methyloleate and methyllinoleatoleate.

Potential Solutions:

- Minimize the use of high cloud point FAME based on palm oil and animal fats in cold temperature environments.
- Maintain mild heating on storage tanks and fuel transfer lines used in B100 FAME and biodiesel fuel applications.
- Determine the impact of biodiesel blending on the final cloud point or CFPP of a diesel fuel blend; reduce or eliminate B100 FAME from diesel blends if the fuel fails to meet seasonal cloud point of CFPP specifications.

8.1.2 Reversion and Increase in Pour Point of Crude Oil or Residual Fuel

8.1.2.1 Possible Cause: Hindrance of Wax Crystal Modifier Performance

Upon treatment with wax crystal modifier, the pour point of some crude oils and residual oils can be effectively reduced by 16°C (30°F) or more. Over time, and upon exposure to changes in temperature during storage, the effect of the wax modifier on the pour point and low-temperature flow of the oil can be lost and may revert upward in temperature. This can be problematic for oils stored in tankage or transported via underground pipeline since the reversion in flow properties can negatively impact the pumpability and low-temperature viscosity of the oil. The oil may no longer flow readily from a tank or flow at the required rate through a pipeline. In extreme cases, the oil may gel and revert back to the base pour point prior to additive addition.

This phenomenon is believed to be due to the influence of certain types of high molecular weight waxes, resins, or asphaltene compounds in the oil on the performance of the wax crystal modifier. Upon initial treatment, wax modifiers and asphaltenes inhibit the formation of a wax lattice network throughout the oil matrix enabling the oil to flow. However, upon continued storage, changes occur, and the wax modifier performance is lost.

One possible reason for this change is due to the inability of the wax modifier to completely overcome the large amount of wax initially present in the oil, and the effect on pour point is lost. A second possibility involves interaction of high molecular weight asphaltenes in the oil with the wax modifier polymer to inhibit the ability of the modifier polymer to fully disperse in the oil. The wax crystal modifier performance is minimized.

What to Do:

- Determine whether the oil treated with a wax crystal modifier will revert in pour point by utilizing either ASTM D5853 Pour Point of Crude Oils method, the British Admiralty Pour Point Reversion test, or the Shell Amsterdam Critical Pour Point method.

Potential Solutions:

- If testing indicates that the oil has reversion tendencies, the addition of higher treat rates of a wax modifier can often overcome the reversion problem.
- Sometimes, reversion in pour point cannot be overcome by the addition of a higher treat rate of a wax crystal modifier. Only dilution with a low viscosity petroleum product or the constant addition of heat will keep the oil fluid at temperatures below the base pour point of the oil.

8.2 Corrosion Problems

8.2.1 Rusting is Identified on Metal Components

8.2.1.1 Possible Cause: Fuel Contains High Concentrations of Water

Water contamination of fuel occurs. Water can originate from fuel processing, atmospheric condensation, or external sources. Water may contain dissolved salts, may be acidic or basic, or may contain solubilized organic compounds. Water-initiated corrosion can result.

What to Do:

- Check the fuel storage tank and distribution system for water.
- Determine the water pH; check for the presence of NaCl in the water.
- Look for the existence of an emulsion at the fuel-water interface.

Potential Solutions:

- Maintain a frequent schedule of draining accumulated water from the bottom of the fuel storage tank.
- Add a ferrous metal corrosion inhibitor to the fuel at a maintenance rate of 5 to 10 ppm.

8.2.1.2 Possible Cause: Microbiologically Influenced Corrosion

Microorganisms can establish growth colonies or plaques on metal surfaces. Corrosion of underlying metal can be severe. Both bacterial and fungal species can initiate this type of corrosion.

What to Do:

- Check tank bottoms and water for the presence of microorganisms and the rate of microbial contamination, both aerobic and anaerobic.
- If the cell count rate is in the range of 100,000 to 1,000,000, biocide treatment is warranted.

Potential Solution:

- Treat the water with a water-soluble biocide approved for use in fuel applications.
- Continue weekly treatment until the microbial growth is under control or water has been drained from the fuel tank.
- An optional method of treatment is to add an oil-soluble, fuel approved biocide into the fuel as it enters tankage.

8.2.2 Ferrous Metal Corrosion Inhibitor Fails to Prevent Rusting of Metal

8.2.2.1 Possible Cause: Caustic Neutralization of Corrosion Inhibitor

During the processing of fuels and fuel components, solutions of sodium hydroxide or potassium hydroxide may be used. These caustic solutions can react with phenols, mercaptans, and naphthenic acids to form water-soluble salts. As a result, these undesirable components are removed from the fuel as the caustic separates from the fuel.

It is possible, however, for some residual caustic to carry over into the finished fuel. Residual caustic that is present may react with a carboxylic acid–based corrosion inhibitor. If this occurs, the ability of the inhibitor to form a protective film on ferrous metal surfaces will be destroyed.

Caustic resistant corrosion inhibitors for fuel have been developed to help overcome this neutralization concern. However, these products must often be used at higher concentrations than carboxylic acid–based inhibitors. Also, some caustic resistant inhibitors tend to emulsify with water later in the fuel distribution channel.

What to Do:

- Check the water in the bottom of the fuel storage vessel for the presence of high levels of sodium or potassium ions by ICAP or AA.
- Check for the presence of sodium or potassium salts of carboxylic acids by IR or FTIR.
- Check the pH of water. If greater than 7, caustic contamination may have occurred.

Potential Solutions:

- Drain water bottoms from the storage tank to remove contaminated water.
- Utilize a caustic resistant corrosion inhibitor.

8.2.2.2 Possible Cause: Acid Initiated Corrosion

Sulfuric acid is used as a catalyst in the production of gasoline alkylate. After processing, this acid must be removed from the finished alkylate. Removal is typically accomplished by gravitational settling, water washing or caustic washing the alkylate. However, if residual sulfuric acid remains in alkylate after settling or washing, use of the alkylate in fuel blending can result in metal corrosion. Filming forming type corrosion inhibitors used in most fuel applications typically do not prevent sulfuric acid initiated corrosion.

8.2.2.3 Possible Cause: Removal of Corrosion Inhibitor Film from Pipeline Wall

In the unlikely event that fuel containing caustic water is shipped from a refinery through a transportation pipeline, removal of the protective corrosion inhibitor film from the pipeline wall by caustic is possible. This removal of inhibitor exposes the pipeline wall to the possibility of corrosion.

What to Do:

- Check outgoing fuel storage tank bottoms for the presence of caustic water by pH determination.

Potential Solution:

- Remove tank bottom water prior to shipment of fuel via pipeline.

8.2.3 Fuel Corrodes Copper, Bronze, Brass, or Silver

8.2.3.1 Possible Cause: Fuel Contains Hydrogen Sulfide or Elemental Sulfur

Copper, copper-containing alloys, and silver are susceptible to attack by H_2S and elemental sulfur. Various cracking reactions and hydroprocessing produce H_2S as a reaction product. If H_2S is not completely removed or stripped from fuel, it can carry into finished products such as gasoline and diesel to create fuel odor and corrosion problems such as copper and silver corrosion.

What to Do:

- Utilize potentiometric titration or length-of-stain detector tube to detect H_2S in fuel.
- Evaluate the fuel for the presence of elemental sulfur by ASTM D5623.

Potential Solutions:

- Caustic process fuel to remove H_2S, mercaptans, and elemental sulfur.
- Treat fuel with a scavenger to remove H_2S.
- Utilize a copper or silver corrosion inhibitor to prevent corrosion by H_2S or elemental sulfur.
- Passing fuel through molecular sieves may remove corrosive sulfur compounds.

8.2.3.2 Possible Cause: Fuel Contains Dissolved Sulfuric Acid

Trace levels of sulfuric acid used in processing fuel can mistakenly carry over into finished products. Copper, bronze, brass, and silver corrosion may result.

What to Do:

- Extract acidic components from fuel with distilled water and check pH.

Potential Solution:

- Lightly caustic wash fuel to remove sulfuric acid. The sodium sulfate salt which forms is water soluble and is removed from the fuel.
- Passing fuel through molecular sieves may remove acidic compounds.

8.3 Emulsion, Haze, or Particulate Contamination

8.3.1 Haze or Emulsion Forms in Fuel

8.3.1.1 Possible Cause: Water Initiated Emulsification of Polar Organics in Fuel

Under certain conditions, fuel detergents, corrosion inhibitors, and other performance enhancing additives can emulsify with water to form water-in-oil emulsions. Conditions of low pH, high pH, or high additive concentration may enhance emulsification of polar organic compounds into fuel.

What to Do:

- Check water for low or high pH.
- Determine whether fuel is treated with performance enhancing additives.

Potential Solutions:

- Utilize an emulsion breaking additive to clear the fuel of haze or emulsion.
- Frequently drain the accumulated water from the bottom of the fuel storage system. This preventive measure alone may eliminate the possibility of water-fuel emulsions.

8.3.1.2 Possible Cause: Presence of Microorganisms

Microorganisms living at the fuel-water interface in fuel storage systems can accumulate to form a distinct, gray to brown colored, rag-like layer of growth. This growth layer can break free and appear as an emulsion in fuel. The products of microbial metabolism are highly polar and can help to stabilize fuel emulsions.

What to Do:

- Check the fuel-water interface for the presence of a microbiological growth layer.

Potential Solutions:

- Treat water with a broad spectrum, water-based biocide. Maintain treatment weekly.
- Frequently drain water bottoms from fuel storage systems.
- Consider the continuous application of an oil-soluble fuel biocide to help minimize growth throughout the fuel distribution and storage system.

8.3.1.3 Possible Cause: Fuel Stream Containing Additives Was Blended into a New Crude Oil Charge and Reprocessed

At times, it may be necessary for refiners to reprocess a fuel stream containing additives. Depending on the additives in the fuel and the crude oil processing steps required, the resulting new fuel stream may be clear or hazy in appearance.

What to Do:

- Determine if a reprocessed fuel stream was used in crude to produce the resulting hazy stream.
- Identify additives that were utilized in the original fuel stream and evaluate their potential interaction in finishing processes such as caustic or amine treatment.

Potential Solutions:

- The addition of a fuel dehazer may provide some performance at clearing fuel of water haze.
- Determine whether clay filtration of the fuel is an economical and practical option at removing haze.

8.3.2 Jet Fuel Fails Particulate Contamination Specification

8.3.2.1 Possible Cause: Water, Rust, Dirt Contamination

During jet fuel processing, intermediate tank storage, common carrier or pipeline transfer of jet fuel, and final storage, water carrying particulate matter may contaminate jet fuel. This contamination may cause failure of the ASTM D2276 or ASTM D5452 particulate matter specification of 1.0 mg/L specified for JP-8 military jet fuel and other jet fuels.

What to Do:

- Collect water from the bottom of a storage tank or vessel holding jet fuel; observe for visible sediment or particulate.
- Analyze water by ICP to identify metal ions that may be present; determine the likely source of metals.

Potential solutions:

- Pass jet fuel through a coalescing filter and/or clay filter to remove suspended water and particulate.

8.3.2.2 Possible Cause: Organic Salt Suspended and Emulsified Particulate in Fuel

During processing of jet fuel, caustic treatment is often utilized to remove naphthenic acids, sulfonic acids, mercaptans, and hydrogen sulfide from the fuel. Sodium salts of these compounds form and separate from the fuel. If not completely removed from the fuel after caustic treatment, the organosodium salts can dislodge rust and particulate matter that may be present on the walls of pipelines and storage tanks. The dislodged compounds may then be suspended or emulsified in the fuel as particulate.

What to Do:

- Determine whether caustic is present in the fuel storage or processing system by checking pH of the storage tank water bottoms.
- If pH is greater than 7, caustic carryover is possible.
- Analyze for the presence of sodium or calcium in the jet fuel. If ions are present, water contamination is likely exist.

Potential Solutions:

- Maintain a schedule of frequent removal of water bottoms from the jet fuel storage system.
- Clay filter jet fuel to remove particulate matter and organosodium salts.

8.3.3 Jet Fuel Fails MSEP or WSIM Specification

8.3.3.1 Possible Cause: Excess Additive in Fuel Enhances Emulsification

Various compounds are used in crude oil processing, fuel refining, and finished jet fuel to improve processing or enable jet fuel to meet specifications. Depending on concentration or additive type, the ability of fuel to rapidly shed water when testing by ASTM D3948 MSEP or WSIM may be affected. Emulsification of water due to a high additive concentration or too much of one additive type may lead to low MSEP values.

What to Do:

- If possible, determine which compounds are used in refining and processing jet fuel and as finished jet fuel additives.
- Verify the rate of each product used, and reduce rate if possible.

Potential Solution:

- Clay filtration of jet fuel may be necessary to remove polar organic compounds that are leading to fuel emulsification and MSEP failure.

8.3.3.2 Possible Cause: Fuel is Contaminated with High Boiling, High Viscosity Components

During transportation or storage, jet fuel may have come in contact with other petroleum products such as diesel fuel, lubricating oil, or residual fuel products. These higher boiling point and higher viscosity products do not filter and shed water rapidly.

What to Do:

- ASTM D86 distillation; check T-90 and residual for high values.
- ASTM D445 viscosity; look for high cSt values.

Potential Solutions:

- This problem is difficult to resolve through finishing processes. Contamination with higher boiling point materials may negatively influence the JFTOT and smoke point values in addition to the MSEP rating.
- Clay filtration and molecular sieve may remove enough of the higher boiling-point polar compounds from jet fuel to improve MSEP.

8.4 Fuel Stability Problems

8.4.1 Diesel Fuel Darkens in Color and Sediment Forms

8.4.1.1 Possible Cause: Fuel Contains Unhydrotreated Cycle Oil

Unhydrotreated light and heavy cycle oil can be blended into diesel fuel. Since cycle oil contains olefinic compounds, the potential for fuel oxidation and darkening is high.

What to Do:

- ASTM D1159 bromine number or D1319 FIA for olefins; if greater than 2 percent, cycle oil may be present in fuel blend.
- If possible, check fuel blend for LCO or HCO.

Potential Solutions:

- Add a distillate fuel stabilizer to control color and sediment.
- Clay filter the fuel to remove color bodies.
- Although outdated in some regions, acid washing to remove olefins and color bodies is a possibility.

8.4.1.2 Possible Cause: Fuel Refined from Crude Containing Polar Aromatic and Resinous Compounds

It has been established that certain nitrogen, oxygen, and sulfur containing compounds contained in crude oil can lead to darkening of refined fuels and oils. Often, these compounds are found in fuels refined from naphthenic crude oil or asphaltic, high sulfur or high nitrogen crudes. Compounds such as indoles, quinolones, and dibenzothiophenes can lead to darkening of fuel.

What to Do:

- Evaluate fuel for nitrogen; if greater than 50 ppm, darkening could be originating from organic nitrogen compounds.
- Determine whether crude source is naphthenic.
- Gas chromatography/mass spectrometry (GC/MS) analysis of the fuel, FCC feed, or crude oil could be performed to identify the presence of polar organics.

Potential Solutions:

- Add a distillate fuel color degradation control additive or stabilizer before fuel enters tankage.
- Clay filter the fuel to remove nitrogen heterocycles.
- Hydroprocess to reduce organically bound nitrogen, sulfur, and oxygen to produce H_2S, ammonia, and water.

8.4.1.3 Possible Cause: Polycyclic Aromatic Hydrocarbons in Fuel

Fuel refiners will sometimes blend low levels of coker gas oil or vacuum gas oil into diesel fuel. These high boiling point fractions may contain high molecular weight polycyclic aromatic compounds which can eventually precipitate to form fuel insoluble sludge and deposits.

What to Do:

- ASTM D86 to look for high residual fraction greater than 1 percent.

Potential Solution:

- Utilize a distillate fuel stabilizer that possesses the ability to perform as a dispersant. By dispersing high boiling point, high molecular weight fractions into fuel, the eventual deposition of these compounds as fuel-insoluble compounds can be minimized.

8.4.1.4 Possible Cause: Naphthenic Acids in Fuel

Cycloparaffinic acidic compounds, or naphthenic acids, in fuel are known to form color bodies and eventually insoluble, high molecular weight reaction products in fuel.

What to Do:

- Look for the presence of these compounds by ASTM D664A or through specific analytical techniques used to identify cyclopentane/cyclohexane acids.

Potential Solution:

- Caustic process fuel to remove naphthenic acids. Naphthenic acid salts are water soluble. Care should be taken to ensure that emulsification does not result.

8.4.2 Jet Fuel Fails Thermal Oxidation Stability Test

8.4.2.1 Possible Cause: Fuel Contains Dissolved Copper

Copper can catalyze fuel degradation under the ASTM D3241 (IP 323) high-temperature test conditions to cause filter screen or tube rating failure.

What to Do:

- Analyze jet fuel for the presence of copper by ICP.

Potential Solutions:

- Utilize a metal chelating additive to help minimize the effect of copper catalyzed degradation of fuel.
- Filter fuel through a clay or a molecular sieve bed to remove copper.

8.4.2.2 Possible Cause: Fuel Contains Heavy Aromatic Compounds

Jet fuel specifications are written to limit the amount of aromatic compounds as well as naphthenes. However, it is still possible that diaromatics or other aromatics with reactive side groups may be present that could lead to tube or filter screen deposits.

What to Do:

- ASTM D1319 (IP 156) fluorescent indicator analysis or ASTM D6379 IP 436 (HPLC) to determine whether fuel contains higher molecular weight aromatics, diaromatics, or naphthenics.

Potential Solutions:

- Reduce jet fuel T90, T95 or FBP temperatures to determine whether fuels pass JFTOT.

- Treat fuel with an antioxidant to determine the impact on screen deposit or tube rating.

8.5 Hydrogen Sulfide Contamination Problems

8.5.1 Contamination Through Fuel Processing

8.5.1.1 Possible Cause: Catalytic Cracking or Thermal Processing of Oil

Catalytic cracking, coking, visbreaking, and other high-temperature refining processes can liberate sulfur from a cracking unit feedstock or from a process stream. Once free, sulfur can react to form low molecular weight compounds such as H_2S, methyl mercaptan, carbon disulfide, and others.

Hydrogen sulfide is a poisonous, flammable gas that can contaminate and dissolve in process fractions such as naphtha and residual oil streams. Hydrogen sulfide is also soluble in water that may be dispersed in these same fractions.

What to Do:

- ASTM D5705 to check for the presence of H_2S in the product vapor.
- Check for the presence of H_2S in the fractions by potentiometric titration, gas chromatography, or other methods.

Potential Solutions:

- Thoroughly blend an oil soluble H_2S scavenger into fuel. Allow the scavenger to react for 30 minutes to several hours. Analyze the fuel for the presence of H_2S.
- In some applications, a H_2S scavenger having both oil and water solubility properties, or a fully water soluble scavenger may be required. The scavengers should be thoroughly blended into the fuel. Allow the scavenger to react for 30 minutes to several hours. Analyze the fuel for the presence of H_2S.
- Vacuum process residual oil to remove H_2S vapors.

8.5.1.2 Possible Cause: Hydroprocessing to Produce Ultralow Sulfur Fuel

Severe hydroprocessing to remove sulfur from naphtha and distillate fractions produces H_2S. Although most H_2S is removed as off-gas or stripped from fuel streams, the possibility exists for low levels of H_2S to remain in fuel fractions.

What to Do:

- ASTM D5705 to check for the presence of H_2S in the fuel vapor.
- Check for the presence of H_2S in the fractions by potentiometric titration, gas chromatography, or other methods.

Potential Solutions:

- Treat fuel with an oil soluble H_2S scavenger.
- Caustic process fuel to remove H_2S.
- Vacuum strip or nitrogen purge to remove H_2S from the fuel, or reprocess the fuel.

8.5.1.3 Possible Cause: Contamination of LPG by Hydrogen Sulfide or Carbonyl Sulfide

What to Do:

- ASTM D2420 lead acetate test method to detect H_2S.
- Gas chromatography method capable of receiving LPG cylinders for H_2S and carbonyl sulfide (COS) analysis.

Potential Solutions:

- Utilize an oil soluble and possibly a water soluble H_2S scavenger.
- Maintain scavenger rates as low as possible to minimize ASTM D2158 residual matter ratings.

8.5.2 Contamination During Fuel Storage

8.5.2.1 Possible Cause: Microbiological Contamination of Fuel in Storage

Sulfate reducing bacteria (SRB) can live in water bottoms of fuel storage tanks. These bacteria can produce growth plaques on metal surfaces and can live in corrosion pits in metal. Hydrogen sulfide is a product of SRB metabolism and can contaminate fuel stored in tanks.

What to Do:

- Check for the presence of anaerobic SRB and H_2S being emitted into tank bottom water.

Potential Solutions:

- Treat water with a biocide capable of killing SRB; maintain regular treatment.
- Drain water from the storage tank on a frequent basis.

CHAPTER 9

Components of Fuel and Fuel Additive Storage and Injection Systems

9.1 Metals

Metals are valued for their properties such as strength, toughness, hardness, and plasticity. Their compressive strength, ductility, and ability to be melted and molded into various shapes greatly expand their usefulness. Alloying with other metals and various elements enhances the performance properties of metals such as corrosion resistance, machinability, and wear resistance.

9.1.1 Steel

Carbon steel is the most common metal used throughout the fuel storage, handling, and transportation systems. The term *carbon steel* describes steel-containing carbon as the principal alloying element added to control hardness and strength. The carbon content can range from approximately 0.05 to about 1.4 percent. Above 1.4 percent, steel becomes brittle and loses cohesiveness and hardness.

Slowly cooled steels are relatively soft. Rapidly quenched steels are strong and hard. Information on the typical uses of steel with varying levels of carbon appears in Table 9.1.

A four- or five-digit numbering system has been established to classify and identify different types of steel. Originally developed by the Society of Automotive Engineers (SAE), this system is now used by the American Iron and Steel Institute (AISI) to classify steel.

In the numbering system, the first digit identifies the type of steel. The second digit is used to identify the approximate content of the primary alloying element. The remaining two digits indicate the

Carbon, %	Typical Use
0.05 to 0.12	Nails, wire, chain
0.10 to 0.20	Soft, tough steel; structural steel, case hardened machine parts, screws
0.20 to 0.30	Better grade of machine and structural steel; gears, shafts, levers
0.30 to 0.40	Responds to heat treatment; connecting rods, shafts, axles
0.40 to 0.50	Crankshafts, gears, axles
0.60 to 0.70	Tool steel; set screws, locomotive wheels, screw drivers
0.70 to 0.80	Tough, hard steel; anvils, hammers, cable wire, wrenches
0.80 to 0.90	Metal punches, rivet sets, cold chisels
0.90 to 1.00	Provides hardness and tensile strength; springs, axes, various dies
1.00 to 1.10	Drills, taps, milling tools, knives
1.10 to 1.20	Used where toughness is the most important consideration; ball bearings, drills, woodworking tools
1.20 to 1.30	Files, knives, tools for milling brass and wood
1.25 to 1.40	Used where a sharp cutting edge is important; razors, saws, wear-resistant machine parts

Table 9.1 Uses for Steel with Varying Carbon Content

typical carbon content percentage. In the case of certain corrosion resistant and heat resistant steels, a three-digit system is used to identify the average carbon content percent.

In addition, a prefix may be used to identify the process used to manufacture the steel. For example:

- A = a basic open-hearth alloy steel
- B = an acid Bessemer carbon steel
- C = a basic open-hearth carbon steel
- D = an acid open-hearth carbon steel
- E = an electric furnace steel

Examples of types of SAE and AISI steel are provided in Table 9.2

9.1.2 Impurities in Steel

Solid impurities such as iron oxides, manganese oxide, silicon, and aluminum can become trapped in the steel and form points of weakness in the steel. Weak points can be the site of eventual wear, fracture, or corrosion.

Steel Type	Identification
Plain carbon steel	10XX
Resulfurized carbon steel	11XX
Resulfurized, rephosphorized carbon steel	12XX
Manganese steel	13XX
High manganese carburizing steel	15XX
Nickel steel	2XXX
5% nickel steel	25XX
Nickel-chromium steel	3XXX
3.5% nickel, 1.5% chromium steel	33XX
Corrosion and heat resistant nickel steel	30XX
Carbon molybdenum steel	40XX
Chromium-nickel-molybdenum steel	43XX
Chromium steel	5XXX
Low chromium steel	51XX
Corrosion and heat resistant chromium steel	51XXX
Chromium-vanadium steel	6XXX
Nickel-chromium-molybdenum steel	86XX and 87XX
Manganese-nickel-chromium-molybdenum steel	94XX

Table 9.2 Types of SAE and AISI Steel

Gaseous impurities such as nitrogen, hydrogen, oxygen, and carbon monoxide are always present. When present as bubbles, they can decrease the plasticity and increase the embrittlement of steel.

9.1.3 Alloying Elements

The addition of low concentrations of elements to steel such as manganese, titanium, or boron can greatly enhance the properties of steel. Improved hardness, strength, machinability, and resistance to corrosion can all be improved by alloying. The effect of various alloying elements is provided in Table 9.3.

9.1.4 Stainless Steel

Stainless steel is quite useful as a construction material due to its resistance to corrosion and scaling at high temperatures. The performance of stainless steel is provided by adding chromium, nickel, molybdenum, or other alloying elements to steel during processing.

A minimum chromium concentration of approximately 11 percent is typical for stainless steel. As more chromium is added, corrosion resistance improves. Concentrations of chromium greater than 20 percent are found in some alloys. Chromium addition leads to the

Alloying Element	Performance Property
Boron	Increases hardness
Chromium	Improves abrasion and wear resistance
Cobalt	Increases hardness
Copper	Improves atmospheric corrosion resistance
Lead	Reduces cutting friction during machining
Manganese	Promotes forgeability
Molybdenum	Improves hardness, heat treatability, and corrosion resistance
Nickel	Increases toughness
Phosphorus	May weaken steel
Sulfur	Forms iron sulfide readily and results in brittle steel
Titanium	Increases strength and heat resistance

Table 9.3 Effect of Alloying Elements on Steel Performance

development of a tight forming oxide film on the surface of metal. This stable film is "self-healing," which means that the film will reform if scratched or broken. This oxide is resistant to attack by acids, bases, organic compounds, and inorganic salts.

Nickel is added to improve resistance to environmental and stress cracking corrosion. Molybdenum provides even greater resistance to corrosion and improves mechanical strength. Copper increases resistance to sulfuric acid attack. Examples of common stainless steel grades and their properties are provided in Table 9.4.

Stainless Steel Type	Alloying Element	Applications
301	Cr and Ni at low levels	General utility steel for structural uses and household utensils
304 and 304L	Ni at higher levels	Improve welding properties and workability in construction; used for piping materials
316 and 316L	Mo for corrosion resistance	Improve tensile strength of constructed materials; used for piping materials

Table 9.4 Common Stainless Steel Grades and Their Properties

9.1.5 Copper and Its Alloys

Copper is softer and more ductile than steel and is utilized frequently in the manufacture of pipes and tubing. Copper has good corrosion resistance but will corrode in the presence of nitric acid and other mineral acids. Organic acids do not corrode copper as readily. Dry ammonia does not corrode copper, but the presence of water in ammonia and ammonium hydroxide will corrode copper. Copper resists corrosion in the presence of caustic solutions, but the addition of zinc will increase corrosion rates. Also carbonate, phosphate, and silicate salts of sodium will corrode copper. The corrosion of copper by ammonia in the presence of water is shown in Eq. (9.1).

$$Cu + 4NH_3 \rightarrow Cu(NH_3)_4^{2+} + 2e^-$$
$$2Cu(NH_3)_4^{2+} + H_2O + 2^{e-} \rightarrow Cu_2O + 2NH_4^+ + 6NH_3 \qquad (9.1)$$

Copper and its nonferrous metal alloys, bronze and brass, are used to manufacture tubing, ferrules, valves, and a variety of fittings. Although their use is somewhat limited in automotive fuel systems, they can be found within fuel storage and distribution systems. Copper steam coils and brass or bronze hardware may be utilized due to their excellent resistance to corrosion and high level of thermal conductivity. Described below are some of the more common alloys of copper and their applications.

9.1.5.1 Brass

Alloys containing copper and zinc are typically classified as *brass*. The term *commercial bronze* may also be used to describe some copper and zinc alloys. Zinc readily alloys with copper in the liquid state and is classified as either alpha brass or alpha-beta brass. Alpha brass containing 5 to 20 percent zinc is termed as *red brass* due to its color. Alpha brass containing 20 to 36 percent zinc is termed as *yellow alpha brass*. Alpha-beta brass contains 36 to 45 percent zinc.

Copper and zinc can also form alloys in the solid state. Solid state solutions with less than 36 percent zinc are identified as *alpha solid solution*, those above 36 percent zinc are classed as *beta solid solution*, and those containing 50 percent or more zinc are termed *gamma solution brass*. Various commercial grades of brass are described and outlined in the following sections:

9.1.5.1.1 Admiralty Brass Composition: 70 percent copper, 29 percent zinc, 1 percent tin, and 0.03 percent arsenic. Admiralty brass has good corrosion resistance, especially in seawater. Arsenic inhibits the loss of zinc from the alloy. It is used for tubing applications in condensers, preheaters, evaporators, and heat exchangers that contact saltwater, oil, steam, and other liquids at operating temperatures below 260°C (500°F).

9.1.5.1.2 Aluminum Brass Composition: 78 percent copper, 20 percent zinc, 2 percent aluminum, and 0.03 percent arsenic. Aluminum brass has better corrosion resistance than admiralty brass due to an ability to "self-heal" when abraded. It is used in applications where high velocities are common such as in marine and land power stations as well as sewage and wastewater treatment plants.

9.1.5.1.3 Cupronickel Alloys Composition: These copper alloys contain from 10 to 30 percent nickel. They make excellent heat exchanger tubes because of their resistance to saltwater corrosion at high temperatures.

9.1.5.1.4 Forging Brass Composition: 60 percent copper, 38 percent zinc, and 2 percent lead. Forging brass is used in hardware and plumbing parts due to its good hot working properties.

9.1.5.1.5 Architectural Bronze Composition: 57 percent copper, 40 percent zinc, and 3 percent lead. Architectural bronze is used in decorative molding, hinges, locks, and industrial forgings.

9.1.5.1.6 Manganese Bronze Composition: This type of commercial alloy is composed of 58.5 percent copper, 39 percent zinc, 1.4 percent iron, 1.0 percent tin, and 0.1 percent manganese. Manganese bronze is strong and has excellent wear resistance. Uses include clutch disks, pump rods, valve stems, and welding rods.

9.1.5.1.7 Naval Bronze or Tobin Bronze Composition: 60 percent copper, 39.95 percent zinc, and 0.05 percent tin. This alloy is resistant to saltwater spray and is used in condenser plates, welding rods, propeller shafts, and marine hardware.

9.1.5.1.8 Red Brass Composition: 85 percent copper and 15 percent zinc. Red brass is commonly found in radiator cores, plumbing pipe, heat exchangers, and valves.

9.1.5.1.9 Muntz Metal Composition: 60 percent copper and 40 percent zinc. This alloy is used in the manufacture of screws, valve stems, brazing rods, condenser tubes, condenser heads, and heat exchanger baffles and plates. Muntz metal is strong, ductile, and corrosion resistant.

9.1.5.2 Bronze

The term *bronze* was originally used to describe copper-tin alloys. It is now used to describe any copper alloy, except copper-zinc alloys, containing up to 25 percent of the principal alloying element. Some examples include the following types of bronze:

9.1.5.2.1 Copper-Tin Bronze Alloys containing principally copper and tin are referred to as true bronze. Tin increases hardness and wear

Tin, %	Use
8	Wire, coins, bronze, sheeting
8 to 12	Gears, machine parts, bearings, marine fittings
12 to 20	Principal metal for bearings
20 to 25	Bells

TABLE 9.5 Common Uses of Copper-Tin Bronze Alloys

resistance more than zinc. It also improves saltwater corrosion resistance. Common uses of copper-tin bronze are described in Table 9.5.

9.1.5.2.2 Aluminum Bronze The concentration of aluminum in this alloy is usually between 5 and 11 percent of the bronze. Other elements such as iron, silicon, nickel, and manganese may be added to increase strength. Aluminum bronze has superior corrosion resistance and wear resistance over tin bronze. Uses include nonsparking hand tools, valve seats, valve guides, bushings, bearings, decorative grills, imitation gold jewelry, and paint pigment.

9.1.5.2.3 Leaded Bronze Lead addition improves machinability and sliding wear properties of bearings. It is usually less than 2 percent of the bronze composition.

9.1.5.2.4 Phosphor Bronze Phosphorus improves casting of molten metal. For this reason it is used in the production of gears and bearings.

9.1.5.2.5 Silicon Bronze The silicon content in this bronze is about 4 percent. Silicon bronze has excellent resistance to corrosion by organic acids and sulfite solutions. Common uses include electrical fittings, boilers, pumps, shafts, and marine hardware.

9.1.6 Aluminum

Aluminum is a soft, ductile, and relatively inexpensive metal. The surface of aluminum readily oxidizes in air and water to form a highly resistant oxide film. This oxide film serves to make aluminum resistant to attack when used in environments containing sulfides, sulfur dioxide, carbon dioxide, and other corrosive gases. It is highly resistant to water initiated corrosion, but it is susceptible to galvanic corrosion by trace amounts of copper, tin, lead, nickel, or carbon steel. The reaction of aluminum in water to form aluminum oxide trihydrate, or Bayerite, is shown in Eq. (9.2).

$$2Al + 6H_2O \rightarrow Al_2O_3 \cdot 3H_2O + 3H_2\uparrow \tag{9.2}$$

Aluminum has a high degree of thermal and electrical conductivity. At low temperatures, the impact strength of aluminum increases, and for this reason, aluminum is commonly used in cryogenic applications.

Alloying Element	Properties Provided
Copper	Minimizes shrinkage; promotes hardness
Magnesium	Improves machinability; improves resistance to corrosion by saltwater and alkaline solutions
Silicon	Improves casting qualities; increases corrosion resistance and impacts toughness
Zinc	Improves mechanical properties

Table 9.6 Common Aluminum Alloying Elements and Their Properties

Alloys are frequently used to improve the strength and hardness of aluminum. These alloys are resistant to corrosion by most organic compounds including hydrocarbon fuels, mercaptans, aldehydes, ethers, ketones, organosulfur, and organonitrogen compounds. Aluminum alloys are also resistant to most organic acids, phenols, halogenated organics, and alcohols. However, aluminum alloys are sometimes corroded by mixtures of organics. Common aluminum alloying elements and their properties are provided in Table 9.6.

Aluminum cladding is an alloying process used to help prevent surface corrosion of other underlying metal components. This process involves hot rolling metal to produce a protective aluminum barrier. Clad aluminum alloys can be found in some heat exchanger tubing applications.

Inorganic acids, bases, and salts with pH values less than 4 and greater than 9 are not compatible with aluminum. Pitting corrosion of aluminum can occur when it comes in contact with saltwater and copper halide salts. The corrosion of aluminum by caustic sodium hydroxide is presented in Eq. (9.3).

$$\mathrm{Al} + \mathrm{NaOH} + \mathrm{H_2O} \rightarrow \mathrm{NaAlO_2} + 1.5\mathrm{H_2}\uparrow \tag{9.3}$$

The effect of pH on corrosion of alloy aluminum is shown in Fig. 9.1.

9.1.7 Nickel

Nickel is used throughout industry because of its excellent corrosion resistance. In addition to its use as a cladding material to provide corrosion resistance to steel tanks and production vessel surfaces, nickel is used as an alloying element in steel production. Nickel is resistant to corrosion by sodium chloride solutions, but it is corroded severely by iron, copper, and mercury salts. Also, nickel has excellent corrosion resistance to most organic acids. Some common nickel alloys are described in the following sections:

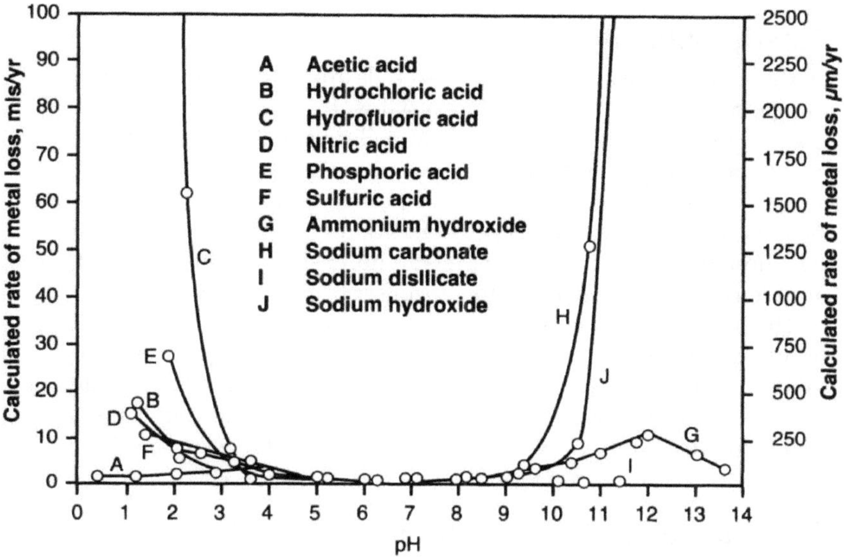

Figure 9.1 Effect of pH on corrosion of alloy aluminum.

9.1.7.1 Nickel-Copper (Monel)

Monel contains approximately 65 percent nickel, 28 percent copper, and other elements including iron, manganese, and cobalt. Monel is very hard, strong, and resistant to corrosion. Uses include pumps, valves, and chemical industry storage and containment vessels.

9.1.7.2 Nickel-Iron

Nickel-iron alloys contain varying levels of nickel. Commercial products include *Invar* and *Permalloy*. The alloys are used in certain high-level technology applications such as transoceanic cabling and other applications whereby machining tolerances are exact, including clocks and variable condensers.

9.1.7.3 Nickel-Iron-Molybdenum

Nickel-iron-molybdenum alloys have a high level of resistance to corrosion, especially at elevated temperatures. Commercial products include *Hastalloy A* and *Hastalloy B*. The compositions of Hastalloy A and Hastalloy B are provided in Table 9.7.

Alloy	% Nickel	% Iron	% Molybdenum	% Others
Hastalloy A	57	20	20	3
Hastalloy B	62	5	30	3

Table 9.7 Composition of Hastalloys A and Hastalloy B

9.2 Plastics and Elastomers

A variety of natural and synthetic materials are used throughout fuel storage, handling, and application systems. Examples include lines, hoses, storage containers, fan blades, impellers, small gears, housings, and various supporting hardware. Some plastics can be degraded by fuels, solvents, additives, and various petroleum based compounds. The most resistant material is polytetrafluoroethylene (PTFE). Ryton and Viton are less resistant, but are still quite stable in fuel systems. Characteristics of PTFE and Ryton are described in Table 9.8.

The primary considerations for the use of plastic materials in fuel and oil systems include resistance to swelling, softening, or embrittlement. Also important are flexibility and stability under a wide range of temperatures. Information on some common elastomers and plastics used in fuel systems is provided in Table 9.9.

9.3 Chemical Storage and Injection Equipment

The equipment used to blend and store fuels and to apply additives into fuel can be designed to fit a specific application. The equipment selection is based upon several factors including the characteristics of a fuel, additive type, operating specifications, and various regulatory issues. A typical system includes the following components:

- Storage tank and site gauge
- Injection/metering pump
- Transfer lines and hoses
- Flowmeter and controller
- Auxiliary equipment such as a heater, insulating blanket, or spill containment enclosure

Elastomer	Characteristics
PTFE	• Trade names Teflon, Kynar, Halar, Tefzel • Suitable for use from –200 to 200°C (–328 to 392°F) • Extreme resistance to fuel combustion by-products and chemical degradation
Ryton	• Resists degradation by ozone • Resists swelling and leaching by fuel components • Is flame retardant • Useful from –40 to 230°C (–40 to 446°F)

TABLE 9.8 Characteristics of PTFE and Ryton

Trade Name	Chemical Type	Typical Properties	Use
Hypalon	Chlorosulfonated polyethylene	Resistant to fuel and temperatures up to 127°C (260°F); flexible at low temperatures	Fuel lines
Lupolen	HDPE	Flexible and resistant to degradation by fuel and elements of the environment	Fuel tanks
Zytel	Polyamine	Stable up to temperatures of 163°C (325°F); resistant to impact, deformation, and fatigue	Fuel rails; filter housings
Viton	Fluoro-elastomer	Resistant to degradation by fuel and temperature up to 260°C (500°F); good compression strength	Fuel line housings; metering or needle valves

Table 9.9 Common Fuel System Elastomers

9.3.1 Chemical Storage Tanks

Tank configuration can be either vertical or horizontal and can vary in size from 500 to over 1,000,000 gal. Most tanks are constructed from carbon steel, although stainless steel, aluminum, fiberglass, and various plastics are sometimes used. Often special requirements or guidelines affect storage tank selection. These can include:

- Vents and vapor recovery systems
- Use of chemical containment area
- Permits
- Location and space
- Exterior painting and coating
- Labeling and identification

Underwriters Laboratories (UL) also approves and rates storage tanks. Factors including tank wall thickness, type of metal used, welding procedure, venting, supports, pressure ratings, and other related topics are specifically defined in the approval codes. Guidelines which relate directly to the storage of flammable and combustible materials are provided under UL Code 142.

Positive Displacement	Nonpositive Displacement
• Delivers fluid as a pulse with each pump stroke or each time the pump chamber outlet opens • Pressure increase is usually not a problem unless the pressure exceeds the rated capacity of the fittings or gaskets, then leakage can occur • Better in series with other pumps; not as sensitive to pressure differences • Usually self-priming	• Provides a smooth, continuous flow • Delivery can be reduced or even halted by high pressure; liquid begins to recirculate within the pump • Not effective in systems where pumps are connected in series; pressure differences can affect efficiency • Not self-priming; the pump must be started with the housing full of fluid and free of air

Table 9.10 Properties and Performance of Positive and Nonpositive Displacement Pumps

9.3.2 Pumps

A variety of positive and nonpositive displacement pumps are available for use in fuel and fuel additive systems. Selection is based on application, required flow, and cost. The primary differences in positive displacement and nonpositive displacement pumps are described in Table 9.10.

9.3.2.1 Nonpositive Displacement Pumps

Nonpositive displacement pumps operate by centrifugal force. Fluid supplied to the center of a rotating impeller is forced to the outer impeller housing by centrifugal force. The housing is spiral in shape and increases in diameter toward the outlet. The force of the rotating impeller moves the fluid out of the housing outlet at a higher pressure than originally delivered.

A propeller pump is another type of nonpositive displacement pump. It operates similar to a fan blade inside of a pipe or tube. As fluid moves past the blade, the energy of the rotating blade moves fluid through the outlet at a higher pressure than introduced.

9.3.2.2 Positive Displacement Pumps

9.3.2.2.1 Vane Pump

A vane pump contains a slotted rotor driven by a rotating drive shaft. Hardened and polished vanes slide in and out of the slots in the rotor as the rotor turns within an elliptical- or circle-shaped cam ring. These spaces between the vanes function as chambers to carry fluid to be pumped.

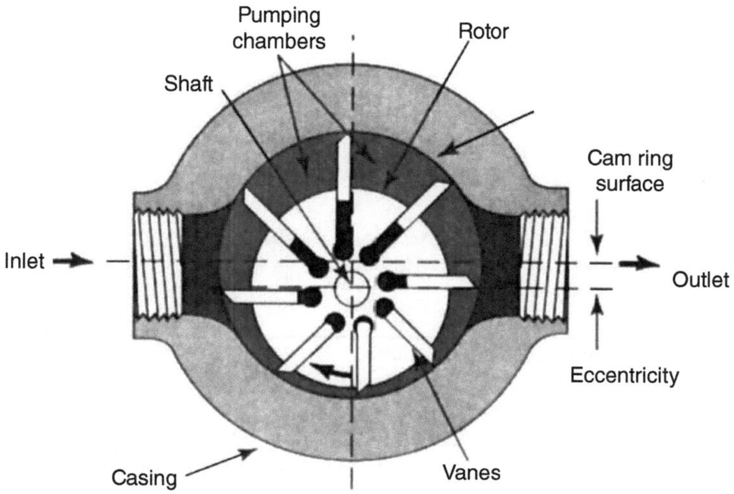

FIGURE 9.2 Vane pump cross section.

As the fluid enters each vane pump chamber through the pump inlet, the area between the sliding pump vanes fills with fluid. As the rotor continues to move, the chambers carry fluid from the inlet to the pump outlet. The fluid is then squeezed or forced from each individual chamber through the outlet at a higher pressure. Vane pumps generally have good efficiency and durability. They are effective within a wide pressure range, displacement volume, and speed. A typical vane pump is shown in Fig. 9.2.

9.3.2.2.2 Gear Pump Gear pumps have a large range of applications. They can be found in automatic transmission systems and in power steering units. The gear pump assembly consists of two interlocking, surface hardened gears which move fluid through the gear teeth at varying rates of speed. One gear is driven by a power source, while the interlocking gear is not. Its rotation is controlled by the powered gear.

As the gear teeth rotate and travel past the pump inlet, a partial vacuum forms. Fluid is carried by the small chambers formed between the teeth. As the teeth mesh near the pump outlet, fluid is delivered at a higher pressure. Flow from a gear pump is continuous and not intermittent.

Because of their design, gear pumps are not typically used to meter small quantities of fluid. This is due to the fact that the pump cannot compensate for pressure changes or flow rate changes readily. Gear pumps are better designed for high-speed and high-flow rates. These pumps are durable and effective at varying pressures. However, their accuracy as a metering pump is limited. A typical gear pump is shown in Fig. 9.3.

Figure 9.3 Gear pump cross section.

9.3.2.2.3 Rotor or Lobe Pump A rotor or lobe pump operates in a fashion similar to a gear pump. External gears control the operation of lobe-shaped impellers within the pump housing. Displacement by this type of pump is much greater than a gear pump, but the potential for wear is also greater. These pumps are typically used to move large volumes of fluid at low pressures. A typical rotor or lobe pump is shown in Fig. 9.4.

9.3.2.2.4 Piston Pump The piston pump consists of a plunger (piston) which moves within a cylinder-like compartment. The path length of the moving piston can be adjusted and/or the piston speed can be adjusted to control flow. These pumps are good at controlling flow at various pumping rates and operate well at high flow rates and pressures. Pump pressures as great as 5000 psi are possible to achieve.

Piston pumps are reliable and durable but are usually higher in cost than other pumps. Examples of piston pump types include radial piston pumps, axial piston pumps, swash plate in-line pumps, wobble plate in-line piston pumps, and bent axis or angle type piston pumps. A piston pump and component parts are shown in Fig. 9.5.

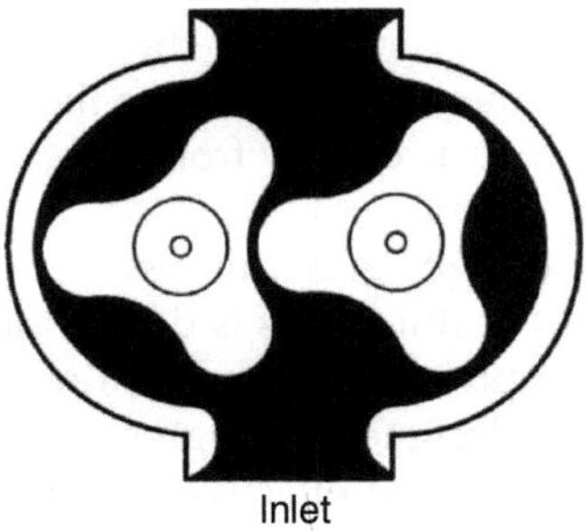

Figure 9.4 Rotor or lobe pump cross section.

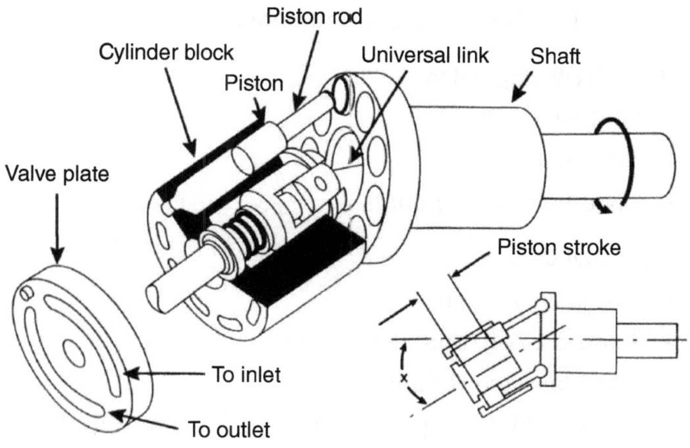

FIGURE 9.5 Piston pump.

9.3.2.2.5 Diaphragm Pump The diaphragm pump consists of a piston that pushes at varying stroke lengths on an oil reservoir. The hydraulic pressure of this oil then moves a flexible, yet rigid disk-like diaphragm. The diaphragm is typically composed of Teflon. The movement of this diaphragm acts to either create a vacuum or exert positive pressure on a fuel or chemical on the opposite side of the diaphragm. This pulsing effect acts to move fluids through at intermittent, controlled rates.

Diaphragm pumps are used for metering small amounts of additive into a fuel or fuel oil. The cost of a diaphragm pump is low compared to other positive displacement pumps. These pumps are excellent metering pumps and are primarily designed for low-pressure, low-flow applications. Also, diaphragm pumps are not recommended for pumping high-viscosity fluids.

9.3.2.2.6 Air-Driven Pump Air-driven pumps usually require a high level of maintenance but are low in cost. Their performance is affected by moisture, temperature changes, and pressure changes.

9.3.3 Flowmeters

Flowmeters sense the flow of a liquid by translating mechanical energy of moving gears or pistons into an electrical signal. This signal is relayed to an electronic sensing device which converts the signal into a flow rate. Factors which influence the selection of a flowmeter include flow rate, viscosity, and compatibility with materials of construction. Flowmeters can be classified as either positive displacement or turbine type meters. A brief description of each flowmeter type is provided in the following sections.

9.3.3.1 Positive Displacement Flowmeters

9.3.3.1.1 Oscillating Piston

- Limited to 150 psi
- Tolerant of dirt and debris
- Good for viscous liquids

9.3.3.1.2 Oval Gear

- Extremely accurate
- Tolerant of dirt and debris in fuel
- Good for viscous fluids
- Good for high-temperature and high-pressure applications

9.3.3.1.3 Piston Type

- Cannot tolerate dirt and debris in fuel stream
- Excellent for viscous fluids
- Effective at high pressures
- Not recommended for temperatures greater than 93.3°C (200°F)

9.3.3.2 Turbine Type Flowmeters

9.3.3.2.1 In-line Turbine

- Not recommended for viscous fluids
- High level of tolerance to dirt and debris
- Good for high-temperature and high-pressure applications
- Very accurate at high and low flow rates

9.3.3.2.2 Insertion Turbine

- Identical to in-line meters but much larger in size

9.3.3.2.3 Insertion Paddlewheel

- Not accurate at low flow rates
- Not recommended for viscous fluids
- Tolerant of dirt and debris
- Low cost

9.3.4 Transfer Lines and Hoses

When transferring fuel and fuel additives from storage to the pump and to its final point of use, different types of lines are utilized. Mild steel, stainless steel, and cross-linked polyethylene hose are the typical materials of construction. These materials withstand the temperature and environmental extremes of outdoor installation and are compatible with fuels and most additives.

Often it is necessary to apply insulation or heat to transfer lines to ensure that products do not gel in the line during cold weather. Both electrical heating and steam heating can be used to "trace" lines. Only explosion proof electrical heat tape and thermostats can be used on fuel systems. Immersion heaters are not recommended for fuel and fuel additive applications. If electrical heat tracing is used, it should be attached only to the bottom of the transfer line. Never wrap a fuel or chemical line with electrical heat tracing. Severe overheating can occur.

Steam tracing is accomplished by winding copper tubing around the length of pipe to be heated. Steam heat is very intense and must be carefully installed to ensure that overheating of piping does not occur. Steam tracing installations are usually permanent, whereby electrical tracing has a shorter life span.

9.3.5 Electrical Service

Positive displacement pumps can be built to run on 440-V, three-phase service, 220-V service, or 110-V service. The site of electrical service may be restricted in some fuel production and distribution areas. Often, explosion proof enclosures and associated equipment must be used when locating equipment in a hazardous area. If explosion proof equipment is required, the cost of the equipment will increase significantly.

9.4 Vehicle Fuel Tanks

The fuel tank must be resistant to corrosion, mechanical impact, temperature fluctuations, and internal pressure increases. Caps must not permit fuel to escape when the vehicle is turning, traveling up and down grades, or when subjected to minor jolts. Also, the tank must be fitted with appropriate relief valves to permit the escape of excessive pressure.

Tank construction materials include steel, aluminum, and plastics. Steel tanks are typically lined internally with a corrosion resistant material. Internal coating materials have included Terne (a tin-lead alloy), zinc containing compounds, and organic coatings. Aluminum tanks are light in weight and resistant to internal corrosion but are sensitive to external corrosion by salts. Plastic tanks are produced from high-density polyethylene and medium-density polyethylene. They are highly corrosion resistant and can be molded into various sizes and shapes.

To help ensure resistance to impact and internal pressure increases, plastic tank wall thickness is greater than steel tank thickness. Plastic tanks are typically greater than 4 mm in thickness, whereas steel tanks are somewhat less than 1 mm in thickness. Plastic tanks, however, are considerably lighter in weight.

9.5 Diesel Fuel Filters

The function of the diesel fuel filter is to prevent contaminants from entering the high-pressure fuel pump and causing damage to sensitive pump parts. Fuel filters are designed to trap organic sludge, wax, and particulate matter. They also help to coalesce water from fuel into larger droplets, which then fall and are collected on the bottom of the filter. Also, some filters are equipped with electronic water sensing devices and bottom drains for removing water.

Although there is no worldwide standard under which fuel filters are manufactured, most contain a hydrophobic paper element with a porosity ranging from approximately 5 to 15 microns with an average porosity of about 8 microns. The porosity of filters used in distributor type fuel pumps is smaller and generally ranges from 4 to 5 microns.

The paper elements can be folded into various forms which appear as pleated, star shaped or V-shaped patterns. These elements are wrapped or coiled around a central porous metal core. Fuels generally flow from the outside shell of the filter inward through the filter media and up through the central core of the filter to the fuel pump feed line. Other filter designs such as stacked disk filters exist but are not common.

9.6 Diesel Fuel Injection Pumps

A typical fuel injection system is composed of the following components: fuel tank, injection pump, mechanical or electronic governor, timing device, fuel supply pump, high-pressure fuel injection lines, and fuel injectors. The high-pressure fuel injection pump is the core component of the fuel injection system. The in-line pump and distributor pump are the most common in fuel systems.

9.6.1 In-Line Injection Pumps

Standard in-line fuel injection pumps are driven by a camshaft following the speed of the engine. For two-cycle engines, the camshaft speed is identical to the crankshaft rpm. For four-cycle engines, the camshaft speed is one-half of the crankshaft rpm. The in-line injection pump parts, such as the camshaft and tappets, are lubricated by the engine oil.

A separate plunger and barrel assembly is utilized for injection of fuel into each cylinder. For example, the in-line pump for a six-cylinder engine contains six individual plunger and barrel assemblies within the pump housing. Plunger movement is controlled by a tappet and spring mechanism driven by the camshaft. The plunger movement in the barrel provides the pumping action needed to inject fuel at high pressure into the fuel lines and through the fuel injectors. The machining tolerance between the plunger and barrel is so tight that no additional seal is needed to prevent fuel from leaking past the plunger as it moves within the barrel sleeve.

The process of fuel injection first involves movement of fuel from the fuel tank by a fuel delivery pump. This low-pressure pump supplies fuel into the fuel delivery port of each plunger and barrel assembly. As the plunger moves past and clears the fuel delivery port, fuel flows into the barrel to fill the chamber above the top of the plunger. Continuous rotation of the camshaft then forces the plunger in the opposite direction, closing off the flow of fuel into the barrel and trapping fuel within the barrel chamber.

The fuel now contained within the barrel is moved by the force of the plunger stroke to the delivery valve held in a closed position by a high-tension spring. As the plunger stroke progresses, the fuel hydraulic pressure increases. Once the hydraulic pressure exceeds the force exerted by the high-tension spring, the fuel delivery valve opens. Immediately, fuel then flows at high pressure through the valve into the fuel injector. The pressure of the injected fuel enables the fuel injector to deliver fuel at high pressure into the combustion chamber.

Honed into the side of the plunger is a groove or slot termed the *helical groove*. This groove is open to the top of the plunger and is filled with fuel. Further down the shaft of the plunger, the groove is aligned with the fuel outlet port, or spill port. As soon as the helical groove opens to the spill port, the pressure of the injected fuel instantly drops, the fuel delivery valve closes, and fuel injection stops.

9.6.2 Distributor Pump or Rotary Pump

A typical distributor pump or rotary fuel pump consists of both low-pressure and high-pressure fuel delivery components. Low-pressure delivery components consist of a vane-type fuel supply pump, fuel pressure control valve, and a fuel overflow return valve. The high-pressure fuel delivery components consist primarily of a distributor-plunger driveshaft mechanism, a rotating and reciprocating metering plunger, and a stationary distributor head through which fuel lines run to individual delivery valves and injectors.

The metering plunger has a series of slots and passages along its length for fuel to enter and exit. The distributor head contains ports and channels that enable fuel to pass into the pump chamber at low pressure then out of the pump chamber at much greater pressure.

Fuel to be injected flows from the pump interior through a fuel inlet port in the distributor housing. The reciprocating plunger retracts and unblocks the fuel inlet port to enable fuel to flow freely into the high-pressure pump chamber. On the return stroke, the plunger then blocks the flow of fuel into the chamber. Fuel in the chamber is then ready for injection.

As the metering plunger rotates and lines up with the correct discharge ports in the distributor head, fuel is forced from the passages through a slot in the plunger body and into the proper injection line via the reciprocating action of the plunger. The fuel is then injected at high pressure into the appropriate line in the distributor head, flows through the fuel injector, and into the combustion chamber.

This fuel distribution and injection process continues four times per plunger revolution for a four-cylinder engine and six times per revolution for a six-cylinder engine. Distributor pumps are not typically utilized on eight-cylinder or larger diesel engines. The amount of metered fuel is controlled by the engine governor. All parts of the distributor pump are contained in one housing. Distributor pumps are lubricated by fuel which fills the housing.

High-pressure diesel pumps are not tolerant of dirt, debris, and organic deposits and can be seriously damaged by fuel contaminants. The barrel and plunger clearance is within a 1- to 2-micron tolerance. This tolerance is necessary to ensure that fuel injection pressures are maintained with minimal leakage past the plunger shaft. For this reason, thorough and complete filtration of fuel is required before the fuel reaches the injection pump.

CHAPTER 10

Safe Shipping and Hazard Information for Common Fuels, Oils, and Solvents

Information is sometimes needed about human and environmental hazards associated with handling and shipping various petroleum products. Safe practices and proper shipping regulations are provided by various transport associations in documents such as the annual publication of the *International Air Transport Association Dangerous Goods Regulations*. For purposes of general awareness, a brief review of hazardous material shipping guidelines and hazard class descriptions are presented in this chapter.

10.1 Hazardous Material Shipping Guidelines

Common carriers require the following information on hazardous materials before shipment:

- Proper shipping name
- Hazard class or division
- UN or ID number
- Packing group
- Subsidiary risk

These requirements are briefly discussed in the sections below.

10.1.1 Proper Shipping Name

A universal description of the material being shipped is the first step that must be taken to ensure that fuels and oils are handled, shipped, stored, and used properly. An extensive listing of proper shipping names for hazardous materials is provided in dangerous goods regulations. Examples of proper shipping names include:

- Corrosive liquid, n.o.s.
- Fuel, aviation, turbine engine
- Petroleum crude oil

The use of the designation "n.o.s." represents the phrase "not otherwise specified" for some materials. The n.o.s. modifier is used if the material to be shipped contains 25 percent or less of the hazardous substance.

10.1.2 Hazard Class or Division

10.1.2.1 Nonhazardous Materials

Many materials may be classified as nonhazardous materials such as residue, scale, sludge, water, and metal parts. These materials may be shipped as nonhazardous materials.

10.1.2.2 Hazardous Materials

Presently, there are nine hazard classes under which hazardous materials are shipped. These are described as follows:

- Class 1: Explosives
- Class 2: Gases
- Class 3: Flammable liquids
- Class 4: Flammable solids; substances liable to spontaneous combustion, substances which in contact with water, emit flammable gases
- Class 5: Oxidizing substances and organic peroxides
- Class 6: Toxic and infectious substances
- Class 7: Radioactive material
- Class 8: Corrosives
- Class 9: Miscellaneous dangerous goods

Most fuels and oils are classified as Class 3: Flammable liquid and combustible liquid. Occasionally, Class 2: Gas samples are shipped.

10.1.3 UN or ID Number

The UN or ID number is specific for each material listed under the hazardous materials registration.

10.1.4 Packing Group

Identification of the correct packing group helps to ensure that the proper shipping container is used to transport hazardous materials.

Selection of the proper packing group also helps to minimize the risk associated with shipping hazardous materials.

Most hazardous materials are assigned to one of the following packing groups according to the level of danger they present:

- Packing Group I: Great danger
- Packing Group II: Medium danger
- Packing Group III: Minor danger

10.1.5 Subsidiary Risk

Some materials may possess more than one hazard risk. As an example, methanol is considered a hazard Class 3: Flammable liquid. It also has a subsidiary risk, which falls in hazard Class 6.1: Toxic substance.

10.1.6 Packaging

Regulatory labels describing the hazard characteristics of a fuel or oil are required on the outside of boxes and containers used for shipping samples. The labels identify the samples as one or more of the following:

- Flammable
- Oxidizer
- Toxic
- Corrosive

10.1.6.1 Hazardous Material Labeling Guidelines

- The hazard labels must be placed on the box in an upright position and must not be covered by any other type of marking or label.
- If a subsidiary risk exists for a material, the subsidiary risk label must be placed near, but not on, the primary label.
- If the proper shipping name and UN number are not included on the label, this information must be printed on the box near the label.
- Also, before air shipment, an orange colored *Cargo Aircraft Only* label must be applied to one side of the shipping container.

10.1.7 International and Domestic Cargo Shipping Carriers

Cargo carriers sometimes have special requirements pertaining to shipment of hazardous materials. To assist customers, most of these carriers provide 24-hour telephone support to help answer questions about cargo shipment.

10.2 Hazard Class Descriptions

Nine air transportation hazard classes have been identified for describing and grouping hazardous materials. This information may prove useful if questions arise on classifying fuels and other hazardous materials for shipment.

10.2.1 Class 1: Explosives

Materials in this class comprise:

1. Explosive substances except those that are too dangerous to transport or those where the predominant hazard is appropriate for another class (a substance that is not itself an explosive but can form an explosive atmosphere of gas, vapor or dust is not included).
2. Explosive articles except devices containing explosive substances in such quantity or of such a character that their inadvertent or accidental ignition or initiation during transport will not cause any effect external to the device either by projection, fire, smoke, heat, or loud noise.
3. Articles and substances not mentioned under (1) and (2) above which are manufactured with a view to producing a practical, explosive, or pyrotechnic effect.

10.2.2 Class 2: Gases

This class is comprised of the following:

10.2.2.1 Compressed Gas

This type of gas when packaged under pressure for transport, is entirely gaseous at –50°C (–58°F). This category includes all gasses with a critical temperature less than or equal to –50°C (–58°F).

10.2.2.2 Liquefied Gas

This type of gas when packaged under pressure for transport, is partially liquid at temperatures above –50°C (58°F). A distinction is made between:

- *High pressure liquefied gas*—a gas with a critical temperature between –50°C (–58°F) and 65°C (149°F).
- *Low pressure liquefied gas*—a gas with a critical temperature above 65°C (149°F).

10.2.2.3 Refrigerated Liquefied Gas

This type of gas when packaged for transport, is made partially liquid because of its low temperature. A cryogenic liquid is the same as a refrigerated liquefied gas.

10.2.2.4 Dissolved Gas

This type of gas when packaged under pressure for transport, is dissolved in a liquid phase solvent.

10.2.3 Class 3: Flammable Liquids

These are liquids, mixtures of liquids, or other liquids containing solids in solution or in suspension (e.g., certain fuels, solvents, oils) that give off a flammable vapor at temperatures not greater than 60°C (140°F) by the closed cup flash point test method or 65.5°C (150°F) by the open cup flash point method.

10.2.4 Class 4: Flammable Solid, Spontaneously Combustible, and Dangerous When Wet

These materials are listed in three classifications as follows:

1. *Flammable solids:* These are solids, other than those classified as explosives, which are readily combustible, or may contribute to fire through friction during transport.
2. *Substances liable to spontaneous combustion:* These are substances that are susceptible to spontaneous heating during transport; substances that heat up in contact with the air and catch fire.
3. *Dangerous when wet:* These substances emit flammable gases when in contact with water.

10.2.5 Class 5: Oxidizing Substances and Organic Peroxides

These materials are classed into two divisions as listed below:

1. Substances that may not combust themselves, but by yielding oxygen, may contribute to the combustion of other materials.
2. Organic substances that contain the R-O-O-R functional group and may be considered derivatives of hydrogen peroxide where one or both of the hydrogen atoms have been replaced by organic radicals. Organic peroxides are thermally unstable and may undergo exothermic, self-accelerating decomposition.

10.2.6 Class 6: Toxic and Infectious Substances

These materials are defined as follows:

1. *Poisonous (toxic) substances:* These are likely to either cause death, serious injury, or harm to human health if they are swallowed, inhaled, or come in contact with the skin.

2. *Infectious substances:* These materials contain viable microorganisms or their toxins that are known, or suspected to cause, disease in animals or humans.

10.2.7 Radioactive Material

Radioactive material refers to any material containing radionucleotides where both the activity and concentration exceed specified values published in shipping regulations. The following radioactive materials are typically not included:

- Radioactive material implanted or incorporated into a person or live animals for diagnosis or treatment.
- Radioactive consumer products that have received regulatory approval following their sale to the end user.
- Natural material and ores containing naturally occurring radionucleotides that are either in their natural state or have only been processed for purposes other than for the extraction of radionucleotides.

10.2.8 Corrosives

These substances, in the event of leakage, can cause severe damage by chemical action when in contact with living tissue. The leakage or escape of these substances from their packaging will materially damage or destroy other goods or the means of transportation.

10.2.9 Miscellaneous Dangerous Goods

This class of substances may present a danger during air transport and substances not covered by another class. These include:

1. *Magnetized material:* This includes any material when packaged for air transport that has a maximum magnetic field strength sufficient to cause a compass deflection of more than 2 degrees at a distance of 2.1 m from any point on the surface of the assembled package. The magnetic field strength at the compass producing a 2-degree deflection is taken to be 0.418 ampere/m or 0.00525 gauss.
2. *Aviation regulated solid or liquid:* This includes any material that has narcotic, noxious, irritating, or other properties that in the event of leakage or spillage could cause extreme discomfort to humans. These materials must not meet any of the definitions for classes 1 through 8.

CHAPTER 11

Fuel Performance Property, Deposit Analysis, and Problem Solving Guide

This chapter contains three different sections that serve as a summary and quick reference guide for use in solving fuel problems.

11.1 Physical, Chemical, and Performance Property Guide

11.1.1 High Viscosity

Problems associated with difficulty in filtering, mixing, and pumping fuel can usually be linked to an increase in fuel viscosity. Contamination of fuel with high-viscosity products or operation at excessively low temperatures can increase the viscosity of fuel and result in fuel pumping problems. Also, if diesel fuel viscosity at −18°C (0°F) is greater than 45 cSt, fuel pumping problems are likely to occur.

11.1.1.1 Possible Effects

- May result in pumping difficulty at low temperatures; wax-related problems may be evident.
- At excessively low temperatures, highly viscous fuel may inadequately fill the pumping chamber of distributor-type fuel injection pumps. Poor fuel distribution and engine operation can result.

- Shearing of high-viscosity diesel fuel by the fuel injection pump can result in excessive heat buildup and distortion of the pump components.
- Increase in the coefficient of friction may result in a possible corresponding decrease in energy efficiency.
- May results in poor atomization of diesel fuel when sprayed through an injector; fuel is sprayed as a stream instead of as small droplets; reduced fuel efficiency and power occurs.
- Flame height in wick-fed kerosene burners is lessened; kerosene consumption is decreased.
- Fuel filter throughput is reduced with increasing fuel viscosity.

11.1.2 High-Sulfur Content

It is known that burning fuel sulfur can form sulfur dioxide and sulfur trioxide compounds. In combination with water, these sulfur oxides can form acidic compounds.

High-sulfur crudes, especially those with a sulfur content greater than 0.5 weight percent complicate refining and are more expensive to convert into fuels. Sulfur and sulfur compounds can be corrosive toward nonferrous metals and can decrease the pH of crude oil. The sulfur content of most crude oil varies from less than 0.1 to more than 5 percent.

Worldwide, low-sulfur and high-sulfur fuel products are being refined. Through hydroprocessing, ultralow-sulfur products are becoming the dominant grade of both naphtha- and distillate-based fuels produced.

11.1.2.1 Possible Effects

- May be corrosive toward copper, brass, and bronze.
- Increase in SO_x during fuel combustion.
- Increase in combustion exhaust odor.
- Potential fuel stability problems especially if sulfur exists in the heterocyclic form.
- May improve fuel lubricity if sulfur exists in the appropriate chemical structure.
- May degrade some elastomers if sulfur exists in the mercaptan form; limited to 30 mass percent in aviation turbine fuel.
- Combustion products can cause corrosive wear of metal.
- Can lead to an increase in combustion chamber deposits; under low temperature, low load, low speed, or stopping and starting driving conditions, condensation can occur; deposits can result.

11.1.3 High-Paraffin Content

Linear, branched, and cyclic paraffins all exist in refined fuel. Fuel performance problems can often be directly related to the type and concentration of paraffin present.

Gums, deposits, and fuel degradation products will typically not be dissolved or held in solution by high-paraffin content fuels. As a result, gums and degradation products will fall from solution and settle onto fuel system parts such as storage tank bottoms and fuel system lines.

11.1.3.1 Possible Effects

- Paraffinic compounds will be a poor solvent for polar compounds such as fuel additives.
- May contribute to lower density and Btu values.
- A high cetane number can be related to a high diesel paraffin content.
- The possibility of wax-related problems at low temperatures can increase.
- Higher smoke point of kerosene and jet fuel.
- Will burn with a yellow flame in kerosene burners.

11.1.4 High Aromatic Content

Monoaromatic, diaromatic, and polycyclic aromatic compounds can all be found in many fuels. Although concentrations vary in different fuels, the presence of these aromatics can produce both desirable and undesirable performance. As an example, a high aromatic content in aviation turbine fuel can cause smoke to form during combustion and lead to carbon deposition in jet engines. A total aromatic content more than 30 percent may lead to deterioration of aircraft fuel system elastomers and lead to fuel leakage.

11.1.4.1 Possible Effects

- Can contribute to low distillate fuel cetane number.
- Increase in heat of combustion or Btu value of heating oil and residual fuel oil.
- Formation of more carbonaceous deposits when burned, increase in smoke and soot.
- Improvement in hydrocarbon solvent properties with increasing aromatic content.
- Increase in gasoline octane number.
- Increase in density, decrease in °API.

- Is directly related to smoke formation during combustion.
- Will burn with a reddish flame in wick-fed kerosene lamps.
- Will burn brilliantly in blue flame, pressure-fed kerosene lamps.
- May cause a reduction in the smoke point of kerosene.

11.1.5 High Acid Number or Neutralization Number

Fuel acid number or neutralization number is a measure of the acidity or basicity of fuel. The presence of acidic compounds can indicate a fuel instability or oxidative degradation problem. Acidic organic compounds in fuel can react with other fuel components to initiate the formation of higher molecular weight, fuel insoluble deposits.

Typically, caustic treatment effectively removes these acidic compounds from fuel. However, after caustic treatment, alkali salts and heavier naphthenic acid salts may remain oil soluble. In fuel, these compounds can act as very effective emulsifying agents. Fuel haze and particulate contamination can be due to these acid salts.

11.1.5.1 Possible Effects

- May adversely affect water separation properties of fuel.
- Can stabilize emulsions.
- Can enhance gum and deposit formation.

11.1.6 Low Viscosity

Low fuel viscosity can be due to the presence of low boiling point, low molecular weight compounds in fuel. Contamination with low boiling point compounds such as solvents, gasoline, and petroleum naphtha can dramatically reduce the viscosity of distillate fuel and residual fuel oil.

A reduction in both the initial boiling point and end point of a fuel can result in a lowering of viscosity. Accompanying a reduction in these distillation parameters is often a consequential increase in the concentration of lower molecular weight, volatile compounds.

If diesel fuel is low in viscosity due to kerosene dilution, fuel lubricity is probably poor. The possibility of wear of high-pressure fuel injection pump parts will increase.

11.1.6.1 Possible Effects

- Leakage of fuel past the diesel fuel injection pump plunger in worn equipment may occur. This may result in a "hot restart" problem whereby fuel leaks past the injection plunger in hot equipment. Restarting may not be possible until the engine cools down.

- Wear of fuel system components can occur due to poor elastohydrodynamic or mixed lubrication properties.
- Soft, nonpenetrating spray from fuel injectors results in poor mixing with air. Combustion quality and power can diminish.
- Fuel readily atomizes into small droplets when sprayed from the fuel injector.

11.1.7 Low Flash Point

The flash point is considered to be an important specification for all finished fuels and oils. The flammability and combustibility characteristics of a material are directly related to the flash point. Also, fuel transportation codes require flammable compounds to be appropriately labeled for safety reasons.

According to OSHA, compounds with flash point values less than 37.8°C (100°F) are considered flammable. DOT and UN codes rate compounds flammable when the flash point is less than 60°C (141°F).

The flash point, vapor pressure, and autoignition values provide important information about the volatility of fuels and solvents. Likewise, the UN number, hazard class, and safety profile provide information about safe shipping and handling of fuels, oil, and solvents.

11.1.7.1 Possible Effects

- High vapor pressure values are usually linked to low flash point compounds.
- Low molecular weight and low flash point contaminants in diesel fuel could lead to early combustion in a diesel engine. This contamination could cause rough running of the engine and formation of exhaust smoke.
- Low molecular weight, low flash point vapors may cause vapor lock. If the Reid vapor pressure of aviation gasoline is no greater than 49 kPa, fuel will be free of vapor locking tendencies under most aircraft use conditions.

11.1.8 °API/Specific Gravity

The American Petroleum Institute (API) established an arbitrary rating system to measure the specific gravity of crude oils and petroleum products. This system assigns water the °API value of 10. Products heavier than water have °API gravity values less than 10. Most petroleum products are lighter in weight than water and have °API values greater than 10. For example, diesel fuel has a typical °API of 30 to 40, and gasoline or petrol has typical °API of 50 to 60. Some asphaltic oils are heavier than water and have °API values less than 10.

The specific gravity of a fuel or oil is a function of the weight per standard volume of the product compared to an equal volume of water. Since aromatic compounds have a greater weight per unit volume than do paraffinic hydrocarbons, the specific gravity of a highly aromatic product would be greater than a paraffinic fuel or oil.

Because fuel and oil weights vary per standard volume, they should always be handled and sold on a weight basis, not volume basis. If sold on a volume basis, the volume should be corrected to 15.6°C (60°F) by utilizing °API volume reduction tables.

11.1.8.1 Possible Effects

- Higher kJ/kg or Btu/lb value for fuel.
- High-density fuel may stratify or settle when mixed with certain heavy fuel oils of a lower density.

11.1.8.2 Possible Reasons for Low °API Fuel

- May contain a high concentration of FCC cycle oil, vacuum gas oil, or coker gas oil.
- High content of aromatic compounds including monoaromatics, diaromatics, and polyaromatics.

11.1.9 Low Electrical Conductivity Values for Fuel

The ability of fuel to conduct electrical charge must be maintained to ensure that static charge does not build as fuel travels through pipelines and accumulates in storage. Fuels containing a higher concentration of polar organic compounds possessing sulfur or nitrogen, such as jet fuel and high-sulfur heating oil, tend to provide good conductivity properties to the fuel.

However, hydroprocessing fuel to remove sulfur has resulted in fuel possessing much lower conductivity values than are often required for fuel to be safely transported and stored. Because of this, electrical conductivity improver is used in fuel to restore conductivity. For example, to ensure diesel fuels provide conductivity, the ASTM D975 Standard Specification for Diesel Fuel Oils requires fuels to meet a specification of 25 pS/m or conductivity units (CU) minimum. On occasion, after treatment with conductivity improver, fuel conductivity values may fall over time. Possible reasons for a continued drop in conductivity are listed below.

11.1.9.1 Possible Causes for a Continued Fall in Fuel Conductivity

- A drop in fuel temperature below 5°C (41°F) can result in a decrease in fuel electrical conductivity values.
- Interaction of fuel with a carboxylic acid–based corrosion inhibitor.

- Interaction with a fuel lubricity improver can result in a decrease in electrical conductivity values.

11.2 Deposit Analysis Guide

A guide for interpreting a deposit analysis is provided in Table 11.1.

Deposit/Component Analysis	Possible Source of Deposit
Sodium or calcium salt of carboxylic acid in deposit	Indicates the presence of a corrosion inhibitor salt or other fatty acid salt
Sodium at high concentration in deposit	Can increase the water solubility of deposits; may indicate the presence of saltwater or caustic water
Vanadium in deposit	Indicates the presence of residual oil; vanadium can cause severe pitting of ferrous metal components exposed to temperatures >1100°F; vanadium can combine with sodium, potassium, or sulfur to cause severe valve corrosion in marine engines
Polymeric amines in deposit	Indicates the presence of dispersant compounds; fuel deposit control additives; fuel stabilizers
Oxyalkylated compounds in deposit	Indicates the presence of water or oil soluble cleaning agents; demulsifying agents
Mg:Ca ratio is in the range of 1:3 to 1:5	Indicates the presence of freshwater
Mg:Ca ratio is in the range of 3:1 to 5:1	Indicates the presence of seawater
Zn, Ca, P, Mg present in deposit	Indicates the presence of a lubricant and lubricant additives
High ash content	Indicates the presence of metals and metal salts that may contribute to deposit formation and wear; could indicate contamination with water containing salts or particulate

Table 11.1 Deposit/Component Analysis and Possible Sources of Deposits

11.3 Fuel Problem Solving Guide

Cetane number of motor diesel fuel is low or below specification	**Possible Causes** • Fuel contains a high percentage of aromatic compounds/low percentage of paraffinic compounds. • Low cetane number component such as LCO has been blended into fuel. • A highly aromatic, low-viscosity kerosene or No. 1 fuel has been blended into motor diesel.
	Problem Solutions and Prevention ✓ Adjust the fuel blend to include more paraffinic components. ✓ Increasing the fuel IBP and T-50 may raise the cetane number. ✓ Add cetane improver to boost the fuel cetane number. Use 250 ppm of cetane improver for each 1 to 2 number boost in diesel cetane number.

Copper and silver are corroding when exposed to fuel	**Possible Causes** • Hydrogen sulfide/elemental sulfur/ mercaptans may be present in the fuel. • Fuel contains dissolved ammonia and water. • Fuel tank bottom water may be contaminated with inorganic acids. • Mercaptans may have attacked copper to form gel-like copper mercaptide.
	Potential Solutions and Prevention ✓ Utilize a scavenger to chemically react with hydrogen sulfide in the fuel or water in the tank. Scavenger rates are based upon the amount of H_2S detected in the fuel vapor or liquid. ✓ Use a fuel soluble copper or silver corrosion inhibitor. ✓ Utilize caustic treatment or amine sweetening to remove corrosive agent.

<table>
<tr><td rowspan="2">Deposits, gums, sediment form in gasoline or diesel fuel</td><td>Possible Causes
• Polymerization, oxidation, or condensation type reaction of olefins and heterocyclic compounds in fuel produced high molecular weight gums sediment.
• Fuel contains or was contaminated with high boiling point, high molecular weight compounds.
• Diesel fuel blend contains nonhydroprocessed cycle oil.
• Inorganic salts of organic compounds may be present in the fuel; waterborne solids may have contaminated fuel through tank or line transfer.
• B100 FAME without antioxidant may have been blended into diesel fuel.</td></tr>
<tr><td>Potential Solutions and Prevention
✓ Look for increased amounts of high molecular weight, high boiling point components in fuel by ASTM D86 or ASTM D1160.
✓ The addition of fuel antioxidant and stabilizer compounds in fuel can help minimize gum and deposit formation.
✓ Minimize the use of high olefin content blendstocks.
✓ Limit exposure of fuel to copper, bronze, or brass in the fuel distribution system; use metal deactivator in fuel for chelating catalytic copper ions.
✓ A high ASTM D86 distillation residue value may indicate a possible existent gum or deposit problem.
✓ The caustic treatment process can often remove color bodies and deposit precursors from fuel.</td></tr>
</table>

Distillate fuel darkens in color/ forms deposits	**Possible Cause** • Fuel contains a high percentage of reactive color bodies such as indoles, pyrroles, quinones, naphthenic acids from crude oil; these compounds can react to form higher molecular weight, insoluble deposits. • Fuel is blended with nonhydroprocessed LCO, coker gas oil, or other cracked distillate. • Fuel is blended with high sulfur or high nitrogen heavy vacuum gas oil (HVGO). • By-products of microbial growth are enhancing fuel degradation and deposit formation. • Fuel degradation is being catalyzed by copper. • B100 FAME used for fuel blending may be oxidizing and darkening in color.
	Problem Solutions and Prevention ✓ Minimize the use of LCO, HVGO, coker gas oil, or other cracked or high boiling fractions. ✓ Limit the use of streams containing high concentrations of heterocyclic compounds that can produce color bodies and potentially react to form deposits. ✓ The addition of fuel stabilizers can help limit distillate fuel and B100 FAME color change and deposit formation. ✓ Caustic treatment and clay filtration can be used to remove color bodies from light distillates such as No. 1 diesel, kerosene, and jet fuel. ✓ Evaluate fuel water bottoms for microorganisms; treat with a biocide if a contamination problem is confirmed. ✓ Minimize the use of copper, brass, and bronze hardware.

Emulsions or haze in fuel	**Possible Causes** • Contamination of fuel with water that contains organic salts has led to emulsion formation. • Accumulated products of microbial metabolism may have enhanced fuel-water emulsion formation. • The presence of high molecular weight, high-viscosity heterocyclic compounds in fuel may stabilize water initiated haze or emulsions. • Fuel blended with B100 FAME at rates greater than 5% may become hazy when exposed to water. • High concentrations of certain fuel additives can enhance the emulsification of water in fuel. • High-acid crude may have been used in fuel refining; if naphthenic acid salts are present in fuel, emulsions can form.
	Problem Solutions and Prevention ✓ Check the pH of tank bottom water to ensure that high pH caustic is not present; caustic can react to stabilize emulsions; drain tank bottom water. ✓ If present, remove tank bottom water contaminated with microorganisms; treat tank bottom water with a biocide. ✓ Mild heating can be used to break some emulsions. ✓ Utilize a demulsifier to break fuel-water emulsions. ✓ Ensure that fuel additive addition rates are within recommended limits and over-treating is avoided; high additive rates may stabilize emulsions in fuel; test fuel ASTM D1094 or D7451 demulsibility. ✓ Limit the exposure of biodiesel blends to water during transport and storage.

Ferrous metal corrosion of tanks, lines, and hardware	**Possible Causes** • Contamination of system with process water, seawater, or condensate water. • Storage and use of fuel that does not possess corrosion inhibiting properties. • Acidic pH water from fuel processing has corroded hardware. • Microbiologically influenced corrosion.
	Potential Solutions and Prevention ✓ Remove as much water as possible from the fuel storage and transportation system. ✓ Check the pH of tank bottom water for the presence of acidic or alkaline water; remove water as soon as possible. ✓ Treat fuel with a ferrous metal corrosion inhibitor. ✓ Check for the presence of sulfate reducing bacteria, hydrocarbon utilizing microbes; treat water with a fuel/water soluble biocide.

Ferrous metal corrosion inhibitor does not prevent corrosion	**Possible Causes** • Scale or deposit on metal is preventing inhibitor from filming metal surface. • Corrosion inhibitor treat rate is too low to prevent corrosion. • Corrosion inhibitor has been stripped from the metal surface by a caustic solution. • Inorganic acids may be present in fuel storage tank water. • Microbiologically influenced corrosion (MIC).
	Problem Solutions and Prevention ✓ Drain water bottom from storage tank frequently. ✓ Ensure that caustic carryover into finished fuel does not occur. ✓ Add higher concentrations of corrosion inhibitor. ✓ Use of a caustic resistant corrosion inhibitor may be required. ✓ Run NACE test and look for emulsification and change in pH of NACE test water after testing.

High gum level/ sediment in gasoline or diesel fuel	**Possible Causes** • High concentration of olefins or nitrogen, sulfur, oxygen containing compounds in fuel. • Fuel may be contaminated with high boiling, high molecular weight compounds. • Diesel fuel blend contains high concentration of cycle oil. • Inorganic salts of organic compounds may be present in fuel; waterborne solids may be in fuel.
	Problem Solution and Prevention ✓ Minimize the use of high-olefin content blend stocks. ✓ Minimize exposure of fuel to copper, bronze, or brass fuel distribution, and storage system hardware. ✓ Use of fuel containing an antioxidant, fuel stabilizing compound, or metal deactivator can minimize the formation of gums and sediment. ✓ Look for increased amounts of high molecular weight, high boiling point components in fuel by ASTM D86 distillation. High boiling compounds may contribute to sediment and gums. ✓ Water washing, caustic washing, or fuel filtration may remove gums and sediment from fuel.

Low induction time in gasoline	**Possible Causes** • High olefin content, possibly due to addition of FCC gasoline to the blend. • Acid catalyst from alkylate may have induced a fuel degradation reaction. • Gasoline may contain methanol; methanol reduces the oxidative stability of gasoline.
	Problem Solutions and Prevention ✓ Addition of an antioxidant to fuel will inhibit the fuel oxidation process and increase induction time; most antioxidants function at rates ranging from 5 to 50 ppm. ✓ Add 2 ppm of a metal deactivator to help prevent copper catalyzed degradation of fuel. ✓ Hydroprocessing can saturate olefins and improve fuel induction time.

Lubricity of diesel fuel fails to meet HFRR wear scar specifications	**Possible Causes** • Fuel was blended with kerosene or a No. 1 distillate fuel to meet low temperature flow and filterability specifications. • Fuel was not treated with a lubricity improver. • An alkyl nitrate cetane improver was added to fuel containing a lubricity improver. Depending upon the cetane improver rate, the HFRR wear scar diameter can increase from 50 to 150 microns in some fuels containing a lubricity improver.
	Problem Solutions and Prevention ✓ Confirm whether or not fuel was treated with lubricity improver. ✓ Minimize the blending rate of kerosene or light distillate into diesel fuel. Low-viscosity kerosene blending may increase the wear scar diameter of a diesel blend. ✓ Utilize a cold flow improver additive to meet low temperature flow and filterability specifications. Substitute cold flow improver for a portion or all of the kerosene or No. 1 diesel blended into the fuel. ✓ If fuel was treated with a cetane improver, increase the lubricity improver addition rate to overcome the impact of cetane improver on the HFRR wear scar.

Pumping and filtration problems with fuel at cold temperatures	**Possible Causes** • Fuel may have a narrow boiling range with a T90 minus T10 temperature difference less than 100°C (180°F). • Fuel is at or near its pour point. • Fuel viscosity may be too great to effectively filter and pump in low-temperature environments. • Fuel may have been stored for several days below its cloud point; wax settling has resulted. • $< 1\%$ *n*-paraffins in fuel $< C_8$ demonstrating few low-viscosity paraffins.
	Problem Solutions and Prevention ✓ Reduce fuel IBP/EP. ✓ Blend with low-viscosity components. ✓ Blend with low molecular weight aromatic components. ✓ Treat fuel with a wax crystal modifier or dispersant. ✓ Heat/insulate fuel system hardware. ✓ Minimize B100 FAME blend component percentage.

Reduced power and/or combustion quality	**Possible Causes** • Fuel may be too cold or too viscous to properly atomize into finely dispersed droplets. • Distillate fuel may be highly diluted with kerosene and not viscous enough to mix with pressurized air when injected into the fuel combustion chamber. • The diesel fuel cetane number may be below 40. • T-50 of diesel fuel may be below 232°C (450°F). • Gasoline may have lost volatile, higher vapor pressure components. • Fuel injector, intake valve, or combustion chamber deposits may be present due to predominantly stop and go, short trip, city driving profile.
	Problem Solutions and Prevention ✓ Check fuel viscosity; viscosity may be too high to flow at cold temperatures. ✓ Minimize the amount of kerosene blended into No. 2 diesel. ✓ Check fuel for an unusually low Btu value; fuel may be diluted with low boiling point, light end components. ✓ ASTM D86 profile of diesel to identify the volatility characteristics of the fuel (IBP, T-50, EP). ✓ Replace aged or low vapor pressure gasoline with new fuel having a higher RVP. ✓ Use fuel containing a detergent to help remove deposits from the fuel intake system. ✓ Use diesel fuel with a cetane number of 40 or higher.

Smoke formation during combustion	**Possible Causes** • Low temperature, low speed, high load operation can enhance smoke formation in diesel engines. • High boiling point aromatics in fuel can burn with more smoke; naphthalene compounds can form dense smoke during combustion. • Fuel injection volume may be too high. • Lubricating oil may be entering the combustion chamber.
	Problem Solutions and Prevention ✓ Use of a lower aromatic content fuel may help reduce smoke. ✓ Pump plunger lift may be too great resulting in an increase in fuel volume injection. ✓ The addition of cetane improver can sometimes help reduce exhaust smoke. ✓ Look for potential source of lube oil leakage such as piston ring wear or valve stem wear.

CHAPTER 12

Synthetic and Alternative Fuels

Synthetic and alternative fuel development and use is a constantly changing field with fuel types entering and exiting the research field and commercial marketplace for a variety of reasons. This chapter contains information on current and some earlier technologies used to develop synthetic and alternative fuels. Technologies ranging from the production of jet fuel from carbon monoxide and hydrogen to the pyrolysis of waste biomass to yield biocrude and synthesis gases will be reviewed.

Synthetic and alternative fuels have been considered as available energy sources to power engines since the early years of combustion engine development. The original engine developed by Rudolph Diesel, for example, was designed to be fueled by powdered coal before settling on liquid fuels. Henry Ford considered ethanol to be a good fuel source since it was renewable and because of its benefit toward agriculture.

Since that time, various alternatives to refined petroleum products have been evaluated and used as fuels for internal combustion engines. Reasons for considering synthetic and alternative fuels stem from two primary motivating factors: (1) the need to ensure an adequate supply of low cost fuel, and (2) the desire to use fuel that burns completely and cleanly without undesirable exhaust emissions.

The terms *synthetic fuel* and *alternative fuel* are closely related but not synonymous. *Synthetic fuel* was coined as a term to include gaseous, liquid, and solid fuels produced from coal, oil shale, tar sands, biomass, gas hydrates, and other sources. *Alternative fuel or nonconventional fuel* appeared later as a popular phrase to refer to fuels such as methanol, ethanol, fuels from other alcohols, biodiesel fuel, solar energy, and other fuels that were not totally based on petroleum and held potential as clean burning, low polluting commercially viable fuels.

12.1 Biomass

Biomass in fuel production could be considered as raw material that is either directly or indirectly derived from biological sources. Biomass-based fuel can come from a number of sources such as grains, legumes, sugar crops, animal waste, municipal waste, forestry products, aquatic plants, and food production waste. A description of biomass raw materials, products and potential use is provided in Table 12.1.

As a fuel source, biomass is considered renewable. Carbon dioxide produced when biomass is burned in subsequently utilized in photosynthesis by plant cells. The conversion of carbon dioxide and water to carbohydrate and oxygen by photosynthetic plant cells is represented by the following equation:

$$nCO_2 + mH_2O \xrightarrow[\text{Chlorophyll}]{\text{Sunlight}} Cn(H_2O)m + nO_2$$

Carbon dioxide is present in the atmosphere at about 0.03 volume percent and also in water as dissolved gas or carbonate. It is estimated that carbon dioxide follows a 350-year life cycle as it passes from air and water, into organic compounds, and then returns.

Raw Material	Specific Products	Use
Legumes, grains, tubers	Soybeans, rapeseed, corn, wheat, rice	Vegetable oils for biodiesel; fermentable starch and sugar for ethanol production
Sugar crops	Sugar cane, sugar beets, sweet sorghum, grapes	Fermentable sugars for ethanol production
Agricultural waste	Cane bagasse, cellulose based materials and pulp waste	Fermentable sugar for ethanol production, on-site combustion or pyrolysis
Animal and municipal waste	Animal renderings, sewage treatment sludge	Fats and oils for biodiesel; anaerobic digestion to methane
Aquatic plants	Kelp, water hyacinth, algae	Combustion or pyrolysis, CO_2 utilization in photosynthesis

TABLE 12.1 Typical Biomass Raw Materials, Products, and Use

Converting biomass into fuel grade products such as gases, alcohols, and oils can be accomplished by processes such as fermentation, anaerobic digestion, thermal conversion, and oil extraction. The fuel grade products produced from these processes can be utilized directly as alternative fuel, or if appropriate, serve as components for blending into existing petroleum based fuel.

12.1.1 Biodiesel

Oils derived from plant sources were some of the first liquid fuels utilized for powering the early diesel engines. One of the first examples was a vegetable oil–fired compression ignition engine introduced by Rudolph Diesel at the 1900 World's Fair in Paris. To compensate for the limited supply of petroleum-based diesel fuel during World War II, vegetable oils were used in some African and Asian countries to power diesel engines.

12.1.1.1 Raw Materials

Since that time, a variety of raw material sources appropriate for use in producing fuels suitable for combustion in a diesel engine are available and include soybeans, rapeseed, canola, palm, jatropha, sunflower, and other plant oils. Processed cooking oil and beef tallow have also been used as raw material sources.

Oils derived from these sources are rich in molecules ranging from C_{10} to C_{24} in carbon number. The most valued oils have high concentrations of molecules in the C_{16} to C_{18} carbon number range. Both saturated and unsaturated compounds are found, but the major portion of the molecules contains one or more double bonds. The molecules are generally linear without branching or cyclization. The properties provide for good combustion in the diesel engine.

In their raw, natural state, these oils exist primarily in the form of a triglyceride or fat. To extract the useful oil from the raw material, the triglyceride must be separated from proteins and sterols also present in the oil. To illustrate this process, soybean oil production will be briefly described.

Soybeans grow on a bushy, annual plant ranging in height from about 75 to 125 cm (30 to 50 in). Stems have a hair-like covering and leaves are trifoliated. Flowers range in color from white to purple with mixed variations. Seedpods usually contain three spherical to oval seeds weighing 0.1 to 0.2 g. Most seeds have a yellow coat containing two oil- and protein-rich cotyledons.

Argentina, Brazil, China, and the United States produce the world's largest supply of soybeans. In the United States, the Midwestern soybean harvest begins in September and October when the moisture content of the soybean is 13 percent. These beans can be stored for about 1 year. Soybeans with less than 12 percent moisture can be stored for 2 years or more. Higher moisture content beans must be aerated and dried in bins before being shipped by truck, rail, or barge for storage in elevators or silos.

Traditional processing then involves the following steps:

1. Cleaning and drying the beans to a 10 to 11 percent moisture content to help loosen the hull or seed coat.
2. Beans are then cracked, dehulled, and conditioned with steam to help facilitate flaking.
3. Conditioned beans are flaked and then mixed with hexane in a Rotocel vat.
4. Oil in the vat is extracted by hexane; the oil-hexane solution is then separated from the residual flake.
5. Hexane is removed from the oil by evaporation, and hexane is reused.
6. Residual flake is treated with steam in the desolventizer-toaster to remove the remaining hexane. Steam also deactivates hard to digest compounds in the flake to improve its value as protein meal.
7. Collected soybean oil is usually degummed to remove phospholipid, wax, and nontriglyceride impurities before further processing.

More efficient procedures utilize expanders and extruders to convert the flakes into collets or porous, sponge like material. Collets are larger and less fragile than soybean flakes aiding in the oil extraction and drainage processes. This technique also minimizes the amount of hexane that needs to be recovered in the desolventizer.

On average, a ton of soybeans yields about 18 percent oil, 72 percent protein meal, 7 percent hulls, and 3 percent loss. A listing of the primary fatty acids found in soybean oil is provided in Table 12.2

12.1.1.2 Production of Fatty Acid Methyl Esters (FAME) or B100 Biodiesel

For natural oils to be suitable as a blending component into petroleum diesel, the fatty acid components of the triglyceride or fat must

Fatty Acid	Carbon Number: Double Bonds	Typical Percent
Palmetic	16:0	11
Stearic	18:0	5
Oleic	18:1	35
Linoleic	18.2	45
Linolenic	18.3	4

Table 12.2 Typical Fatty Acid Composition of Soybean Oil

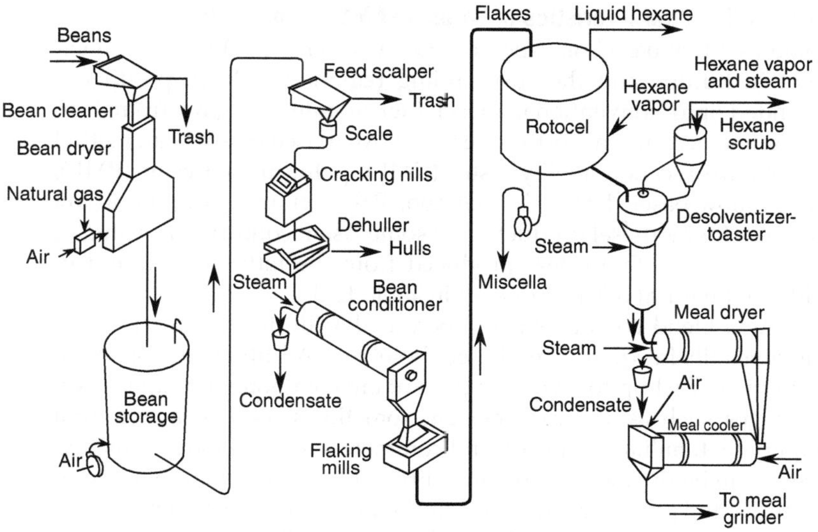

FIGURE 12.1 Schematic for processing soy beans into oil and meal.

be converted into fuel soluble, biodiesel grade methyl esters. The process of transesterification is used for this purpose. Through transesterification, the triglyceride is cleaved to yield glycerol and the individual fatty acid methyl esters or FAME compounds. The typical yield of biodiesel from soy oil is approximately 1-gal of biodiesel from 7.3 lb of soybean oil. The basic process of FAME production is provided in Fig. 12.1.

Most biodiesel transesterification processes are base catalyzed and have the operational advantages of low temperature, low pressure, and high conversion rates. Steps involved typically include the following:

1. Degummed soy oil is reacted with an alkaline catalyst such as sodium or potassium methoxide in the presence of an excess of methyl alcohol.
2. Reaction temperature is maintained at approximately 65°C (150°F) and a pressure of 20 psi until transesterification is complete.
3. Glycerol, free fatty acids, and unreacted methyl alcohol are separated from the methyl ester product.
4. The methyl ester is purified by removal of residual methyl alcohol and any other low boiling point compounds before use in biodiesel fuel.

12.1.1.3 Characteristics of Various FAME Compounds

Biodiesel fuel based on rapeseed methyl ester (RME) is most prevalent in Europe and the surrounding region. The RME product is blended into conventional diesel fuel at rates ranging from 5 to 7 volume percent. In North America, FAME based on soybean methyl ester (SME), canola methyl ester (CME), palm methyl ester (PME), and tallow-based FAME is common. Palm oil and jatropha oil are sources for biodiesel production in some Asian countries. A listing of the various methyl esters produced from the fatty acids found in different natural oils is provided in Table 12.3.

As a blend component, 100 percent FAME, termed B100, mixes quite readily into petroleum-based diesel fuel. Addition rates of FAME into diesel fuel up to 5 volume percent are common with higher percentages such as 10 or 20 percent sometimes used in agricultural equipment, military vehicles, and commercial fleets. Biodiesel methyl esters are biodegradable, low in sulfur content, free of aromatic compounds, and possess oxygen in their chemical structure. When blended into petroleum diesel, B100 FAME can typically improve both the lubricity and cetane number of the fuel but may degrade the water demulsibility properties and the long-term storage stability of the fuel blend. At subzero temperatures, the handling characteristics of biodiesel become more difficult to control than conventional petroleum diesel. A comparison of the cloud point and pour point temperatures of FAME produced from different oil sources is presented in Table 12.4.

FAME Compounds Detected by GC/MS	FAME Type, Area %				
	Soy	RME	Palm	Jatropha	Tallow
16:0 Methyl palmate	8.9	9.0	44.7	14.2	24.5
16:1 Methyl palmitoleate	—	—	—	—	1.6
18:0 Methyl stearate	6.2	3.3	11.7	5.5	13.6
18:1 Methyl oleate	38.1	59.2	37.5	52.2	43.8
18:2 Methyl linoleate	43.6	26.2	5.3	27.1	12.8
18:3, 20:0, 20:1 + others	3.2	2.3	0.8	1.0	3.7

Table 12.3 FAME Compounds Formed from Different Natural Oils

Cold Temperature Property	No 2. S15 Diesel	Canola B100	Palm B100	Soy B100	Tallow B100
Cloud point, °C (°F)	−10 (14)	−2 (28)	17 (62)	−1 (30)	6 (42)
Pour point, °C (°F)	−15 (5)	−7 (20)	13 (55)	−4 (25)	2 (35)

Table 12.4 Comparison of No. 2 S15 Diesel and B100 FAME Cloud and Pour Point Values

12.1.1.4 Biodiesel Grades

Biodiesel specification ASTM D6751 Standard Specification for Biodiesel Blendstock (B100) for Middle Distillate Fuels classifies four grades of B100 for use as a blend component with petroleum diesel into the following:

Grade No. 1-B S15: A special purpose biodiesel fuel with good low temperature flow properties that is sensitive to the presence of partially reacted glycerides and contains a maximum sulfur content of 15 ppm.

Grade No. 1-B S500: A special purpose biodiesel fuel with good low temperature flow properties that is sensitive to the presence of partially reacted glycerides and contains a maximum sulfur content of 500 ppm.

Grade No. 2-B S15: A general purpose biodiesel fuel intended for use in applications requiring a blend containing a maximum sulfur content of 15 ppm.

Grade No. 2-B S500: A general purpose biodiesel fuel intended for use in applications requiring a blend containing a maximum sulfur content of 500 ppm.

A common terminology also exists to describe blends containing both diesel fuel and FAME. The blend nomenclature is shown in Table 12.5.

12.1.1.5 Biodiesel Performance Properties

The cold temperature filterability and the storage stability of biodiesel have emerged as two important performance properties that B100 FAME must maintain. To help assure its use and application under these conditions, the following B100 FAME evaluation procedures have been developed and implemented.

12.1.1.5.1 ASTM D7501 Cold Soak Filtration Test After B100 FAME is produced, the product may be stored or transported prior to immediate use. If the FAME is exposed to cold temperatures during storage

Common Name	Volume % Diesel	Volume % FAME
B0	100	0
B5	95	5
B10	90	10
B20	80	20
B50	50	50
B100	0	100

Table 12.5 Nomenclature to Describe Biodiesel Blends

or transportation, insoluble compounds may agglomerate or precipitate from the cold oil. Upon use, the compounds may be trapped on filter surfaces and halt the flow of fuel to a pump or engine.

- The Cold Soak Filteration Test procedure used to evaluate filter blocking tendency involves cooling 300 mL of B100 FAME to a temperature of 4.5 ± 0.5°C (40 ± 1°F) for 16 hours, then warming to room temperature and filtering the oil at a controlled vacuum. For Grade B-1 fuel, the oil must be filtered within 200 seconds. For Grade B-2 fuel, the oil must be filtered within 360 seconds. For B100 intended for application at temperatures of −12°C (10.4°F) or below, a maximum cold soak filtration limit of 200 seconds must be met.

12.1.1.5.2 EN 14112 and EN 15751 for B100 FAME (accelerated oxidation method) Depending upon the proximity of the biodiesel production facility to a fuel blending terminal, produced B100 FAME may be transported by a short haul tanker or by long distance rail car to a fuel blending terminal. Exposure of unsaturated methyl oleate and methyl linoleate compounds in FAME to air during transportation and storage can result in oxidation and degradation of FAME. Reaction products such as carboxylic acids, aldehydes, diols, and ketones can form in addition to the development of higher molecular weight compounds. The quality of FAME is impacted.

- To determine the resistance of FAME to degradation during transportation, storage, and use, accelerated oxidation testing is completed. Testing involves heating 7.5 g samples of B100 FAME or diesel blends containing FAME to a temperature of 110°C and passing air through the FAME samples at a rate of 10 L/h. As oxygen in air reacts with FAME, volatile, low molecular weight carboxylic acids will form. Vapor containing these volatile acids leaves the oil sample, passes through a section of tubing, and into water. As acids accumulate, water conductivity increases. The conductivity is recorded and charted over time until an inflection point is reached and conductivity rapidly accelerates. At this point, testing ends and the oxidation lifetime or oxidation stability is measured from the recorded and charted conductivity results. Accepted oxidation stability results ranging from 3 to 6 hours and greater are specified by regional standards. Antioxidants are occasionally added to FAME to help meet oxidation stability requirements.

12.1.1.6 Biodiesel and Petroleum Diesel Property Comparison

The physical, chemical, and performance properties of B100 FAME and biodiesel blends can be different from petroleum-based diesel fuel. Some of these properties are presented and briefly discussed in Table 12.6.

FAME Property	Observation and Comparison
Cetane number	The cetane number of B100 FAME is typically higher than conventional diesel fuel due to the presence of large amounts of C_{16} and C_{18} carbon number compounds; cetane numbers of 52 and higher are common; minimum specification is 47
Demulsibility	Very poor; biodiesel and biodiesel blends do not shed water as effectively as conventional diesel fuel; fuel haze, gelling, and low temperature handling problems can develop if biodiesel is contaminated with water during storage and transportation
Density, g/cm^3 @ 15.5°C (60°F)	Density is typically higher than conventional diesel fuel; values of 0.88 to 0.89 are expected; blends of biodiesel with conventional diesel fuel will produce densities closer to values of 0.82 to 0.86
Lubricity	HFRR lubricity wear scar diameter measurements for B100 FAME are typically lower than conventional petroleum diesel fuel; HFRR values for diesel improve in biodiesel fuel blends.
Monoglyceride, % mass	Monoglyceride presence is indicative of incomplete transesterification and removal of fatty acids from triglyceride; limited to 0.4% mass in Grade No. 1-B FAME
Oxygen Content	B100 FAME contains approximately 10.5% oxygen and aids in fuel combustion to help lower exhaust emissions, especially carbon monoxide
Sulfur, % mass	Sulfur in FAME is usually quite low and is specified in Grades No. 1-B S15 and No. 2-B S500 at respective ranges of 0.0015 to 0.05% mass.
Total acid number, mg KOH/g	Total acid number or TAN for FAME is specified in ASTM D6751 as 0.50 or less.

Table 12.6 Biodiesel and Petroleum Diesel Property Comparison

12.1.1.7 Biodiesel Lubricity

The lubricity properties of B100 FAME and biodiesel blends are typically outstanding. Presented in Fig. 12.2 are the ASTM D6079 HFRR wear scar diameter results for 100 percent nonadditized Grade No. 2 S15 ULSD and blends of Grade No. 2-B S15 biodiesel containing 2 volume percent FAME produced from corn oil, palm oil, soy oil, rapeseed oil, and tallow. In each case, 2 volume percent FAME in the blend produced low wear scar diameter ratings that fall below the U.S. specification of 520 microns and at or below the 460- to 400-micron international specifications.

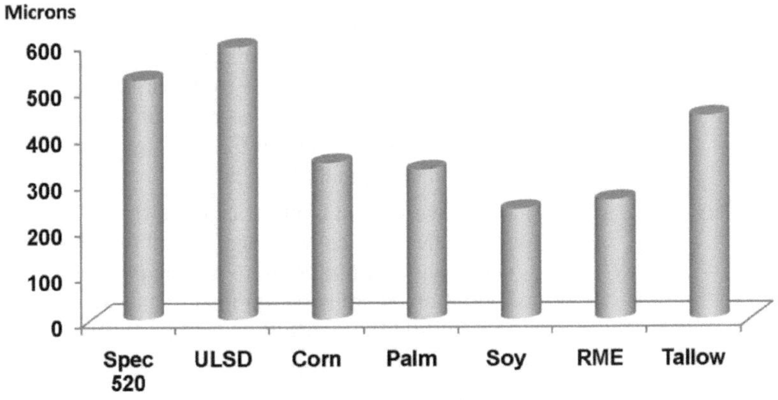

FIGURE 12.2 HFRR lubricity ratings of ULSD and B2 blends.

12.1.2 Fuel Grade Ethanol

Fuel grade ethanol has been produced and utilized worldwide. Brazil, the Scandinavian region, Japan, New Zealand, the European Union, the United States, and many other countries have found ethanol to be an effective alternative fuel for transportation vehicles. For example, ethanol fuels were marketed in the United States during the 1920s and 1930s, and by 1938 an ethanol facility in Atchison, Kansas was producing 18 million gal of ethanol annually. After World War II, interest in fuel ethanol production declined significantly, primarily due to the low cost of petroleum-based fuels.

During the world energy crisis of the 1970s, however, interest in ethanol as a transportation fuel source was once again spurred. Brazil developed widespread production of ethanol motor fuels. Products produced included "Alcool," a hydrated ethanol, and "Gasolina" a 78:22 ratio blend of gasoline and ethanol. In the United States "Gasohol," a 90:10 blend of gasoline and ethanol was marketed. Other countries also developed policies to maintain a fuel grade ethanol industry by conversion of biomass. Also, clean air and reformulated fuel regulations helped in promoting the use of ethanol as a viable alternative fuel.

12.1.2.1 Fermentation and Ethanol Production

Ethanol can be readily produced by industrial techniques such as the sulfonation-hydrolysis process and the direct catalytic hydration process as outlined in Eqs. (12.1) and (12.2).

$$\underset{\text{Ethylene}}{2CH_2{=}CH_2} + H_2SO_4 \rightarrow \underset{\text{Diethyl sulfate}}{(CH_3CH_2)_2SO_4}$$

$$\underset{\text{Diethyl sulfate}}{(CH_3CH_2)_2SO_4} + 2H_2O \rightarrow \underset{\text{Ethanol}}{2CH_3CH_2OH} + H_2SO_4 \qquad (12.1)$$

$$CH_2{=}CH_2 + H_2O \xrightarrow[\text{Catalyst}]{} CH_3CH_2OH \tag{12.2}$$

Biomass fermentation, however, yields the greatest supply of fuel grade ethanol. Fermentation is the anaerobic decomposition of organic compounds, basically carbohydrate, into alcohols by the action of microbial enzymes. Ethyl alcohol, or ethanol, is the desired product of carbohydrate fermentation, but other alcohols, acids, ketones, and aldehydes can form as by-products. One common example of ethanol production from biomass is the fermentation of corn. Both dry mill and wet mill processes are used in processing corn into ethanol.

12.1.2.1.1 Dry Mill Processing Traditionally, dry-milling is the predominant process and first involves grinding or milling corn kernels into meal that is flour-like in consistency. At this stage, the starch and cellulose components remain together in the milled corn, but the non-starch cellulose is later separated. The meal is then mixed with water, and the corn-water slurry mixture, termed "mash," is boiled.

Various enzymes are involved in the fermentation process to convert starch into ethanol. Enzymes include:

- *Diastase:* Catalyzes the hydrolysis or saccharification of starch into maltose. Diastase contains α-amylase and β-amylase produced during the germination of barley grain. The α-amylase also helps prevent gelation of starch.
- *Glucoamylase:* A pure enzyme that performs similar to diastase.
- *Maltase:* Catalyzes the hydrolysis of maltose into glucose.
- *Zymase:* Converts glucose into ethanol.
- *Invertase:* Converts sucrose into fermentable glucose.

For fuel grade ethanol production, the mash is maintained at a temperature range of about 50 to 60°C (122 to 140°F) and the hydrolyzing enzymes diastase or glucoamylase are added to convert the starch polysaccharide to maltose, a disaccharide. This process takes about an hour to complete, and the final product of hydrolysis is termed "wort." Solids and cellulose are filtered from the wort before continuing.

Prior to fermentation, the wort is cooled to temperatures below 30°C (85°F), and the pH is adjusted to approximately 5. Yeasts such as *Saccharomyces cerevisiae, Saccharomyces carlsbergensis,* or *Candida brassicae* are added, and fermentation proceeds for 2 to 3 days under batch processing conditions. Yeast produces the enzymes maltase, zymase, and invertase that respectively convert maltose into glucose, glucose into ethanol, and sucrose into fermentable glucose.

Since fermentation takes place in a dilute aqueous solution, the reaction continues until the alcohol concentration approaches about 14 percent. At higher alcohol concentrations, the process becomes self-inhibitory. By-products of fermentation to produce ethanol can include higher molecular weight alcohols, glycerine, ethers, and carbon dioxide. Usually no more than 10 percent starch is converted to these compounds. After fermentation, the product is termed *beer* and is distilled. Atmospheric distillation, vacuum distillation, and membrane separation techniques can be used to recover ethanol from the final fermented product. After distillation, water is removed from 190-proof ethanol by passing ethanol through a molecular sieve unit to yield 200-proof ethanol.

The distillate bottoms material, termed *stillage*, is centrifuged for further processing where a soluble product is separated from coarse grain solids. The following products are then produced:

- The soluble product undergoes concentration through evaporation resulting in condensed distillers solubles, CDS or "syrup."
- In combination, the syrup and coarse grain solids can be mixed to produce a wet distillers grains product or dried to yield dried distillers grains with solubles, or DDGS, for use in livestock and poultry feed.
- Recovered syrup can also be further processed into components such as corn distillers oil and raw material for biodiesel production.

12.1.2.1.2 Wet Mill Processing In the wet milling process, corn is steeped in water to separate compounds such as starch, gluten, fibrous material, and the oil-containing germ from the corn kernel. One process involves steeping corn in a warm, dilute solution of sulfurous acid for about 1 to 2 days to facilitate the separation process. At this point, the steeped solution containing a complex blend of materials can be processed by different routes such as filtration, grinding, and screening to separate the corn germ and fibrous material and by centrifuging to separate insoluble gluten from water soluble starch. At this point, different processing routes are taken such as:

- The oil-containing corn germ is extracted to produce corn oil.
- Steep water and fiber are dried together to produce corn gluten livestock feed.
- The steeped solution containing concentrated starch is fermented to produce ethanol.
- The gluten is dried and processed into corn protein gluten meal used in poultry feed.
- The starch containing solution is dried and sold as various grades of corn starch.
- The starch solution is processed and refined into corn syrup.

12.1.2.2 Ethanol Fuel Grades

As a transportation fuel, ethanol can be used directly, blended with conventional fuel, or used as an oxygenate in reformulated fuels. After production, fuel grade ethanol is denatured with 1.96 to 2.5 volume percent of an allowed denaturant including natural gasoline, gasoline blendstocks, or unleaded gasoline. No other denaturants are approved due to potential detrimental effects on fuel stability, elastomers, and piston scuffing. Requirements for denatured fuel grade ethanol are described in ASTM D4806 Standard Specification for Denatured Fuel Ethanol for Blending with Gasolines for Use as Automotive Spark-Ignition Engine Fuel. Common grades of ethanol blended fuels are briefly described.

12.1.2.2.1 Reformulated Gasoline Reformulated gasoline can contain ethanol. Although reformulated gasoline must contain a minimum of 2.7 weight percent oxygen, an EPA waiver for ethanol and certain other alcohols allows as much as 3.7 weight percent oxygen in fuel or about 10.5 weight percent ethanol.

12.1.2.2.2 Ethanol Flex-Fuel This fuel is defined by ASTM D5798 Standard Specification for Fuel Ethanol Blends for Flexible-Fuel Automotive Spark-Ignition Engines. Fuel blended to this specification contains 51 to 83 volume percent ethanol. This fuel was formerly defined as Ed75-Ed85 or commonly known as E85.

12.1.2.2.3 E5, E10, E15 These blends contain conventional gasoline plus ethanol at respective volume percent rates of 5, 10, and 15 percent. In Europe, E5 blends are most common, whereby in the United States, E10 blends are most prevalent. Blends of E15 are approved for use in vehicles of model year 2001 and newer.

12.1.2.2.4 Diesel-Ethanol Blends Diesel fuel blends formulated with ethanol volumes ranging from 7 to 30 percent and coadditive volumes averaging 5 percent have been proposed and evaluated. This technology has not yet reached the commercial stage of development.

12.1.2.3 Ethanol Handling and Storage

Since ethanol is a relatively pure, highly polar molecule, it can provide performance different from a traditional, non-polar hydrocarbon fuel that contains numerous organic molecules. Because of this fact, handling, storage, and use of ethanol and ethanol blended fuels can also differ from conventional fuels.

When storing or transporting ethanol or ethanol blended fuels, contact with water from any source should be avoided. Water present in the bottom of a gasoline storage tank, and on the interior walls of pipelines, transfer lines, or flexible hoses may be solubilized by fuel ethanol. Likewise, any compound or particulate matter dissolved in the water may also become soluble in the ethanol blended fuel. These soluble compounds can create problems with fuel filter plugging or fuel discoloration.

Water contamination of gasoline can also cause ethanol in fuel to separate from gasoline-ethanol blends. If water at a rate of 0.5 to 1 percent contacts a gasoline-ethanol blend, the water will become soluble in the ethanol leading to an increase the specific gravity of ethanol-water solution. This contamination results in separation of the heavier ethanol-water solution as a distinct phase from the lighter specific gravity gasoline. The problem of phase separation in gasoline-ethanol blends is more pronounced at cold temperatures.

12.1.2.4 Ethanol pHe

As an outcome of the 1977 Clean Air Act, the blending of oxygenates such as ethanol, methyltertiarybutyl ether (MTBE) and other oxygen containing organics into fuel was outlined.

Simultaneously, research was being completed by a major engine manufacturer to determine the cause of engine failure on vehicles fueled on gasoline-ethanol blends. Engine problems such as cylinder bore wear and fuel pump wear were reported. At the time, total fuel acidity by ASTM 1613 was being measured, but this was not providing the full picture of the impact of ethanol on fuel blends. Research turned toward evaluating fuel grade ethanol being produced and its properties.

Measurement of ethanol pH was considered a valid option since products of fermentation such as dissolved CO_2, water, carbonic acid, acetic acid, and lactic acid were identified in some fuel grade ethanol. Testing using a known concentration of acid in both water and ethanol revealed that pH measurements differed. Ethanol produced a lower pH value upon exposure of the pH electrode to ethanol for 30 seconds. This method of pH measurement was considered an effective way to measure the acid strength of ethanol and a better method to predict the corrosion potential of ethanol containing low concentrations of strong acids.

Much work followed expanding the industry's knowledge about fuel grade ethanol and its performance in automotive engines. Efforts resulted in the 1999 publication of ASTM D6423 Standard Test Method for Determination of pHe of Ethanol, Denatured Ethanol, and Fuel Ethanol.

12.1.2.5 Ethanol Compatibility

Ethanol compatibility with tank and piping materials such as carbon steel, stainless steel, and fiberglass is good. Zinc linings and epoxy linings are not stable in ethanol systems and are not recommended for use. Teflon and nylon are compatible elastomers. In E10 storage, Kalrez, Ryton, Viton, and other elastomers may be used.

Very fine filters are recommended for ethanol dispensers to ensure that waterborne solids do not contaminate vehicle filters or fuel injectors. Dispensing hoses delivering ethanol and other alcohol fuels to vehicles may be green or blue in color to signify their compatibility with alcohol fuels. Also, blending ethanol with conventional fuels helps to minimize compatibility problems with existing fuel system components.

Commercial use of ethanol fuel is regulated by NFPA 30, Flammable and Combustible Material Code. For ethanol storage and dispensing systems, the NFPA 30A Automotive and Marine Service Station Code would apply.

12.1.2.6 Ethanol Use and Performance

Flexible fueled vehicles (FFVs) or variable fueled vehicles (VFVs) have been developed to operate on either gasoline or alcohol based fuels. These vehicles are equipped with on-board sensors and controls to adjust the spark advance and the fuel injector timing to correct for differences in fuel performance.

Specific topics related to fuel ethanol performance in vehicles are summarized below.

12.1.2.6.1 Cold Weather Starting The cold temperature starting properties for ethanol fuels is poor unless blended with gasoline or some other starting fluid such as dimethyl ether or isopentane. The minimum starting temperature of neat ethanol is about 15.6°C (60°F). In ethanol flex fuel and E10 blends, starting problems are minimized.

12.1.2.6.2 Autoignition Temperature The elevated autoignition temperature of ethanol makes its use as a direct substitute for conventional diesel fuel in compression ignition engines less likely. The autoignition temperature of ethanol is about 704°C (1300°F), or more than two times higher than the temperature for conventional diesel fuel. Newly designed, higher compression diesel engines have been developed to operate on alcohol-based fuels and have limited commercial use.

12.1.2.6.3 Octane Number The octane number ratings for ethanol can be higher than gasoline making ethanol a good blend component. Ethanol has a research octane number (RON) of 109 and a motor octane number (MON) of 90. When together as a gasoline-ethanol blend, the octane numbers can become even higher depending on the properties of the base gasoline.

12.1.2.6.4 Ethanol pHe The pHe values for fuel grade ethanol are specified to be a minimum of 6.5 and a maximum of 9.0. Below pHe 6.5, fuel injectors can fail due to corrosive wear, excessive cylinder wear can occur, and fuel pumps can malfunction. Above pHe 9.0,

Property	Test Method	Limit
Water content, vol %, (mass) %, max	E203 or E1064	1.0 (1.26)
pHe	D6423	6.5 to 9.0
Sulfur, mass ppm, max	D2622, D3120, D5453	30
Existent sulfate, mass ppm, max	D7318, D7319, D7328	4
Copper, mg/kg, max	D1688	0.1
Solvent washed gum, mg/100 mL, max	D381	5.0

Table 12.7 Selected Performance Requirements for Denatured Fuel Ethanol

plastic pump parts can fail. The impact of ethanol pHe on engine components is minimized in E10.

12.1.2.6.5 Exhaust Emissions The reduced exhaust emission levels of carbon dioxide, particulate, and hydrocarbons are a primary advantage of alcohol-based fuel combustion. The increased oxygen content available from ethanol participates in driving combustion to completion yielding carbon dioxide and water. Since ethanol contains no sulfur, particulate emissions are minimized.

12.1.2.6.6 Vapor Pressure The Reid vapor pressure of gasoline increases when ethanol is blended into fuel as an oxygenate. At a 10 percent ethanol blend volume, the vapor pressure of gasoline may increase by 1.0 to 5.0 psi.

Selected performance requirements for fuel ethanol listed in ASTM D4806 Standard Specification for Denatured Fuel Ethanol for Blending with Gasolines for Use as Automotive Spark-Ignition Fuel are provided in Table 12.7.

12.2 Synthesis Gas and Gas-to-Liquid Products

The ability to synthesize fuel products from a variety of sources other than petroleum has resulted in the expansion of existing technologies and the development of new technologies in fuel production. Some of these fuel products contain little or no sulfur, aromatic, or heterocyclic compounds. Many are gases that can be used directly as fuel or are further processed to yield hydrogen rich, low particulate emission, liquid fuel products.

12.2.1 Synthesis Gas

In 1792 William Murdoch, a Scottish engineer and partner of James Watt, distilled coal in an iron oven and used the coal gas to fuel lamps in his home. This discovery led to the development of the coal gas industry. By 1812, distilled coal gas was being used to fuel street lamps in London. Use of coal gas quickly spread throughout Europe and remained as an important source of domestic and industrial heating and lighting into the twentieth century.

Distilled coal gas contained about 50 percent hydrogen, 20 to 30 percent methane, with the remainder being principally carbon monoxide. The calorific value of coal gas served as the benchmark for the "town gas" industry that was to follow. The solid coke byproduct produced from coal gas processing was a source of further fuel development during later years of the nineteenth century. Complete gasification of coke was achieved by blasting hot coke with alternating amounts of steam and air to produce a "blue water gas." This fuel burned with a distinct blue flame and was composed of about 50 percent hydrogen, 40 percent carbon monoxide, with the remainder being carbon dioxide and nitrogen.

In 1868, Siemens suggested the use of an underground gasification process to derive gas from coal remaining in the ground after mining. A similar process was being developed in Russia by Demetri Mendeleev. The process involved blowing air into a coal seam through an injection well and then igniting the system. Through a second production well, the resulting coal gas was withdrawn and used. This type of underground coal gasification process was utilized commercially in parts of Europe and Asia for several years.

Underground coal gasification processes have since given way to more favorable and highly developed aboveground processes. Reactors for use in aboveground coal gasification are usually one of three designs: packed bed, fluidized bed, or entrained or concurrent flow. Depending on the process, gasification temperatures can be controlled by reaction temperature, reaction pressure, and the feed rates of carbon, oxygen, and steam. Worldwide interest in coal gasification processes continues even after its replacement by the natural gas industry.

A major step in synthetic fuel chemistry took place in 1925 when Professor Franz Fischer, founding director of the Kaiser Wilhelm Institute of Coal Research and Dr. Hans Tropsch, head of the department, applied for a patent involving production of a liquid fuel from a carbon monoxide and hydrogen containing synthesis gas derived from coal.

Synthesis gas is defined as a gas mixture containing primarily carbon monoxide and hydrogen generated from the gasification or pyrolysis of carbon containing materials such as coal, petroleum coke, natural gas, biomass, and other carbon sources. The combustible synthesis gas can be used as a fuel source, in production of chemical intermediates, or in gas-to-liquid (GTL) fuel synthesis.

Equations representing production of synthesis gas are presented in Eq. (12.3).

$$\begin{aligned}
CH_n + O_2 &\rightarrow CO_2 + (H)_n \\
CH_n + CO_2 &\rightarrow 2CO + (H)_n \\
CH_n + H_2O &\rightarrow CO + H_2 + (H)_n \\
(CH_2O)_n &\rightarrow {}_nCO + {}_nH_2
\end{aligned} \tag{12.3}$$

Steam reforming of natural gas or methane is another common method of producing commercial hydrogen and synthesis gas. This process is often referred to as steam-methane reforming and is a continuous, catalytic process forming carbon monoxide and hydrogen from steam and methane. Both endothermic and exothermic reactions exist. Equation (12.4) represents an endothermic steam-methane reaction.

$$CH_4 + H_2O \rightarrow CO + 3H_2 \quad \Delta H_{298K} = +97{,}400 \text{ Btu/lb} \tag{12.4}$$

12.2.2 Fischer-Tropsch Process and Gas-to-Liquid Fuels

Coal and methane derived synthesis gas processes have been further developed into controlled gasification techniques followed by catalytic reactions to yield a valuable mixture of Fischer-Tropsch type hydrocarbon products. Fischer-Tropsch synthesis is a process involving the addition of methylene groups to a growing chain of carbon atoms. Simplified Fischer-Tropsch gas-to-liquid reactions equations to produce liquid hydrocarbons and oxygenated compounds are listed in Eq. (12.5).

$$\begin{aligned}
CO + 2H_2 &\rightarrow (-CH_2-) + H_2O \\
nCO + 2nH_2 &\rightarrow C_nH_{2n} + nH_2O \\
CO + 2H_2 &\rightarrow (\text{hydrocarbons, alcohols, acids}) + H_2O
\end{aligned} \tag{12.5}$$

After initial production in Germany, the Fischer-Tropsch gas to liquid synthesis process was expanded into commercial scale manufacture by the South African Coal and Gas Corporation. Production at the Sasol facility in Sasolburg, South Africa, began in the mid-1950s and yielded a variety of Fischer-Tropsch liquid fuels from synthesis gas. Several Sasol production facilities are operational today producing a range of hydrocarbon fuels in addition to other products from coal and natural gas derived synthesis gas.

Fischer-Tropsch reaction products consist of a mixture of paraffinic and olefinic gases, plus higher molecular weight hydrocarbons with varying percentages of paraffins, aromatics, and oxygenated

Reaction Product	Molar Formula	Mass Percent
Gases		
Methane	CH_4	11
Ethylene	C_2H_4	4
Ethane	C_2H_6	6
Propene	C_3H_6	11
Propane	C_3H_8	2
Butene	C_4H_8	8
Liquids		
C_5 to C_7	$C_{5.5}H_{11}$	8
Light oils	C_8H_{16}	33
Heavy oils	$C_{20}H_{24}$	6
Alcohols	$C_{2.4}H_{6.8}O$	9
Acids	$C_{2.4}H_{4.8}O_2$	2
Total		100

TABLE 12.8 Typical Fischer-Tropsch Reaction Products

compounds. Both gaseous and liquid products are produced. Low carbon number alcohols and acids are the primary oxygenates formed. A listing of representative products from the Sasol gas-to-liquid process is provided in Table 12.8.

12.2.2.1 Low Temperature Fischer-Tropsch Reaction

Low temperature Fischer-Tropsch (LTFT) reactions are carried out in temperatures ranging from approximately 200 to 240°C (392 to 464°F). Catalyzed reactions in this temperature range tend to produce higher carbon number compounds in a 1:2 ratio of gasoline to diesel. Carbon number range compounds produced by LTFT reactions are displayed in Fig. 12.3.

12.2.2.2 High Temperature Fischer-Tropsch Reaction

High temperature Fischer-Tropsch (HTFT) reactions are carried out in temperatures ranging from approximately 300 to 350°C (572 to 662°F). Catalyzed reactions in this temperature range tend to produce lower carbon number products in a 2:1 ratio of gasoline to diesel. Branching is also increased at higher temperatures. Fischer-Tropsch diesel products are comparable to No. 2 S15 diesel in several properties. Examples of carbon number range compounds produced by HTFT reactions are displayed below in Fig. 12.4.

Fuel product yield depends on various factors such as reactor design, synthesis gas composition, catalyst properties, reaction

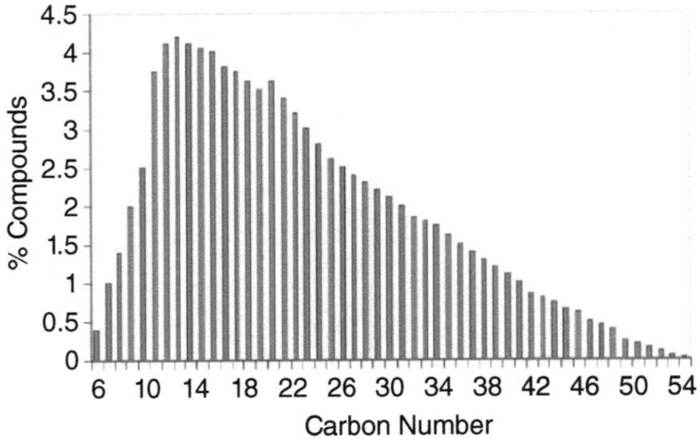

FIGURE 12.3 Low temperature Fischer-Tropsch (LTFT) compounds produced.

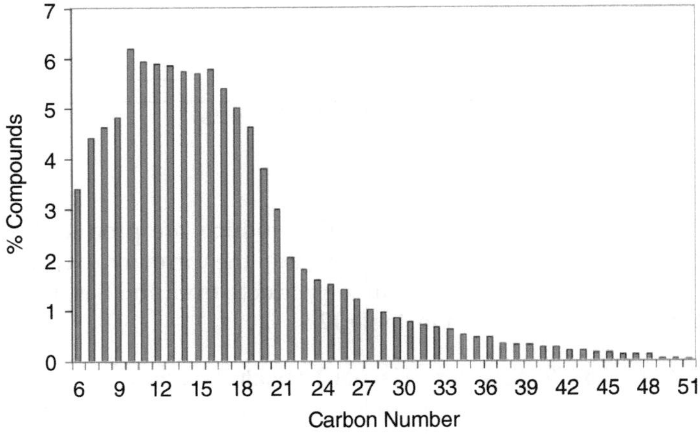

FIGURE 12.4 High temperature Fischer-Tropsch (HTFT) compounds produced.

pressure, and temperature. Iron, cobalt, nickel, and ruthenium are effective catalysts in producing Fischer-Tropsch products. A general comparison of catalyst properties is provided below:

Cobalt

- High conversion rates and long lifetime
- More sensitive than iron to poisoning by sulfur in feedstock

Iron

- Flexibility in use and wide availability
- Shorter lifetime than cobalt

Nickel

- High catalytic activity
- Excessive methane production

Ruthenium

- Simplest system producing high molecular weight hydrocarbons
- Limited supply and highest cost

Fischer-Tropsch reactors are designed to ensure effective temperature control and heat removal to help maintain product quality and long catalyst lifetimes. Fixed bed, fluidized bed and slurry type reactors are the most common. Fixed bed and slurry bed reactors are used in LTFT reactions, and fluidized bed reactors are used in HTFT reactions. A fixed bed reactor consists of numerous tubes containing a stationary catalyst bed through which the reaction mixture flows. In a slurry reactor, catalyst particles are suspended in the moving reaction liquid and synthesis gas components. Catalyst in the fluidized bed reactor is in powdered form and is suspended in a stream of upward flowing gas or liquid.

12.2.3 Synthetic Natural Gas

Synthetic natural gas or biogas production is accomplished by utilizing anaerobic digestion to convert biomass into methane. Anaerobic digestion is a multistage process of biomass decomposition involving hydrolysis, acidogenesis, acetogenesis, and methanogenesis of complex organic compounds by symbiotic bacterial species. Bacteria are grouped into the following classes:

- Hydrolytic bacteria that break apart complex organic matter into compounds such as fatty acids, sugars, amino acids.
- Acidogenic or fermentative bacteria convert these compounds into alcohols, organic acids, hydrogen, carbon dioxide, and ammonia.
- Acetogenic bacteria further metabolize compounds into acetate, carbon dioxide, and hydrogen.
- Methanogenic bacteria produce methane and carbon dioxide.

12.2.3.1 Production of Synthetic Natural Gas

The production of synthetic natural gas involves blending pulverized biomass material with water and any added nutrient material to form a mixture known as a *substrate*. The substrate is then held in a digester where bacteria begin to propagate and establish conditions necessary for biogas generation. The bacteria required to establish the process of anaerobic digestion are typically present naturally in the biomass material. Development of a thriving population of microorganisms is termed *acclimation*. Sludge from a previous process batch can be used to help inoculate or seed the digester.

The bacteria are mostly mesophiles, effective between temperatures of 20 to 45°C (68 to 112°F), and thermophiles, effective between temperatures of 45 to 65°C (112 to 149°F). Digester operating temperatures can be adjusted to control bacterial population and reaction time. Mesophilic processing may require digester residence times from 10 to 15 days. Thermophilic processing may require a shorter 5 to 10 day digester residence time. Addition of nutrients containing phosphorus, nitrogen, and trace elements, plus maintenance of pH between 6.6 and 7.6, aids in the digestion process.

To ensure that the anaerobic bacteria thrive, careful control over substrate concentration, pH, temperature, nutrients, and toxins must be maintained. For example, acetate concentrations above 2 g/L and ammonia concentrations above 3 g/L will significantly diminish methane production. Oxygen, certain metals, and antibiotic contaminants can be harmful. A two-stage anaerobic digester is provided in Fig. 12.5.

Raw biogas produced from anaerobic digestion contains primarily methane, carbon dioxide, ammonia, hydrogen sulfide, and water vapor. Yield of methane from conversion of biomass is about 30 weight percent or 272 kg (600 lb) for each ton of biomass digested. Inorganic gas contaminants can be removed from biogas by liquid or solid absorption media or by catalytic processing. Also, it is possible to burn biogas as an on-site fuel after removal of hydrogen sulfide. The remaining high water content supernatant liquid can sometimes be

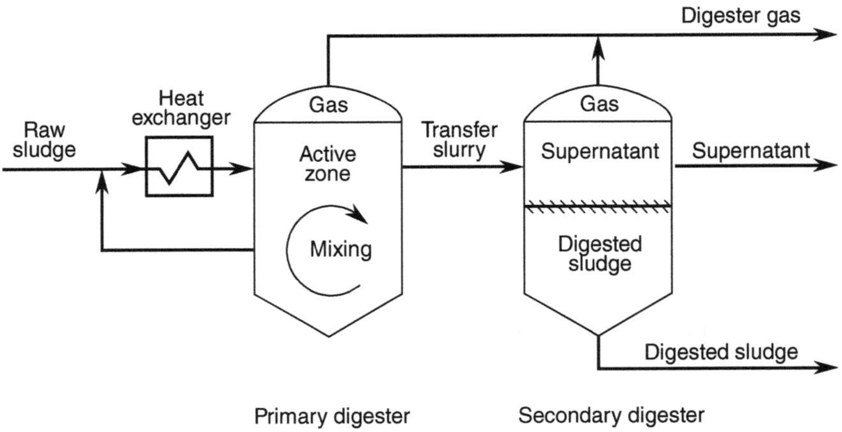

FIGURE 12.5 Two-stage anaerobic digester.

Component	Percent of Composition, wt%
Methane	55 to 65
Carbon dioxide	35 to 45
Ammonia	0 to 1
Hydrogen sulfide	0 to 1
Nitrogen	0 to 3
Hydrogen	0 to 1
Oxygen	0 to 2
Water	Variable

TABLE 12.9 Typical Composition of Biogas

recycled after proper treatment, or if appropriate, used for irrigation. Residual sludge can have a high nitrogen content and may serve as fertilizer.

12.2.3.2 Biogas Composition

The typical composition of biogas is outlined in Table 12.9.

12.2.4 Synthesis Gas to Methanol

Methanol, along with nitromethane and isooctane, has been used as a racing fuel for many years. When used in high compression ratio engines, methanol has good antiknock properties and can be injected at very rich fuel:air ratios such as 4:1. Methanol has undergone extensive evaluation as an alternative fuel source. Its high oxygen content and characteristic production of very low levels of exhaust emissions when used as an internal combustion engine fuel make methanol an attractive alternative fuel.

Methanol was first produced commercially in 1830 by the pyrolysis of wood to produce "wood alcohol." Nearly a century later, a process was developed by BASF to produce synthetic methanol from coal synthesis gas. The first synthetic methanol production processes were introduced in 1923 by BASF and by DuPont 1927. In the late 1940s, natural gas replaced coal synthesis gas as the primary feedstock for methanol production.

Today, steam reforming of natural gas is the primary process used in methanol production. A steam reformer is a large process furnace possessing a series of externally heated, catalyst filled reaction tubes. Special high strength metallurgy is used in the construction of reformer tubes to help ensure longer life and better efficiency. The principal reaction taking place in the steam reformer produces a synthesis gas represented by Eq. (12.6).

$$C_nH_m + nH_2O \leftrightarrow nCO + (0.5m + n)H_2 \tag{12.6}$$

12.2.4.1 Feedstock Pretreatment

Residual sulfur compounds present in the natural gas feed into the steam reformer must be removed before synthesis gas production. Sulfur severely degrades and poisons the CuZn reforming catalyst. In this step, natural gas is heated with steam to 260°C (500°F) and hydroprocessed through a CoMo or a NiMo catalyst bed. Sulfur is converted into H2S and then absorbed onto a ZnO catalyst bed. This process reduces sulfur to less than 0.1 ppm. After desulfurization, the feedstock is heated to 500 to 550°C (930 to 1020°F) and passed through a special nickel-alumina catalyst bed.

12.2.4.2 Methanol Synthesis

Once past the nickel-alumina catalyst, the synthesis gas is compressed to about 250 psig and passed through a CuZn catalyst to yield crude methanol, water, and low quantities of ethanol, acetone, and other impurities. Noncondensable synthesis gases from the process such as hydrogen, carbon monoxide, and carbon dioxide are recycled across the CuZn catalyst bed.

Depending on the grade of methanol being produced, up to four separate distillation steps may be required to produce methanol Grade AA, the purest grade of methanol. For fuel grade methanol production, one distillation step is usually required.

The specification for fuel grade methanol is described in ASTM D5797 Standard Specification for Fuel Methanol (M70-M85) for Automotive Spark Engines. Currently, the production of methanol fueled engines and the commercial availability of fuel grade methanol is not widespread.

12.3 Oil Sands or Tar Sands

Oil sand or tar sand reserves are found in Venezuela, Madagascar, the Uinta Basin of Utah, and extensively in the Athabasca deposits in Alberta, Canada. In the late 1940s, Sidney Ellis and Karl Clark developed processes to enable valuable oil to be extracted from oil sands. Pilot plant operations initiated in the 1950s led to commercial scale construction of a synthetic oil facility in the Athabasca region in 1965 by Great Canadian Oil Sands. Ltd. The plant went into operation in 1968. Syncrude Canada, Ltd., began operating a commercial facility in 1978. Synthetic oil production from oil sand processing has developed into the largest industry of its type in the world.

The oil sands from the Athabasca deposits are a mixture of sand grains, water, and high-viscosity crude hydrocarbon called bitumen. At room temperature, bitumen is a semisolid, but will convert to a viscous liquid at about 150°C (302°F). Sand grains are about 0.075 to 0.25 mm in diameter with small amounts of attached clay. The grains are all in direct contact and wet with a thin, continuous sheet-like layer of water. Bitumen fills the void between the wetted sand grains

Component	Mass %
Carbon	83.2
Hydrogen	10.2
Nitrogen	0.4
Sulfur	4.6
Oxygen	1.6

TABLE 12.10 Elemental Composition of Athabasca Oil Sand

and forms a continuous phase through the pores. Bitumen is about 17 to 18 percent of the mass. The elemental composition of Athabasca oil sand bitumen is provided in Table 12.10.

12.3.1 Bitumen Recovery from Surface Mining

To obtain one barrel of recoverable, synthetic crude oil from oil sand, about 1.6 tons of oil sand must be mined and processed. It is estimated that 10 percent of the energy obtained from the final synthetic crude is required for oil sand mining. Raw oil sand is collected in open pit mines using massive bucket wheel excavators and electric drag lines. Once mined, methods such as hot water extraction, cold water extraction, solvent extraction, direct coking, and fluid coking are employed to recover the oil sand hydrocarbon for fuel processing. Hot water processing and direct coking are described in the following sections.

12.3.1.1 Hot Water Processing

Hot water processing requires mixing the sand with water at a temperature of about 90°C (194°F) to form a pulp containing 65 to 85 percent solids. Sodium hydroxide and other conditioners can be added to control pH and aid in separation. Conditioning is carried out in a rotating, steam-heated drum. Large quantities of water are added and the layers disintegrate. The condensed pulp is pumped to separation cells where the sand settles and the bitumen oil froths to the surface on the rising bubbles. The sand is separated, and the bitumen oil mixture is collected for further upgrading.

12.3.1.2 Direct Coking

Direct coking is a pyrolysis process utilizing a fluidized bed reactor to distill and remove bitumen from the oil sand matrix. By direct coking, oil sand is pyrolyzed by heating in a coker to about 482°C (900°F). At this temperature, the volatile synthetic oil separates from the sand matrix and is recovered from the coker as a condensed liquid.

Coke deposits form on the sand as a result of thermal cracking of bitumen. The coke bearing sand is then fluidized with air and transferred to a burner operating at 760°C (1400°F) to drive coke from the sand. This produces flue gas and clean, hot sand. A portion of this hot sand supplies heat for additional coking.

12.3.2 Bitumen Recovery from In-Situ Processing

In regions where oil sand deposits are covered by dense rock or other formations, in-situ recovery may be more practical. Steam injection and underground combustion processes may be used for oil recovery.

12.3.2.1 Steam Injection or Flooding

Recovery of oil sand bitumen by steam flooding involves either cyclic or continuous injection of steam at high pressure into a well. In cyclic injection, steam is injected into the oil sand formation for a period of time ranging from weeks to months. The steam spreads or floods the formation creating heat and pressure. The steam is then turned off, and the well is sealed for several weeks. When the well is reopened, a mixture of water and bitumen can be withdrawn.

12.3.2.2 Steam Assisted Gravity Drainage

Steam assisted gravity drainage (SAGD) is a continuous process and is accomplished by drilling two wells into the oil sand deposit, an upper injection well and a lower, interconnecting production well. The two wells are at elevation differences to allow liquid to flow by gravity from the injection to the production wells. Steam is injected into the upper well to heat the oil sand and significantly reduce the viscosity of the bitumen. A steam chamber develops in the injection well which grows laterally and vertically. The reduced viscosity bitumen and condensed water flow to the lower production well and are collected.

In the SAGD process, the well zone grows only by gravity leaving the injection and production wells to operate at approximately the same pressure as the reservoir. This eliminates pressure differences that could lead to channeling or fracturing. Gases produced by the process include methane, carbon dioxide, and low levels of hydrogen sulfide.

12.3.3 Upgrading Bitumen to Produce Synthetic Fuels

The processes used to upgrade bitumen to liquid, synthetic fuels typically involve two conversion steps as follows:

- Primary conversion or primary upgrading incorporates delayed coking and fluid coking to cracking bitumen into a lighter crude product for further processing.
- Secondary upgrading involves the use of catalytic hydroprocessing to remove sulfur, nitrogen, and olefins from the synthetic crude.

Property	Synthetic Crude	Brent Crude
Gravity, °API	31.8	38.6
Sulfur, wt%	0.1	0.3
C_4 – minus	3.4	2.9
C_5—177°C (350°F)	15.0	28.6
177 to 343°C (350 to 650°F)	44.0	29.6
343 to 566°C (650 to 1050°F)	37.6	29.8
566°C (1050°F) and above	0.00	9.1

TABLE 12.11 Comparison of Hydroprocessed Synthetic Crude and Brent Crude

Hydroprocessed synthetic crude oil typically has less naphtha and a higher content of distillate and vacuum gas oil than conventional crude. A comparison of selected properties of hydroprocessed synthetic crude oil and Brent crude is presented in Table 12.11.

Through hydroprocessing, reactive olefins formed during coking are saturated and converted to paraffins. Sulfur and nitrogen present in heterocyclic compounds are processed into hydrogen sulfide and ammonia. The synthetic crude can also be fractionated into separate streams such as naphtha, kerosene, and gas oil and independently hydroprocessed. Refined gas oil or diesel fuel is aromatic in character and contains a higher percentage of cycloparaffins than most conventional crude oil. The fuel is low in cetane number, high in density, and typically has good cold temperature handling properties.

12.4 Gaseous Fuels

Availability of raw materials and the production of low levels of exhaust emissions when fuels are combusted are two primary reasons for interest in gaseous fuels. Environmental benefits of using compressed gases as alternative fuel sources can include:

- inherently low exhaust emission levels of nonmethane hydrocarbons such as benzene.
- low fuel emission levels on cold starting; engine starting is not as significantly impacted by temperature.
- minimal fugitive emission levels from storage and refueling facilities due to tightly sealed tanks and handling hardware.

Results of a California study using flexible fuel vehicles capable of operating on either gasoline or compressed natural gas (CNG) found significantly lower emissions after 50,000 miles for CNG fueled vehicles. Test values are provided in Table 12.12.

Fuel	Emissions, g/mile			
	NMOG*	CO	NO_x	HCHO†
Gasoline	0.20	1.2	0.19	n/a‡
CNG	0.021	0.4	0.04	0.0002

*Nonmethane organic gases.
†Formaldehyde.
‡Not available.

Table 12.12 Exhaust Emission Values after 50,000 Miles Comparing Gasoline and CNG in a Flexible Fueled Engine

12.4.1 Combustion Properties

The combustion properties of gaseous fuels flowing through engine system orifices can vary depending upon the gas composition. Pure gases will possess different combustion properties than mixed gases. Stoichiometric air-fuel control systems utilized on compressed gas engines must be able to account for these differences in fuel quality and make accurate adjustments in fuel metering. This is accomplished by determining the *Wobbe index* of the fuel being metered. The Wobbe index (W) is determined by the following equation:

$$W = H/\rho$$

where H is the volumetric heating value of the gas and ρ is the specific gravity of the gas.

A change in the Wobbe index of gaseous fuels will result in a nearly proportional change in the rate of energy flow, and thus in the fuel-air ratio. A lower Wobbe index provides a richer mixture. Stoichiometric engine control systems are able to compensate for reasonable variations in fuel Wobbe index and provide appropriate air-fuel metering adjustments. Without this adjustment mechanism, problems such as low power, knock, increased exhaust emissions, or overheating could result. Wobbe index values for compressed gases typically range between 48.5 and 52.9 MJ/m^3 (1300 and 1420 Btu/standard cubic foot (scf).

12.4.2 LPG and CNG Vehicles

Liquefied petroleum gas (LP gas or LPG) refers to mixtures of three and four carbon number hydrocarbons such as propane, propene, butane, butene, and isobutane. Low amounts of ethane may also be present. Natural gas liquids and petroleum refining processes are the primary sources of LPG. The LPG from petroleum refining often contains significant quantities of olefinic compounds. Important uses include heating residential and commercial buildings, fuel for cooking, chemical feedstock, and natural gasoline blend stock. For use as a motor vehicle fuel, LPG containing mostly propane is required.

In many respects, LPG engines resemble conventional spark ignition engines. Engine conversion kits utilize a digital stoichiometric air-fuel ratio control system with feedback via an oxygen sensor to minimize emissions while maintaining good fuel economy. Vehicle storage tanks are cylindrical with rounded ends and are capable of holding fuel at pressures of 240 to 250 psig. Tanks are much stronger than those used to hold conventional liquid fuels but less sophisticated than CNG storage tanks. Carbon steel or aluminum can be used for tank construction. Because propane can dissolve plasticizers used in elastomer production, only propane resistant elastomers such as neoprene, nitrile polymer, or Teflon should be used. Except for the need for a gastight connection, LPG can be dispensed in the same manner as gasoline or diesel fuel. To ensure that adequate vapor space is available for gas expansion, automatic limiters are incorporated to maintain an 80 percent maximum fill rate.

For CNG fueled engines, sophisticated vehicle storage and pumping systems are required to maintain cryogenic temperatures needed to maintain safe use of the fuel. Fueling hoses and dispensing units must be carefully maintained to ensure complete and accurate pressure regulation, leak detection, flow regulation, and vapor recovery. In vehicles, CNG is stored at 3000 to 3600 psig in double-walled, vacuum-insulated cryogenic containers and then vaporized prior to combustion. The vaporizer utilizes engine heat to expand the cold liquid into a gaseous form suitable for injection into the fuel combustion system. Some designs utilize an expansion tank or buffer to provide additional space for containment of cold LPG when engine demand for fuel is suddenly reduced. This buffer also stores expanded gas for use in engine starting. Stainless steel is the recommended material for tank construction. Welded joints, compression fittings, and Teflon gaskets are used in assembly because of their strength and compatibility under cryogenic conditions.

A listing of the number of LPG and CNG vehicles sold in the United States from the years 2000 to 2010 is provided in App. 9.

12.5 Additional Synthetic and Alternative Fuel Processes

The continued development of synthetic and alternative processes to produce fuel has led to a number of routes and discoveries that have potential for producing viable fuels and other raw materials. Processes, such as, ethanol production from lignocellulose biomass and the thermal conversion of biomass have both reached various levels of commercial success. These two processes are described in the following text.

12.5.1 Cellulosic Ethanol

Producing ethanol by fermenting sugars and starch found in plants such as corn kernels, fruits, and seeds is a common and well-known process. However, producing ethanol on a large scale from the lignocellulose containing components of plants such as husks, wood pulp, crop waste, and materials including cardboard and other paper products has been a technical challenge. Today, however, commercial scale processes have been developed and are in place to produce bioethanol from lignocellulose biomass.

12.5.1.1 Lignocellulose

The composition of lignocellulose consists primarily of cellulose, hemicellulose, and lignin. The percentage of each component in lignocellulose can vary depending on the source plant properties.

12.5.1.1.1 Cellulose Cellulose is composed of β-linked, D-glucose molecules linked together in a linear polymeric form. Polymer chain length varies and is dependent on the properties of the plant source. For example the number of D-glucose repeating units on a cellulose polymer can range from approximately one-hundred to several thousand molecules. Individual cellulose polymers are tightly bound together through hydrogen bonding.

12.5.1.1.2 Hemicellulose Hemicellulose is composed of several branched biopolymers such as xylan, xyloglucan, arabinoxylan, and glucomannan. These polymers contain mostly five-carbon pentose sugars such as xylose and mannose plus acidic sugar derivatives. They are typically shorter in chain length than cellulose polymers.

12.5.1.1.3 Lignin Lignin is composed of a very large and highly cross-linked polymer consisting of aromatic phenyl propanoid monomer units. Both cellulose and hemicellulose bind to lignin through covalent and hydrogen bonding forming the basis for cellulose microfibrils.

12.5.1.2 Pretreatment of Lignocellulose

To free glucose from the cellulose and hemicellulose bound within lignocellulose biomass, pretreatment and hydrolysis steps must be completed. Physiochemical pretreating utilizes compounds that alter the physical or chemical properties of biomass. Treatments release the bound cellulose and hemicellulose polymers. Several pretreatment methods are briefly outlined below:

Mechanical methods: Crushing, grinding, shredding, and other methods are used to reduce biomass particle size.

Steam and hot water: Hydrolysis and separation of bound polymer chains occurs with these treatments.

Organic solvents: Low molecular weight solvents, such as methanol and ethanol, improve the hydrolysis and extractability of lignin from bound polymer chains.

Ammonia: Causes swelling and changes the crystal structure of cellulose. This helps in removal of lignin from bound hemicellulose and cellulose.

Chemical treatment: Numerous chemical treatment processes are available to aid in removal of lignin and the release of bound cellulose and hemicellulose. Treatments can include the use of aqueous acids and bases, ozone, sulfur dioxide, and other compounds.

Enzyme treatment: Some fungi produce extracellular enzymes that help separate lignin and hemicellulose from cellulose.

12.5.1.3 Enzymatic Hydrolysis of Lignocellulose

Hydrolysis of the 1,4-β-D-glucoside linkage between glucose molecules in cellulose has been a major challenge in the development of large-scale cellulosic ethanol production. Research has overcome this challenge, and a group of enzymes that are active in depolymerization of cellulose and hemicellulose to yield glucose and xylose has been identified. These enzymes are identified as cellulase. The cellulase enzymes are produced by bacteria, fungi, and protozoa including organisms such as Trichoderma species, Penicillium funiculosum, Clostridium globosum, and Monospora species. Three major classes of cellulase enzymes and their function are outlined in Table 12.13.

Due to the complexity of the components found in hemicellulose, a series of enzymes must be used in hemicellulose hydrolysis. Various xylanase, xylosidase, mannosidase, and esterase enzymes are required in depolymerization. Enzymatic depolymerization of hemicellulose is an ordered process affecting specific bonds on

Enzyme	Function
Endocellulase	Cleaves internal bonds of the cellulose polymer to create new chains of a shorter length
Exocellulase	Cleaves these new, shorter chains from the ends to form units containing 2 and 4 glucose molecules
Cellobiase (β-glucosidase)	Hydrolyzes the 2 and 4 glucose-containing fragments into D-glucose molecules

Table 12.13 Cellulase Enzymes

individual polymer molecules. Hemicellulose enzymes are produced by a variety of organisms such as microorganisms, crustaceans, and insects. Some of the microorganisms involved include *Aspergillus niger* producing endo-1,4-β-xylanase, *Clostridium acetobutylicum* producing β-1,4-xylosidase, and *Pyrococcus furiosus* producing β-mannosidase.

12.5.1.4 Acid Hydrolysis of Lignocellulose

Acid hydrolysis is one of the oldest and most common methods for treatment of cellulose and hemicellulose. Both concentrated and dilute sulfuric acid methods are utilized. Sulfuric acid breaks apart hydrogen bonds within cellulose polymers, disrupts the crystalline order of the polymer, and cleaves the β-linkage between the cellulose bonds to free glucose for fermentation.

12.5.1.5 Recent Developments in Catalytic Lignocellulose Hydrolysis

Ionic liquid hydrolysis and solid acid hydrolysis are two recent developments in lignocellulosic biomass hydrolysis. Catalytic liquid hydrolysis involves the use of liquids such as various alkyl 3-methylimidazolium salts in combination with catalytic amounts of sulfuric acid to effectively hydrolyze cellulose biomass. Examples of solid substrates incorporating acidic catalyst sites can include carbonaceous solids, metal oxides, and exchange resins. Catalytic sites are typically activated through acid sulfonation.

12.5.1.6 Contaminant Removal and Fermentation

Before progressing with fermentation, contaminants from cellulosic ethanol processing must be removed. Contaminants can include organic acids, aldehydes, phenols, and metal ions. Methods of removal can include physical methods of separation such as evaporation or absorption, chemical methods including neutralization or ion exchange, and biological methods such as enzymatic detoxification.

After removal of contaminants, sugars released from lignocellulose biomass can be converted to ethanol using processes similar to those used to ferment starch and sugar-based raw materials. Yeasts such as *Saccharomyces cerevisia*, other organisms including the bacteria *Zymomonas mobilis*, or recently developed recombinant microbes can be used in fermentation to yield cellulosic ethanol.

12.5.2 Thermal Conversion of Biomass

Thermal conversion processes such as pyrolysis, gasification, and combustion of biomass have a long history of use in generating useful products and energy. These processes are compared and contrasted in Table 12.14.

Conversion Process Condition	Type of Conversion Process		
	Pyrolysis	Gasification	Combustion
Temperature	<600°C (<1112°F)	800 to 1100°C (1472 to 2012°F)	>1500°C (>2732°F)
Environment	No oxygen	Oxygen, hydrogen, steam	Air
Products	Solid char produced at low temperatures; liquid and gases at higher temperatures	High yield of hydrogen, carbon monoxide, carbon dioxide, methane, ethane, nitrogen, and water	Recoverable heat can be used to generate steam and produce electricity
Other factors	High quantity and quality of liquid and gas produced in a hydrogen atmosphere	Higher yield and quality of gas produced than with pyrolysis; more costly than pyrolysis	No synthetic fuel gas produced

TABLE 12.14 Biomass Thermal Conversion Process Comparison

12.5.2.1 Wood Pyrolysis

The pyrolysis and gasification of wood can be used to illustrate thermal conversion processes and demonstrate the impact heating can have on products produced.

- When heated to temperatures between 71 and 79.4°C (160 and 175°F), a watery distillate is produced called *pyroligneous acid.* The liquid contains between 5 to 10 percent acetic acid and 1.5 to 3 percent methyl alcohol, or wood alcohol.
- At higher pyrolysis temperatures, viscous, unstable wood pyrolysis oil is collected. This oil contains useful tall oil and creosote oil. Wood pyrolysis oil has a high oxygen content, low sulfur content, and can serve as an on-site burner fuel to meet energy needs.

At much higher gasification temperatures, valuable synthesis gases are formed, such as hydrogen, carbon monoxide, carbon dioxide, plus methane and higher molecular weight hydrocarbons.

APPENDIX 1

Heat of Combustion of Fuels—Approximate Btu-Gravity Relation

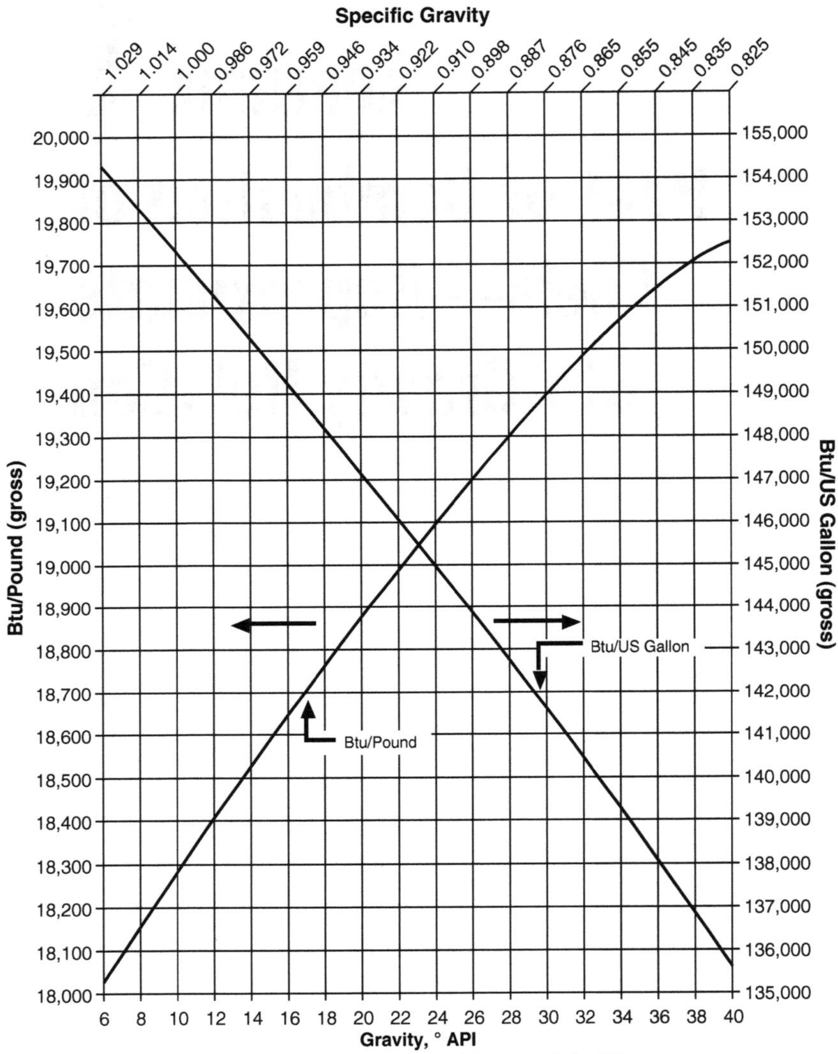

Multiply Btu/pound by 0.556 to obtain kilogram-calories/kilogram.
Multiply Btu/gallon by 0.0066 to obtain kilogram-calories/liter.

APPENDIX 2

Factors for Converting Volumes to 60°F

Observed Temperature °F	Group Number and API Gravity Range at 60°F: Group 0, 0–14.9° API	Group 1, 15.0–34.9° API	Group 2, 35.0–50.9° API	Group 3, 51.0–63.9° API
	Factor for Reducing Volume to 60°F			
0	1.0211	1.0241	1.0298	1.0362
1	1.0208	1.0237	1.0293	1.0356
2	1.0204	1.0233	1.0288	1.0350
3	1.0201	1.0229	1.0283	1.0344
4	1.0197	1.0225	1.0278	1.0338
5	1.0194	1.0221	1.0273	1.0332
6	1.0190	1.0217	1.0268	1.0326
7	1.0186	1.0213	1.0263	1.0320
8	1.0183	1.0209	1.0258	1.0314
9	1.0179	1.0205	1.0253	1.0308
10	1.0176	1.0201	1.0248	1.0302
11	1.0172	1.0197	1.0243	1.0296
12	1.0169	1.0193	1.0238	1.0290
13	1.0165	1.0189	1.0233	1.0284
14	1.0162	1.0185	1.0228	1.0278
15	1.0158	1.0181	1.0223	1.0272
16	1.0155	1.0177	1.0218	1.0266
17	1.0151	1.0173	1.0214	1.0260
18	1.0148	1.0168	1.0209	1.0253
19	1.0144	1.0164	1.0204	1.0247
20	1.0141	1.0160	1.0199	1.0241
21	1.0137	1.0156	1.0194	1.0235
22	1.0133	1.0152	1.0189	1.0229
23	1.0130	1.0148	1.0184	1.0223
24	1.0126	1.0144	1.0179	1.0217
25	1.0123	1.0140	1.0174	1.0211
26	1.0119	1.0136	1.0169	1.0205
27	1.0116	1.0132	1.0164	1.0199
28	1.0112	1.0128	1.0159	1.0193
29	1.0109	1.0124	1.0154	1.0187
30	1.0105	1.0120	1.0149	1.0181
31	1.0102	1.0116	1.0144	1.0175
32	1.0098	1.0112	1.0139	1.0169
33	1.0095	1.0108	1.0134	1.0163
34	1.0091	1.0104	1.0129	1.0157
35	1.0088	1.0100	1.0124	1.0151
36	1.0084	1.0096	1.0119	1.0145
37	1.0081	1.0092	1.0114	1.0139
38	1.0077	1.0088	1.0109	1.0132
39	1.0074	1.0084	1.0104	1.0127
40	1.0070	1.0080	1.0099	1.0122
41	1.0067	1.0076	1.0094	1.0115
42	1.0063	1.0072	1.0089	1.0109
43	1.0060	1.0068	1.0084	1.0103
44	1.0056	1.0064	1.0079	1.0097
45	1.0053	1.0060	1.0075	1.0091
46	1.0049	1.0056	1.0070	1.0085
47	1.0046	1.0052	1.0065	1.0079
48	1.0042	1.0048	1.0060	1.0073
49	1.0038	1.0044	1.0055	1.0067

Observed Temperature °F	Group Number and API Gravity Range at 60°F: Group 0, 0–14.9° API	Group 1, 15.0–34.9° API	Group 2, 35.0–50.9° API	Group 3, 51.0–63.9° API
	Factor for Reducing Volume to 60°F			
50	1.0035	1.0040	1.0050	1.0061
51	1.0031	1.0036	1.0045	1.0054
52	1.0028	1.0032	1.0040	1.0048
53	1.0024	1.0028	1.0035	1.0042
54	1.0021	1.0024	1.0030	1.0036
55	1.0017	1.0020	1.0025	1.0030
56	1.0014	1.0016	1.0020	1.0024
57	1.0010	1.0012	1.0015	1.0018
58	1.0007	1.0008	1.0010	1.0012
59	1.0003	1.0004	1.0005	1.0006
60	1.0000	1.0000	1.0000	1.0000
61	0.9997	0.9996	0.9995	0.9994
62	0.9993	0.9992	0.9990	0.9988
63	0.9990	0.9988	0.9985	0.9982
64	0.9986	0.9984	0.9980	0.9976
65	0.9983	0.9980	0.9975	0.9970
66	0.9979	0.9976	0.9970	0.9964
67	0.9976	0.9972	0.9965	0.9958
68	0.9972	0.9968	0.9960	0.9951
69	0.9969	0.9964	0.9955	0.9945
70	0.9965	0.9960	0.9950	0.9939
71	0.9962	0.9956	0.9945	0.9933
72	0.9958	0.9952	0.9940	0.9927
73	0.9955	0.9948	0.9935	0.9921
74	0.9951	0.9944	0.9930	0.9915
75	0.9948	0.9940	0.9925	0.9909
76	0.9944	0.9936	0.9920	0.9903
77	0.9941	0.9932	0.9916	0.9897
78	0.9937	0.9929	0.9911	0.9891
79	0.9934	0.9925	0.9906	0.9885
80	0.9930	0.9921	0.9901	0.9879
81	0.9927	0.9917	0.9896	0.9873
82	0.9923	0.9913	0.9891	0.9866
83	0.9920	0.9909	0.9886	0.9860
84	0.9916	0.9905	0.9881	0.9854
85	0.9913	0.9901	0.9876	0.9848
86	0.9909	0.9897	0.9871	0.9842
87	0.9906	0.9893	0.9866	0.9836
88	0.9902	0.9889	0.9861	0.9830
89	0.9899	0.9885	0.9856	0.9824
90	0.9896	0.9881	0.9851	0.9818
91	0.9892	0.9877	0.9846	0.9812
92	0.9889	0.9873	0.9841	0.9806
93	0.9885	0.9869	0.9836	0.9799
94	0.9882	0.9865	0.9831	0.9793
95	0.9878	0.9861	0.9826	0.9787
96	0.9875	0.9857	0.9821	0.9781
97	0.9871	0.9854	0.9816	0.9775
98	0.9868	0.9850	0.9811	0.9769
99	0.9864	0.9846	0.9806	0.9763

Observed Temperature °F	Group 0 0–14.9° API	Group 1 15.0–34.9° API	Group 2 35.0–50.9° API	Group 3 51.0–63.9° API
	Factor for Reducing Volume to 60°F			
100	0.9861	0.9842	0.9801	0.9757
101	0.9857	0.9838	0.9796	0.9751
102	0.9854	0.9834	0.9791	0.9745
103	0.9851	0.9830	0.9786	0.9738
104	0.9847	0.9826	0.9781	0.9732
105	0.9844	0.9822	0.9776	0.9726
106	0.9840	0.9818	0.9771	0.9720
107	0.9837	0.9814	0.9766	0.9714
108	0.9833	0.9810	0.9761	0.9708
109	0.9830	0.9806	0.9756	0.9702
110	0.9826	0.9803	0.9751	0.9696
111	0.9823	0.9799	0.9746	0.9690
112	0.9819	0.9795	0.9741	0.9683
113	0.9816	0.9791	0.9736	0.9677
114	0.9813	0.9787	0.9731	0.9671
115	0.9809	0.9783	0.9726	0.9665
116	0.9806	0.9779	0.9721	0.9659
117	0.9802	0.9775	0.9717	0.9653
118	0.9799	0.9771	0.9712	0.9647
119	0.9795	0.9767	0.9707	0.9641
120	0.9792	0.9763	0.9702	0.9634
121	0.9788	0.9760	0.9697	0.9628
122	0.9785	0.9756	0.9692	0.9622
123	0.9782	0.9752	0.9687	0.9616
124	0.9778	0.9748	0.9682	0.9610
125	0.9775	0.9744	0.9677	0.9604
126	0.9771	0.9740	0.9672	0.9598
127	0.9768	0.9736	0.9667	0.9592
128	0.9764	0.9732	0.9662	0.9585
129	0.9761	0.9728	0.9657	0.9579
130	0.9758	0.9725	0.9652	0.9573
131	0.9754	0.9721	0.9647	0.9567
132	0.9751	0.9717	0.9642	0.9561
133	0.9747	0.9713	0.9637	0.9555
134	0.9744	0.9709	0.9632	0.9549
135	0.9740	0.9705	0.9627	0.9542
136	0.9737	0.9701	0.9622	0.9536
137	0.9734	0.9697	0.9617	0.9530
138	0.9730	0.9693	0.9612	0.9524
139	0.9727	0.9690	0.9607	0.9518
140	0.9723	0.9686	0.9602	0.9512
141	0.9720	0.9682	0.9597	0.9506
142	0.9716	0.9678	0.9592	0.9499
143	0.9713	0.9674	0.9587	0.9493
144	0.9710	0.9670	0.9582	0.9487
145	0.9706	0.9666	0.9577	0.9481
146	0.9703	0.9662	0.9572	0.9475
147	0.9699	0.9659	0.9567	0.9469
148	0.9696	0.9655	0.9562	0.9462
149	0.9693	0.9651	0.9557	0.9456

Observed Temperature °F	Group 0 0–14.9° API	Group 1 15.0–34.9° API	Group 2 35.0–50.9° API	Group 3 51.0–63.9° API
	Factor for Reducing Volume to 60°F			
150	0.9689	0.9647	0.9552	0.9450
151	0.9686	0.9643	0.9547	0.9444
152	0.9682	0.9639	0.9542	0.9438
153	0.9679	0.9635	0.9537	0.9432
154	0.9675	0.9632	0.9532	0.9426
155	0.9672	0.9628	0.9527	0.9419
156	0.9669	0.9624	0.9522	0.9413
157	0.9665	0.9620	0.9517	0.9407
158	0.9662	0.9616	0.9512	0.9401
159	0.9658	0.9612	0.9507	0.9395
160	0.9655	0.9609	0.9502	0.9389
161	0.9652	0.9605	0.9497	0.9382
162	0.9648	0.9601	0.9492	0.9376
163	0.9645	0.9597	0.9487	0.9370
164	0.9641	0.9593	0.9482	0.9364
165	0.9638	0.9589	0.9477	0.9358
166	0.9635	0.9585	0.9472	0.9351
167	0.9631	0.9582	0.9467	0.9345
168	0.9628	0.9578	0.9462	0.9339
169	0.9624	0.9574	0.9457	0.9333
170	0.9621	0.9570	0.9452	0.9327
171	0.9618	0.9566	0.9447	0.9321
172	0.9614	0.9562	0.9442	0.9314
173	0.9611	0.9559	0.9437	0.9308
174	0.9607	0.9555	0.9432	0.9302
175	0.9604	0.9551	0.9428	0.9296
176	0.9601	0.9547	0.9423	0.9290
177	0.9597	0.9543	0.9418	0.9283
178	0.9594	0.9539	0.9413	0.9277
179	0.9590	0.9536	0.9408	0.9271
180	0.9587	0.9532	0.9403	0.9265
181	0.9584	0.9528	0.9398	0.9259
182	0.9580	0.9524	0.9393	0.9252
183	0.9577	0.9520	0.9388	0.9246
184	0.9574	0.9517	0.9383	0.9240
185	0.9570	0.9513	0.9378	0.9234
186	0.9567	0.9509	0.9373	0.9228
187	0.9563	0.9505	0.9368	0.9221
188	0.9560	0.9501	0.9363	0.9215
189	0.9557	0.9498	0.9358	0.9209
190	0.9553	0.9494	0.9353	0.9203
191	0.9550	0.9490	0.9348	0.9197
192	0.9547	0.9486	0.9343	0.9190
193	0.9543	0.9482	0.9338	0.9184
194	0.9540	0.9478	0.9333	0.9178
195	0.9536	0.9475	0.9328	0.9172
196	0.9533	0.9471	0.9323	0.9166
197	0.9530	0.9467	0.9318	0.9159
198	0.9526	0.9463	0.9313	0.9153
199	0.9523	0.9460	0.9308	0.9147

Observed Temperature °F	Group Number and API Gravity Range at 60°F			
	Group 0 0–14.9° API	Group 1 15.0–34.9° API	Group 2 35.0–50.9° API	Group 3 51.0–63.9 API
	Factor for Reducing Volume to 60°F			
200	0.9520	0.9456	0.9303	0.9141
201	0.9516	0.9452	0.9298	
202	0.9513	0.9448	0.9293	
203	0.9509	0.9444	0.9288	
204	0.9506	0.9441	0.9283	
205	0.9503	0.9437	0.9278	
206	0.9499	0.9433	0.9273	
207	0.9496	0.9429	0.9268	
208	0.9493	0.9425	0.9263	
209	0.9489	0.9422	0.9258	
210	0.9486	0.9418	0.9253	
211	0.9483	0.9414	0.9248	
212	0.9479	0.9410	0.9243	
213	0.9476	0.9407	0.9238	
214	0.9472	0.9403	0.9233	
215	0.9469	0.9399	0.9228	
216	0.9466	0.9395	0.9223	
217	0.9462	0.9391	0.9218	
218	0.9459	0.9388	0.9213	
219	0.9456	0.9384	0.9208	
220	0.9452	0.9380	0.9203	
221	0.9449	0.9376	0.9198	
222	0.9446	0.9373	0.9193	
223	0.9442	0.9369	0.9188	
224	0.9439	0.9365	0.9183	
225	0.9436	0.9361	0.9178	
226	0.9432	0.9358	0.9173	
227	0.9429	0.9354	0.9168	
228	0.9426	0.9350	0.9163	
229	0.9422	0.9346	0.9158	
230	0.9419	0.9343	0.9153	
231	0.9416	0.9339	0.9148	
232	0.9412	0.9335	0.9143	
233	0.9409	0.9331	0.9138	
234	0.9405	0.9328	0.9133	
235	0.9402	0.9324	0.9128	
236	0.9399	0.9320	0.9123	
237	0.9395	0.9316	0.9118	
238	0.9392	0.9313	0.9113	
239	0.9389	0.9309	0.9108	
240	0.9385	0.9305	0.9103	
241	0.9382	0.9301	0.9098	
242	0.9379	0.9298	0.9093	
243	0.9375	0.9294	0.9088	
244	0.9372	0.9290	0.9083	
245	0.9369	0.9286	0.9078	
246	0.9365	0.9283	0.9073	
247	0.9362	0.9279	0.9068	
248	0.9359	0.9275	0.9063	
249	0.9356	0.9272	0.9058	
250	0.9352	0.9268	0.9053	

Observed Temperature °F	Group Number and API Gravity Range at 60°F		
	Group 0 0–14.9° API	Group 1 15.0–34.9° API	Group 2 35.0–50.9° API
	Factor for Reducing Volume to 60°F		
251	0.9349	0.9264	
252	0.9346	0.9260	
253	0.9342	0.9257	
254	0.9339	0.9253	
255	0.9336	0.9249	
256	0.9332	0.9245	
257	0.9329	0.9242	
258	0.9326	0.9238	
259	0.9322	0.9234	
260	0.9319	0.9231	
261	0.9316	0.9227	
262	0.9312	0.9223	
263	0.9309	0.9219	
264	0.9306	0.9216	
265	0.9302	0.9212	
266	0.9299	0.9208	
267	0.9296	0.9205	
268	0.9293	0.9201	
269	0.9289	0.9197	
270	0.9286	0.9194	
271	0.9283	0.9190	
272	0.9279	0.9186	
273	0.9276	0.9182	
274	0.9273	0.9179	
275	0.9269	0.9175	
276	0.9266	0.9171	
277	0.9263	0.9168	
278	0.9259	0.9164	
279	0.9256	0.9160	
280	0.9253	0.9157	
281	0.9250	0.9153	
282	0.9246	0.9149	
283	0.9243	0.9146	
284	0.9240	0.9142	
285	0.9236	0.9138	
286	0.9233	0.9135	
287	0.9230	0.9131	
288	0.9227	0.9127	
289	0.9223	0.9124	
290	0.9220	0.9120	
291	0.9217	0.9116	
292	0.9213	0.9113	
293	0.9210	0.9109	
294	0.9207	0.9105	
295	0.9204	0.9102	
296	0.9200	0.9098	
297	0.9197	0.9094	
298	0.9194	0.9091	
299	0.9190	0.9087	

Observed Temperature °F	Group 0 0–14.9° API	Group 1 15.0–34.9° API
	Factor for Reducing Volume to 60°F	
300	0.9187	0.9083
301	0.9184	0.9080
302	0.9181	0.9076
303	0.9177	0.9072
304	0.9174	0.9069
305	0.9171	0.9065
306	0.9167	0.9061
307	0.9164	0.9058
308	0.9161	0.9054
309	0.9158	0.9050
310	0.9154	0.9047
311	0.9151	0.9043
312	0.9148	0.9039
313	0.9145	0.9036
314	0.9141	0.9032
315	0.9138	0.9029
316	0.9135	0.9025
317	0.9132	0.9021
318	0.9128	0.9018
319	0.9125	0.9014
320	0.9122	0.9010
321	0.9118	0.9007
322	0.9115	0.9003
323	0.9112	0.9000
324	0.9109	0.8996
325	0.9105	0.8992
326	0.9102	0.8989
327	0.9099	0.8985
328	0.9096	0.8981
329	0.9092	0.8978
330	0.9089	0.8974
331	0.9086	0.8971
332	0.9083	0.8967
333	0.9079	0.8963
334	0.9076	0.8960
335	0.9073	0.8956
336	0.9070	0.8952
337	0.9066	0.8949
338	0.9063	0.8945
339	0.9060	0.8942
340	0.9057	0.8938
341	0.9053	0.8934
342	0.9050	0.8931
343	0.9047	0.8927
344	0.9044	0.8924
345	0.9040	0.8920
346	0.9037	0.8916
347	0.9034	0.8913
348	0.9031	0.8909
349	0.9028	0.8906

Observed Temperature °F	Group 0 0–14.9° API	Group 1 15.0–34.9° API
	Factor for Reducing Volume to 60°F	
350	0.9024	0.8902
351	0.9021	0.8899
352	0.9018	0.8895
353	0.9015	0.8891
354	0.9011	0.8888
355	0.9008	0.8884
356	0.9005	0.8881
357	0.9002	0.8877
358	0.8998	0.8873
359	0.8995	0.8870
360	0.8992	0.8866
361	0.8989	0.8863
362	0.8986	0.8859
363	0.8982	0.8856
364	0.8979	0.8852
365	0.8976	0.8848
366	0.8973	0.8845
367	0.8969	0.8841
368	0.8966	0.8838
369	0.8963	0.8834
370	0.8960	0.8831
371	0.8957	0.8827
372	0.8953	0.8823
373	0.8950	0.8820
374	0.8947	0.8816
375	0.8944	0.8813
376	0.8941	0.8809
377	0.8937	0.8806
378	0.8934	0.8802
379	0.8931	0.8799
380	0.8928	0.8795
381	0.8924	0.8792
382	0.8921	0.8788
383	0.8918	0.8784
384	0.8915	0.8781
385	0.8912	0.8777
386	0.8908	0.8774
387	0.8905	0.8770
388	0.8902	0.8767
389	0.8899	0.8763
390	0.8896	0.8760
391	0.8892	0.8756
392	0.8889	0.8753
393	0.8886	0.8749
394	0.8883	0.8746
395	0.8880	0.8742
396	0.8876	0.8738
397	0.8873	0.8735
398	0.8870	0.8731
399	0.8867	0.8728

Observed Temperature °F	Group 0 0–14.9° API	Group 1 15.0–34.9° API
	Factor for Reducing Volume to 60°F	
400	0.8864	0.8724
401	0.8861	0.8721
402	0.8857	0.8717
403	0.8854	0.8714
404	0.8851	0.8710
405	0.8848	0.8707
406	0.8845	0.8703
407	0.8841	0.8700
408	0.8838	0.8696
409	0.8835	0.8693
410	0.8832	0.8689
411	0.8829	0.8686
412	0.8826	0.8682
413	0.8822	0.8679
414	0.8819	0.8675
415	0.8816	0.8672
416	0.8813	0.8668
417	0.8810	0.8665
418	0.8806	0.8661
419	0.8803	0.8658
420	0.8800	0.8654
421	0.8797	0.8651
422	0.8794	0.8647
423	0.8791	0.8644
424	0.8787	0.8640
425	0.8784	0.8637
426	0.8781	0.8633
427	0.8778	0.8630
428	0.8775	0.8626
429	0.8772	0.8623
430	0.8768	0.8619
431	0.8765	0.8616
432	0.8762	0.8612
433	0.8759	0.8609
434	0.8756	0.8605
435	0.8753	0.8602
436	0.8749	0.8599
437	0.8746	0.8595
438	0.8743	0.8592
439	0.8740	0.8588
440	0.8737	0.8585
441	0.8734	0.8581
442	0.8731	0.8578
443	0.8727	0.8574
444	0.8724	0.8571
445	0.8721	0.8567
446	0.8718	0.8564
447	0.8715	0.8560
448	0.8712	0.8557
449	0.8709	0.8554

Observed Temperature °F	Group Number and API Gravity Range at 60°F	
	Group 0 0–14.9° API	Group 1 15.0–34.9° API
	Factor for Reducing Volume to 60°F	
450	0.8705	0.8550
451	0.8702	0.8547
452	0.8699	0.8543
453	0.8696	0.8540
454	0.8693	0.8536
455	0.8690	0.8533
456	0.8687	0.8529
457	0.8683	0.8526
458	0.8680	0.8522
459	0.8677	0.8519
460	0.8674	0.8516
461	0.8671	0.8512
462	0.8668	0.8509
463	0.8665	0.8505
464	0.8661	0.8502
465	0.8658	0.8498
466	0.8655	0.8495
467	0.8652	0.8492
468	0.8649	0.8488
469	0.8646	0.8485
470	0.8643	0.8481
471	0.8640	0.8478
472	0.8636	0.8474
473	0.8633	0.8471
474	0.8630	0.8468

Observed Temperature °F	Group Number and API Gravity Range at 60°F	
	Group 0 0–14.9° API	Group 1 15.0–34.9° API
	Factor for Reducing Volume to 60°F	
475	0.8627	0.8464
476	0.8624	0.8461
477	0.8621	0.8457
478	0.8618	0.8454
479	0.8615	0.8451
480	0.8611	0.8447
481	0.8608	0.8444
482	0.8605	0.8440
483	0.8602	0.8437
484	0.8599	0.8433
485	0.8596	0.8430
486	0.8593	0.8427
487	0.8590	0.8423
488	0.8587	0.8420
489	0.8583	0.8416
490	0.8580	0.8413
491	0.8577	0.8410
492	0.8574	0.8406
493	0.8571	0.8403
494	0.8568	0.8399
495	0.8565	0.8396
496	0.8562	0.8393
497	0.8559	0.8389
498	0.8556	0.8386
499	0.8552	0.8383
500	0.8549	0.8379

APPENDIX 3

Reduction of Observed API Gravity to API Gravity at 60°F

Observed Temperature °F	API Gravity at Observed Temperature																			
	5	6	7	8	9	10	11	12	13	14	15	16	17	18	19	20	21	22	23	24
	Corresponding API Gravity at 60°F																			
0	7.9	8.9	10.0	11.0	12.1	13.1	14.2	15.2	16.3	17.3	18.3	19.4	20.4	21.5	22.5	23.6	24.6	25.7	26.7	27.8
1	7.8	8.9	9.9	11.0	12.0	13.1	14.1	15.2	16.2	17.2	18.3	19.3	20.4	21.4	22.5	23.5	24.6	25.6	26.7	27.7
2	7.8	8.8	9.9	10.9	12.0	13.0	14.1	15.1	16.1	17.2	18.2	19.3	20.3	21.4	22.4	23.5	24.5	25.6	26.6	27.7
3	7.7	8.8	9.8	10.9	11.9	13.0	14.0	15.0	16.1	17.1	18.2	19.2	20.3	21.3	22.4	23.4	24.4	25.5	26.5	27.6
4	7.7	8.7	9.8	10.8	11.9	12.9	13.9	15.0	16.0	17.1	18.1	19.2	20.2	21.2	22.3	23.3	24.4	25.4	26.5	27.5
5	7.6	8.7	9.7	10.8	11.8	12.8	13.9	14.9	16.0	17.0	18.1	19.1	20.1	21.2	22.2	23.3	24.3	25.4	26.4	27.5
6	7.6	8.6	9.7	10.7	11.8	12.8	13.8	14.9	15.9	17.0	18.0	19.0	20.1	21.1	22.2	23.2	24.3	25.3	26.4	27.4
7	7.5	8.6	9.6	10.7	11.7	12.7	13.8	14.8	15.9	16.9	17.9	19.0	20.0	21.1	22.1	23.2	24.2	25.2	26.3	27.3
8	7.5	8.5	9.6	10.6	11.6	12.7	13.7	14.8	15.8	16.8	17.9	18.9	20.0	21.0	22.0	23.1	24.1	25.2	26.2	27.3
9	7.4	8.5	9.5	10.6	11.6	12.6	13.7	14.7	15.7	16.8	17.8	18.9	19.9	20.9	22.0	23.0	24.1	25.1	26.2	27.2
10	7.4	8.4	9.5	10.5	11.5	12.6	13.6	14.7	15.7	16.7	17.8	18.8	19.8	20.9	21.9	23.0	24.0	25.0	26.1	27.1
11	7.3	8.4	9.4	10.5	11.5	12.5	13.6	14.6	15.6	16.7	17.7	18.8	19.8	20.8	21.9	22.9	23.9	25.0	26.0	27.1
12	7.3	8.3	9.4	10.4	11.4	12.5	13.5	14.5	15.6	16.6	17.7	18.7	19.7	20.8	21.8	22.8	23.9	24.9	26.0	27.0
13	7.2	8.3	9.3	10.3	11.4	12.4	13.5	14.5	15.5	16.6	17.6	18.6	19.7	20.7	21.7	22.8	23.8	24.9	25.9	26.9
14	7.2	8.2	9.3	10.3	11.3	12.4	13.4	14.4	15.5	16.5	17.5	18.6	19.6	20.7	21.7	22.7	23.8	24.8	25.8	26.9
15	7.1	8.2	9.2	10.2	11.3	12.3	13.3	14.4	15.4	16.5	17.5	18.5	19.6	20.6	21.6	22.7	23.7	24.7	25.8	26.8
16	7.1	8.1	9.2	10.2	11.2	12.3	13.3	14.3	15.4	16.4	17.4	18.5	19.5	20.5	21.6	22.6	23.6	24.7	25.7	26.8
17	7.0	8.1	9.1	10.1	11.2	12.2	13.2	14.3	15.3	16.3	17.4	18.4	19.4	20.5	21.5	22.5	23.6	24.6	25.6	26.7
18	7.0	8.0	9.1	10.1	11.1	12.2	13.2	14.2	15.3	16.3	17.3	18.3	19.4	20.4	21.4	22.5	23.5	24.5	25.6	26.6
19	6.9	8.0	9.0	10.0	11.1	12.1	13.1	14.2	15.2	16.2	17.3	18.3	19.3	20.4	21.4	22.4	23.5	24.5	25.5	26.6
20	6.9	7.9	9.0	10.0	11.0	12.0	13.1	14.1	15.1	16.2	17.2	18.2	19.3	20.3	21.3	22.4	23.4	24.4	25.5	26.5
21	6.8	7.9	8.9	9.9	11.0	12.0	13.0	14.1	15.1	16.1	17.1	18.2	19.2	20.2	21.3	22.3	23.3	24.4	25.4	26.4
22	6.8	7.8	8.8	9.9	10.9	11.9	13.0	14.0	15.0	16.1	17.1	18.1	19.1	20.2	21.2	22.2	23.3	24.3	25.3	26.4
23	6.7	7.8	8.8	9.8	10.9	11.9	12.9	13.9	15.0	16.0	17.0	18.1	19.1	20.1	21.1	22.2	23.2	24.2	25.3	26.3
24	6.7	7.7	8.7	9.8	10.8	11.8	12.9	13.9	14.9	15.9	17.0	18.0	19.0	20.1	21.1	22.1	23.1	24.2	25.2	26.2
25	6.6	7.7	8.7	9.7	10.8	11.8	12.8	13.8	14.9	15.9	16.9	17.9	19.0	20.0	21.0	22.1	23.1	24.1	25.1	26.2
26	6.6	7.6	8.6	9.7	10.7	11.7	12.8	13.8	14.8	15.8	16.9	17.9	18.9	19.9	21.0	22.0	23.0	24.1	25.1	26.1
27	6.5	7.6	8.6	9.6	10.7	11.7	12.7	13.7	14.8	15.8	16.8	17.8	18.9	19.9	20.9	21.9	23.0	24.0	25.0	26.0
28	6.5	7.5	8.5	9.6	10.6	11.6	12.7	13.7	14.7	15.7	16.8	17.8	18.8	19.8	20.9	21.9	22.9	23.9	25.0	26.0
29	6.4	7.5	8.5	9.5	10.6	11.6	12.6	13.6	14.6	15.7	16.7	17.7	18.7	19.8	20.8	21.8	22.8	23.9	24.9	25.9
30	6.4	7.4	8.4	9.5	10.5	11.5	12.5	13.6	14.6	15.6	16.6	17.7	18.7	19.7	20.7	21.8	22.8	23.8	24.8	25.9
31	6.3	7.4	8.4	9.4	10.5	11.5	12.5	13.5	14.5	15.6	16.6	17.6	18.6	19.7	20.7	21.7	22.7	23.7	24.8	25.8
32	6.3	7.3	8.4	9.4	10.4	11.4	12.4	13.5	14.5	15.5	16.5	17.6	18.6	19.6	20.6	21.6	22.7	23.7	24.7	25.7
33	6.3	7.3	8.3	9.3	10.4	11.4	12.4	13.4	14.4	15.5	16.5	17.5	18.5	19.5	20.6	21.6	22.6	23.6	24.6	25.7
34	6.2	7.2	8.3	9.3	10.3	11.3	12.3	13.4	14.4	15.4	16.4	17.4	18.5	19.5	20.5	21.5	22.5	23.6	24.6	25.6
35	6.2	7.2	8.2	9.2	10.2	11.3	12.3	13.3	14.3	15.3	16.4	17.4	18.4	19.4	20.4	21.5	22.5	23.5	24.5	25.5
36	6.1	7.1	8.2	9.2	10.2	11.2	12.2	13.3	14.3	15.3	16.3	17.3	18.3	19.4	20.4	21.4	22.4	23.4	24.5	25.5
37	6.1	7.1	8.1	9.1	10.1	11.2	12.2	13.2	14.2	15.2	16.3	17.3	18.3	19.3	20.3	21.3	22.4	23.4	24.4	25.4
38	6.0	7.0	8.1	9.1	10.1	11.1	12.1	13.1	14.2	15.2	16.2	17.2	18.2	19.2	20.3	21.3	22.3	23.3	24.3	25.4
39	6.0	7.0	8.0	9.0	10.0	11.1	12.1	13.1	14.1	15.1	16.1	17.2	18.2	19.2	20.2	21.2	22.2	23.3	24.3	25.3

Observed Temperature °F	API Gravity at Observed Temperature																			
	5	6	7	8	9	10	11	12	13	14	15	16	17	18	19	20	21	22	23	24
	Corresponding API Gravity at 60°F																			
40	5.9	6.9	8.0	9.0	10.0	11.0	12.0	13.0	14.1	15.1	16.1	17.1	18.1	19.1	20.1	21.2	22.2	23.2	24.2	25.2
41	5.9	6.9	7.9	8.9	9.9	11.0	12.0	13.0	14.0	15.0	16.0	17.0	18.1	19.1	20.1	21.1	22.1	23.1	24.2	25.2
42	5.8	6.8	7.9	8.9	9.9	10.9	11.9	12.9	13.9	15.0	16.0	17.0	18.0	19.0	20.0	21.0	22.1	23.1	24.1	25.1
43	5.8	6.8	7.8	8.8	9.8	10.9	11.9	12.9	13.9	14.9	15.9	16.9	17.9	19.0	20.0	21.0	22.0	23.0	24.0	25.0
44	5.7	6.7	7.8	8.8	9.8	10.8	11.8	12.8	13.8	14.9	15.9	16.9	17.9	18.9	19.9	20.9	21.9	23.0	24.0	25.0
45	5.7	6.7	7.7	8.7	9.7	10.8	11.8	12.8	13.8	14.8	15.8	16.8	17.8	18.8	19.9	20.9	21.9	22.9	23.9	24.9
46	5.6	6.7	7.7	8.7	9.7	10.7	11.7	12.7	13.7	14.7	15.8	16.8	17.8	18.8	19.8	20.8	21.8	22.8	23.8	24.9
47	5.6	6.6	7.6	8.6	9.6	10.7	11.7	12.7	13.7	14.7	15.7	16.7	17.7	18.7	19.7	20.8	21.8	22.8	23.8	24.8
48	5.5	6.6	7.6	8.6	9.6	10.6	11.6	12.6	13.6	14.6	15.6	16.7	17.7	18.7	19.7	20.7	21.7	22.7	23.7	24.7
49	5.5	6.5	7.5	8.5	9.5	10.6	11.6	12.6	13.6	14.6	15.6	16.6	17.6	18.6	19.6	20.6	21.6	22.7	23.7	24.7
50	5.5	6.5	7.5	8.5	9.5	10.5	11.5	12.5	13.5	14.5	15.5	16.5	17.6	18.6	19.6	20.6	21.6	22.6	23.6	24.6
51	5.4	6.4	7.4	8.4	9.4	10.5	11.5	12.5	13.5	14.5	15.5	16.5	17.5	18.5	19.5	20.5	21.5	22.5	23.5	24.6
52	5.4	6.4	7.4	8.4	9.4	10.4	11.4	12.4	13.4	14.4	15.4	16.4	17.4	18.5	19.5	20.5	21.5	22.5	23.5	24.5
53	5.3	6.3	7.3	8.3	9.3	10.4	11.4	12.4	13.4	14.4	15.4	16.4	17.4	18.4	19.4	20.4	21.4	22.4	23.4	24.4
54	5.3	6.3	7.3	8.3	9.3	10.3	11.3	12.3	13.3	14.3	15.3	16.3	17.3	18.3	19.3	20.3	21.4	22.4	23.4	24.4
55	5.2	6.2	7.2	8.2	9.2	10.2	11.3	12.3	13.3	14.3	15.3	16.3	17.3	18.3	19.3	20.3	21.3	22.3	23.3	24.3
56	5.2	6.2	7.2	8.2	9.2	10.2	11.2	12.2	13.2	14.2	15.2	16.2	17.2	18.2	19.2	20.2	21.2	22.2	23.2	24.2
57	5.1	6.1	7.1	8.1	9.1	10.1	11.2	12.2	13.2	14.2	15.2	16.2	17.2	18.2	19.2	20.2	21.2	22.2	23.2	24.2
58	5.1	6.1	7.1	8.1	9.1	10.1	11.1	12.1	13.1	14.1	15.1	16.1	17.1	18.1	19.1	20.1	21.1	22.1	23.1	24.1
59	5.0	6.0	7.0	8.0	9.0	10.0	11.1	12.1	13.1	14.1	15.1	16.1	17.1	18.1	19.1	20.1	21.1	22.1	23.1	24.1
60	5.0	6.0	7.0	8.0	9.0	10.0	11.0	12.0	13.0	14.0	15.0	16.0	17.0	18.0	19.0	20.0	21.0	22.0	23.0	24.0
61	5.0	6.0	7.0	8.0	9.0	10.0	10.9	11.9	12.9	13.9	14.9	16.0	16.9	17.9	18.9	20.0	21.0	21.9	22.9	23.9
62	4.9	5.9	6.9	7.9	8.9	9.9	10.8	11.9	12.9	13.9	14.9	15.9	16.9	17.9	18.9	19.9	20.9	21.9	22.9	23.9
63	4.9	5.9	6.9	7.9	8.9	9.9	10.8	11.8	12.8	13.8	14.8	15.8	16.8	17.8	18.8	19.8	20.8	21.8	22.8	23.8
64	4.8	5.8	6.8	7.8	8.8	9.8	10.8	11.8	12.8	13.8	14.8	15.8	16.8	17.8	18.8	19.8	20.8	21.8	22.8	23.8
65	4.8	5.8	6.8	7.8	8.8	9.8	10.7	11.7	12.7	13.7	14.7	15.7	16.7	17.7	18.7	19.7	20.7	21.7	22.7	23.7
66	4.7	5.7	6.7	7.7	8.7	9.7	10.7	11.7	12.7	13.7	14.7	15.7	16.7	17.7	18.7	19.7	20.7	21.6	22.6	23.6
67	4.7	5.7	6.7	7.7	8.7	9.7	10.6	11.6	12.6	13.6	14.6	15.6	16.6	17.6	18.6	19.6	20.6	21.6	22.6	23.6
68	4.6	5.6	6.6	7.6	8.6	9.6	10.6	11.6	12.6	13.6	14.6	15.6	16.6	17.6	18.5	19.5	20.5	21.5	22.5	23.5
69	4.6	5.6	6.6	7.6	8.6	9.6	10.6	11.5	12.5	13.5	14.5	15.5	16.5	17.5	18.5	19.5	20.5	21.5	22.5	23.5
70	4.5	5.5	6.5	7.5	8.5	9.5	10.5	11.5	12.5	13.5	14.5	15.5	16.5	17.4	18.4	19.4	20.4	21.4	22.4	23.4
71	4.5	5.5	6.5	7.5	8.5	9.5	10.4	11.4	12.4	13.4	14.4	15.4	16.4	17.4	18.4	19.4	20.4	21.4	22.3	23.3
72	4.5	5.4	6.4	7.4	8.4	9.4	10.4	11.4	12.4	13.4	14.4	15.4	16.3	17.3	18.3	19.3	20.3	21.3	22.4	23.3
73	4.4	5.4	6.4	7.4	8.4	9.4	10.3	11.3	12.3	13.3	14.3	15.3	16.3	17.3	18.3	19.3	20.2	21.2	22.2	23.2
74	4.4	5.4	6.3	7.3	8.3	9.3	10.3	11.3	12.3	13.3	14.3	15.2	16.2	17.2	18.2	19.2	20.2	21.2	22.2	23.2
75	4.3	5.3	6.3	7.3	8.3	9.3	10.2	11.2	12.2	13.2	14.2	15.2	16.2	17.2	18.2	19.1	20.1	21.1	22.1	23.1
76	4.3	5.3	6.3	7.3	8.2	9.2	10.2	11.2	12.2	13.2	14.2	15.1	16.1	17.1	18.1	19.1	20.1	21.1	22.0	23.0
77	4.2	5.2	6.2	7.2	8.2	9.2	10.2	11.1	12.1	13.1	14.1	15.1	16.1	17.1	18.0	19.0	20.0	21.0	22.0	23.0
78	4.2	5.2	6.2	7.1	8.1	9.1	10.1	11.1	12.1	13.1	14.0	15.0	16.0	17.0	18.0	19.0	20.0	20.9	21.9	22.9
79	4.1	5.1	6.1	7.1	8.1	9.1	10.1	11.0	12.0	13.0	14.0	15.0	16.0	17.0	17.9	18.9	19.9	20.9	21.9	22.9
80	4.1	5.1	6.1	7.1	8.0	9.0	10.0	11.0	12.0	13.0	13.9	14.9	15.9	16.9	17.9	18.9	19.8	20.8	21.8	22.8
81	4.1	5.0	6.0	7.0	8.0	9.0	10.0	10.9	11.9	12.9	13.9	14.9	15.9	16.8	17.8	18.8	19.8	20.8	21.8	22.7
82	4.0	5.0	6.0	7.0	7.9	8.9	9.9	10.9	11.9	12.9	13.8	14.8	15.8	16.8	17.8	18.8	19.7	20.7	21.7	22.7
83	4.0	5.0	5.9	6.9	7.9	8.9	9.9	10.8	11.8	12.8	13.8	14.8	15.8	16.7	17.7	18.7	19.7	20.7	21.6	22.6
84	3.9	4.9	5.9	6.9	7.8	8.8	9.8	10.8	11.8	12.8	13.7	14.7	15.7	16.7	17.7	18.6	19.6	20.6	21.6	22.6
85	3.9	4.9	5.8	6.8	7.8	8.8	9.8	10.7	11.7	12.7	13.7	14.7	15.6	16.6	17.6	18.6	19.6	20.5	21.5	22.5
86	3.8	4.8	5.8	6.8	7.8	8.7	9.7	10.7	11.7	12.7	13.6	14.6	15.6	16.6	17.5	18.5	19.5	20.5	21.5	22.4
87	3.8	4.8	5.7	6.7	7.7	8.7	9.7	10.6	11.6	12.6	13.6	14.6	15.5	16.5	17.5	18.5	19.5	20.4	21.4	22.4
88	3.7	4.7	5.7	6.7	7.7	8.6	9.6	10.6	11.6	12.5	13.5	14.5	15.5	16.5	17.4	18.4	19.4	20.4	21.3	22.3
89	3.7	4.7	5.7	6.6	7.6	8.6	9.6	10.5	11.5	12.5	13.5	14.5	15.4	16.4	17.4	18.4	19.3	20.3	21.3	22.3

Observed Temperature °F	API Gravity at Observed Temperature																			
	5	6	7	8	9	10	11	12	13	14	15	16	17	18	19	20	21	22	23	24
	Corresponding API Gravity at 60°F																			
90	3.7	4.6	5.6	6.6	7.6	8.5	9.5	10.5	11.5	12.4	13.4	14.4	15.4	16.4	17.3	18.3	19.3	20.3	21.2	22.2
91	3.6	4.6	5.6	6.5	7.5	8.5	9.5	10.4	11.4	12.4	13.4	14.3	15.3	16.3	17.3	18.3	19.2	20.2	21.2	22.2
92	3.6	4.5	5.5	6.5	7.5	8.4	9.4	10.4	11.4	12.3	13.3	14.3	15.3	16.2	17.2	18.2	19.2	20.1	21.1	22.1
93	3.5	4.5	5.5	6.5	7.4	8.4	9.4	10.3	11.3	12.3	13.3	14.2	15.2	16.2	17.2	18.1	19.1	20.1	21.1	22.0
94	3.5	4.5	5.4	6.4	7.4	8.4	9.3	10.3	11.3	12.2	13.2	14.2	15.2	16.1	17.1	18.1	19.1	20.0	21.0	22.0
95	3.4	4.4	5.4	6.4	7.3	8.3	9.3	10.2	11.2	12.2	13.2	14.1	15.1	16.1	17.1	18.0	19.0	20.0	20.9	21.9
96	3.4	4.4	5.3	6.3	7.3	8.3	9.2	10.2	11.2	12.1	13.1	14.1	15.1	16.0	17.0	18.0	18.9	19.9	20.9	21.9
97	3.4	4.3	5.3	6.3	7.2	8.2	9.2	10.2	11.1	12.1	13.1	14.0	15.0	16.0	17.0	17.9	18.9	19.9	20.8	21.8
98	3.3	4.3	5.3	6.2	7.2	8.2	9.1	10.1	11.1	12.0	13.0	14.0	15.0	15.9	16.9	17.9	18.8	19.8	20.8	21.7
99	3.3	4.2	5.2	6.2	7.1	8.1	9.1	10.1	11.0	12.0	13.0	13.9	14.9	15.9	16.8	17.8	18.8	19.7	20.7	21.7
100	3.2	4.2	5.2	6.1	7.1	8.1	9.0	10.0	11.0	11.9	12.9	13.9	14.9	15.8	16.8	17.8	18.7	19.7	20.7	21.6
101	3.2	4.1	5.1	6.1	7.1	8.0	9.0	10.0	10.9	11.9	12.9	13.8	14.8	15.8	16.7	17.7	18.7	19.6	20.6	21.6
102	3.1	4.1	5.1	6.0	7.0	8.0	8.9	9.9	10.9	11.8	12.8	13.8	14.7	15.7	16.7	17.6	18.6	19.6	20.5	21.5
103	3.1	4.1	5.0	6.0	7.0	7.9	8.9	9.9	10.8	11.8	12.8	13.7	14.7	15.7	16.6	17.6	18.6	19.5	20.5	21.5
104	3.0	4.0	5.0	5.9	6.9	7.9	8.8	9.8	10.8	11.7	12.7	13.7	14.6	15.6	16.6	17.5	18.5	19.5	20.4	21.4
105	3.0	4.0	4.9	5.9	6.9	7.8	8.8	9.8	10.7	11.7	12.7	13.6	14.6	15.6	16.5	17.5	18.4	19.4	20.4	21.3
106	3.0	3.9	4.9	5.9	6.8	7.8	8.8	9.7	10.7	11.6	12.6	13.6	14.5	15.5	16.5	17.4	18.4	19.4	20.3	21.3
107	2.9	3.9	4.8	5.8	6.8	7.7	8.7	9.7	10.6	11.6	12.6	13.5	14.5	15.5	16.4	17.4	18.3	19.3	20.3	21.2
108	2.9	3.8	4.8	5.8	6.7	7.7	8.7	9.6	10.6	11.5	12.5	13.5	14.4	15.4	16.4	17.3	18.3	19.2	20.2	21.2
109	2.8	3.8	4.8	5.7	6.7	7.6	8.6	9.6	10.5	11.5	12.5	13.4	14.4	15.3	16.3	17.3	18.2	19.2	20.2	21.1
110	2.8	3.7	4.7	5.7	6.6	7.6	8.6	9.5	10.5	11.5	12.4	13.4	14.3	15.3	16.3	17.2	18.2	19.1	20.1	21.1
111	2.7	3.7	4.7	5.6	6.6	7.6	8.5	9.5	10.4	11.4	12.4	13.3	14.3	15.2	16.2	17.2	18.1	19.1	20.0	21.0
112	2.7	3.7	4.6	5.6	6.5	7.5	8.5	9.4	10.4	11.4	12.3	13.3	14.2	15.2	16.1	17.1	18.1	19.0	20.0	20.9
113	2.7	3.6	4.6	5.5	6.5	7.5	8.4	9.4	10.3	11.3	12.3	13.2	14.2	15.1	16.1	17.1	18.0	19.0	19.9	20.9
114	2.6	3.6	4.5	5.5	6.5	7.4	8.4	9.3	10.3	11.3	12.2	13.2	14.1	15.1	16.0	17.0	18.0	18.9	19.9	20.8
115	2.6	3.5	4.5	5.5	6.4	7.4	8.3	9.3	10.2	11.2	12.2	13.1	14.1	15.0	16.0	16.9	17.9	18.9	19.8	20.8
116	2.5	3.5	4.4	5.4	6.4	7.3	8.3	9.2	10.2	11.2	12.1	13.1	14.0	15.0	15.9	16.9	17.8	18.8	19.8	20.7
117	2.5	3.4	4.4	5.4	6.3	7.3	8.2	9.2	10.2	11.1	12.1	13.0	14.0	14.9	15.9	16.8	17.8	18.8	19.7	20.7
118	2.4	3.4	4.4	5.3	6.3	7.2	8.2	9.1	10.1	11.1	12.0	13.0	13.9	14.9	15.8	16.8	17.7	18.7	19.6	20.6
119	2.4	3.4	4.3	5.3	6.2	7.2	8.1	9.1	10.1	11.0	12.0	12.9	13.9	14.8	15.8	16.7	17.7	18.6	19.6	20.5
120	2.4	3.3	4.3	5.2	6.2	7.1	8.1	9.1	10.0	11.0	11.9	12.9	13.8	14.8	15.7	16.7	17.6	18.6	19.5	20.5
121	2.3	3.3	4.2	5.2	6.1	7.1	8.1	9.0	10.0	10.9	11.9	12.8	13.8	14.7	15.7	16.6	17.6	18.5	19.5	20.4
122	2.3	3.2	4.2	5.1	6.1	7.0	8.0	9.0	9.9	10.9	11.8	12.8	13.7	14.7	15.6	16.6	17.5	18.5	19.4	20.4
123	2.2	3.2	4.1	5.1	6.0	7.0	8.0	8.9	9.9	10.8	11.8	12.7	13.7	14.6	15.6	16.5	17.5	18.4	19.4	20.3
124	2.2	3.1	4.1	5.0	6.0	7.0	7.9	8.9	9.8	10.8	11.7	12.7	13.6	14.6	15.5	16.5	17.4	18.4	19.3	20.3
125	2.1	3.1	4.0	5.0	6.0	6.9	7.9	8.8	9.8	10.7	11.7	12.6	13.6	14.5	15.5	16.4	17.4	18.3	19.3	20.2
126	2.1	3.0	4.0	5.0	5.9	6.9	7.8	8.8	9.7	10.7	11.6	12.6	13.5	14.5	15.4	16.4	17.3	18.3	19.2	20.2
127	2.1	3.0	4.0	4.9	5.9	6.8	7.8	8.7	9.7	10.6	11.6	12.5	13.5	14.4	15.4	16.3	17.3	18.2	19.2	20.1
128	2.0	3.0	3.9	4.9	5.8	6.8	7.7	8.7	9.6	10.6	11.5	12.5	13.4	14.4	15.3	16.3	17.2	18.2	19.1	20.0
129	2.0	2.9	3.9	4.8	5.8	6.7	7.7	8.6	9.6	10.5	11.5	12.4	13.4	14.3	15.3	16.2	17.2	18.1	19.0	20.0
130	1.9	2.9	3.8	4.8	5.7	6.7	7.6	8.6	9.5	10.5	11.4	12.4	13.3	14.3	15.2	16.2	17.1	18.0	19.0	19.9
131	1.9	2.8	3.8	4.7	5.7	6.6	7.6	8.5	9.5	10.4	11.4	12.3	13.3	14.2	15.2	16.1	17.0	18.0	18.9	19.9
132	1.8	2.8	3.7	4.7	5.6	6.6	7.5	8.5	9.4	10.4	11.3	12.3	13.2	14.2	15.1	16.1	17.0	17.9	18.9	19.8
133	1.8	2.7	3.7	4.6	5.6	6.5	7.5	8.4	9.4	10.3	11.3	12.2	13.2	14.1	15.1	16.0	16.9	17.9	18.8	19.8
134	1.8	2.7	3.7	4.6	5.6	6.5	7.4	8.4	9.3	10.3	11.2	12.2	13.1	14.1	15.0	15.9	16.9	17.8	18.8	19.7
135	1.7	2.7	3.6	4.6	5.5	6.5	7.4	8.4	9.3	10.2	11.2	12.1	13.1	14.0	15.0	15.9	16.8	17.8	18.7	19.7
136	1.7	2.6	3.6	4.5	5.5	6.4	7.4	8.3	9.3	10.2	11.1	12.1	13.0	14.0	14.9	15.8	16.8	17.7	18.7	19.6
137	1.6	2.6	3.5	4.5	5.4	6.4	7.3	8.3	9.2	10.2	11.1	12.0	13.0	13.9	14.9	15.8	16.7	17.7	18.6	19.5
138	1.6	2.5	3.5	4.4	5.4	6.3	7.3	8.2	9.2	10.1	11.0	12.0	12.9	13.9	14.8	15.7	16.7	17.6	18.6	19.5
139	1.5	2.5	3.4	4.4	5.3	6.3	7.2	8.2	9.1	10.1	11.0	11.9	12.9	13.8	14.8	15.7	16.6	17.6	18.5	19.4

Observed Temperature °F	API Gravity at Observed Temperature																			
	25	26	27	28	29	30	31	32	33	34	35	36	37	38	39	40	41	42	43	44
	Corresponding API Gravity at 60°F																			
0	28.9	30.0	31.0	32.0	33.1	34.2	35.3	36.3	37.4	38.5	39.6	40.7	41.8	42.9	44.0	45.1	46.2	47.3	48.4	49.5
1	28.8	29.8	30.9	32.0	33.0	34.1	35.2	36.3	37.3	38.4	39.5	40.6	41.7	42.8	43.9	45.0	46.1	47.2	48.3	49.4
2	28.7	29.8	30.8	31.9	33.0	34.0	35.1	36.2	37.3	38.3	39.4	40.5	41.5	42.7	43.8	44.9	46.0	47.1	48.2	49.3
3	28.7	29.7	30.8	31.8	32.9	34.0	35.0	36.1	37.2	38.3	39.3	40.4	41.5	42.6	43.7	44.8	45.9	47.0	48.1	49.2
4	28.6	29.6	30.7	31.8	32.8	33.9	35.0	36.0	37.1	38.2	39.3	40.3	41.4	42.5	43.6	44.7	45.8	46.9	48.0	49.1
5	28.5	29.6	30.6	31.7	32.8	33.8	34.9	36.0	37.0	38.1	39.2	40.3	41.3	42.4	43.5	44.6	45.7	46.8	47.9	49.0
6	28.5	29.5	30.6	31.6	32.7	33.7	34.8	35.9	36.9	38.0	39.1	40.2	41.3	42.3	43.4	44.5	45.6	46.7	47.8	48.9
7	28.4	29.4	30.5	31.6	32.6	33.7	34.7	35.8	36.9	37.9	39.0	40.1	41.2	42.3	43.3	44.4	45.5	46.6	47.7	48.8
8	28.3	29.4	30.4	31.5	32.5	33.6	34.7	35.7	36.8	37.9	38.9	40.0	41.1	42.2	43.3	44.3	45.4	46.5	47.6	48.7
9	28.3	29.3	30.4	31.4	32.5	33.5	34.6	35.7	36.7	37.8	38.9	39.9	41.0	42.1	43.2	44.3	45.3	46.4	47.5	48.6
10	28.2	29.2	30.3	31.3	32.4	33.5	34.5	35.6	36.6	37.7	38.8	39.8	40.9	42.0	43.1	44.2	45.2	46.3	47.4	48.5
11	28.1	29.2	30.2	31.3	32.3	33.4	34.4	35.5	36.6	37.6	38.7	39.8	40.8	41.9	43.0	44.1	45.2	46.2	47.3	48.4
12	28.1	29.1	30.2	31.2	32.3	33.3	34.4	35.4	36.5	37.6	38.6	39.7	40.8	41.8	42.9	44.0	45.1	46.2	47.2	48.3
13	28.0	29.0	30.1	31.1	32.2	33.2	34.3	35.4	36.4	37.5	38.5	39.6	40.7	41.7	42.8	43.9	45.0	46.1	47.1	48.2
14	27.9	29.0	30.0	31.1	32.1	33.2	34.2	35.3	36.3	37.4	38.5	39.5	40.6	41.7	42.7	43.8	44.9	46.0	47.1	48.1
15	27.9	28.9	30.0	31.0	32.1	33.1	34.2	35.2	36.3	37.3	38.4	39.4	40.5	41.6	42.7	43.7	44.8	45.9	47.0	48.0
16	27.8	28.8	29.9	30.9	32.0	33.0	34.1	35.1	36.2	37.2	38.3	39.4	40.4	41.5	42.6	43.6	44.7	45.8	46.9	48.0
17	27.7	28.8	29.8	30.9	31.9	33.0	34.0	35.1	36.1	37.2	38.2	39.3	40.4	41.4	42.5	43.6	44.6	45.7	46.8	47.9
18	27.7	28.7	29.7	30.8	31.8	32.9	33.9	35.0	36.0	37.1	38.1	39.2	40.3	41.3	42.4	43.5	44.5	45.6	46.7	47.8
19	27.6	28.6	29.7	30.7	31.8	32.8	33.9	34.9	36.0	37.0	38.1	39.1	40.2	41.2	42.3	43.4	44.4	45.5	46.6	47.7
20	27.5	28.6	29.6	30.7	31.7	32.7	33.8	34.8	35.9	36.9	38.0	39.0	40.1	41.2	42.2	43.3	44.4	45.4	46.5	47.6
21	27.5	28.5	29.5	30.6	31.6	32.7	33.7	34.8	35.8	36.9	37.9	39.0	40.0	41.1	42.1	43.2	44.3	45.3	46.4	47.5
22	27.4	28.4	29.5	30.5	31.6	32.6	33.6	34.7	35.7	36.8	37.8	38.9	39.9	41.0	42.1	43.1	44.2	45.2	46.3	47.4
23	27.3	28.4	29.4	30.5	31.5	32.5	33.6	34.6	35.7	36.7	37.8	38.8	39.9	40.9	42.0	43.0	44.1	45.2	46.2	47.3
24	27.3	28.3	29.3	30.4	31.4	32.5	33.5	34.5	35.6	36.6	37.7	38.7	39.8	40.8	41.9	42.9	44.0	45.1	46.1	47.2
25	27.2	28.2	29.3	30.3	31.4	32.4	33.4	34.5	35.5	36.6	37.6	38.7	39.7	40.8	41.8	42.9	43.9	45.0	46.0	47.1
26	27.1	28.2	29.2	30.2	31.3	32.3	33.4	34.4	35.4	36.5	37.5	38.6	39.6	40.7	41.7	42.8	43.8	44.9	46.0	47.0
27	27.1	28.1	29.1	30.2	31.2	32.3	33.3	34.3	35.4	36.4	37.5	38.5	39.5	40.6	41.6	42.7	43.7	44.8	45.9	46.9
28	27.0	28.0	29.1	30.1	31.1	32.2	33.2	34.3	35.3	36.3	37.4	38.4	39.5	40.5	41.6	42.6	43.7	44.7	45.8	46.8
29	27.0	28.0	29.0	30.0	31.1	32.1	33.1	34.2	35.2	36.3	37.3	38.3	39.4	40.4	41.5	42.5	43.6	44.6	45.7	46.7
30	26.9	27.9	28.9	30.0	31.0	32.0	33.1	34.1	35.1	36.2	37.2	38.3	39.3	40.4	41.4	42.4	43.5	44.5	45.6	46.6
31	26.8	27.9	28.9	29.9	30.9	32.0	33.0	34.0	35.1	36.1	37.1	38.2	39.2	40.3	41.3	42.4	43.4	44.5	45.5	46.6
32	26.8	27.8	28.8	29.8	30.9	31.9	32.9	34.0	35.0	36.0	37.1	38.1	39.1	40.2	41.2	42.3	43.3	44.4	45.4	46.5
33	26.7	27.7	28.7	29.8	30.8	31.8	32.9	33.9	34.9	36.0	37.0	38.0	39.1	40.1	41.1	42.2	43.2	44.3	45.3	46.4
34	26.6	27.7	28.7	29.7	30.7	31.8	32.8	33.8	34.9	35.9	36.9	38.0	39.0	40.0	41.1	42.1	43.1	44.2	45.2	46.3
35	26.6	27.6	28.6	29.6	30.7	31.7	32.7	33.8	34.8	35.8	36.8	37.9	38.9	39.9	41.0	42.0	43.1	44.1	45.1	46.2
36	26.5	27.5	28.6	29.6	30.6	31.6	32.7	33.7	34.7	35.7	36.8	37.8	38.8	39.9	40.9	41.9	43.0	44.0	45.1	46.1
37	26.4	27.5	28.5	29.5	30.5	31.6	32.6	33.6	34.6	35.7	36.7	37.7	38.8	39.8	40.8	41.9	42.9	43.9	45.0	46.0
38	26.4	27.4	28.4	29.4	30.5	31.5	32.5	33.5	34.6	35.6	36.6	37.6	38.7	39.7	40.7	41.8	42.8	43.8	44.9	45.9
39	26.3	27.3	28.4	29.4	30.4	31.4	32.4	33.5	34.5	35.5	36.5	37.6	38.6	39.6	40.7	41.7	42.7	43.8	44.8	45.8
40	26.2	27.3	28.3	29.3	30.3	31.4	33.4	32.4	34.4	35.4	36.5	37.5	38.5	39.6	40.6	41.6	42.6	43.7	44.7	45.7
41	26.2	27.2	28.2	29.2	30.3	31.3	32.3	33.3	34.3	35.4	36.4	37.4	38.4	39.5	40.5	41.5	42.6	43.6	44.6	45.7
42	26.1	27.1	28.2	29.2	30.2	31.2	32.2	33.3	34.3	35.3	36.3	37.3	38.4	39.4	40.4	41.4	42.5	43.5	44.5	45.6
43	26.1	27.1	28.1	29.1	30.1	31.1	32.2	33.2	34.2	35.2	36.2	37.3	38.3	39.3	40.3	41.4	42.4	43.4	44.4	45.5
44	26.0	27.0	28.0	29.0	30.1	31.1	32.1	33.1	34.1	35.2	36.2	37.2	38.2	39.2	40.3	41.3	42.3	43.3	44.4	45.4
45	25.9	26.9	28.0	29.0	30.0	31.0	32.0	33.0	34.1	35.1	36.1	37.1	38.1	39.2	40.2	41.2	42.2	43.2	44.3	45.3
46	25.9	26.9	27.9	28.9	29.9	30.9	32.0	33.0	34.0	35.0	36.0	37.0	38.1	39.1	40.1	41.1	42.1	43.2	44.2	45.2
47	25.8	26.8	27.8	28.8	29.9	30.9	31.9	32.9	33.9	34.9	35.9	37.0	38.0	39.0	40.0	41.0	42.1	43.1	44.1	45.1
48	25.7	26.8	27.8	28.8	29.8	30.8	31.8	32.8	33.8	34.9	35.9	36.9	37.9	38.9	39.9	41.0	42.0	43.0	44.0	45.0
49	25.7	26.7	27.7	28.7	29.7	30.7	31.7	32.8	33.8	34.8	35.8	36.8	37.8	38.8	39.9	40.9	41.9	42.9	43.9	44.9

Observed Temperature °F	API Gravity at Observed Temperature																			
	25	26	27	28	29	30	31	32	33	34	35	36	37	38	39	40	41	42	43	44
	Corresponding API Gravity at 60°F																			
50	25.6	26.6	27.6	28.6	29.7	30.7	31.7	32.7	33.7	34.7	35.7	36.7	37.8	38.8	39.8	40.8	41.8	42.8	43.8	44.9
51	25.6	26.6	27.6	28.6	29.6	30.6	31.6	32.6	33.6	34.6	35.7	36.7	37.7	38.7	39.7	40.7	41.7	42.7	43.8	44.8
52	25.5	26.5	27.5	28.5	29.5	30.5	31.5	32.6	33.6	34.6	35.6	36.6	37.6	38.6	39.6	40.6	41.6	42.7	43.7	44.7
53	25.4	26.4	27.4	28.5	29.5	30.5	31.5	32.5	33.5	34.5	35.5	36.5	37.5	38.5	39.5	40.6	41.6	42.6	43.6	44.6
54	25.4	26.4	27.4	28.4	29.4	30.4	31.4	32.4	33.4	34.4	35.4	36.4	37.5	38.5	39.5	40.5	41.5	42.5	43.5	44.5
55	25.3	26.3	27.3	28.3	29.3	30.3	31.3	32.3	33.4	34.4	35.4	36.4	37.4	38.4	39.4	40.4	41.4	42.4	43.4	44.4
56	25.2	26.3	27.3	28.3	29.3	30.3	31.3	32.3	33.3	34.3	35.3	36.3	37.3	38.3	39.3	40.3	41.3	42.3	43.3	44.3
57	25.2	26.2	27.2	28.2	29.2	30.2	31.2	32.2	33.2	34.2	35.2	36.2	37.2	38.2	39.2	40.2	41.2	42.2	43.3	44.3
58	25.1	26.1	27.1	28.1	29.1	30.1	31.1	32.1	33.1	34.1	35.1	36.1	37.1	38.2	39.2	40.2	41.2	42.2	43.2	44.2
59	25.1	26.1	27.1	28.1	29.1	30.1	31.1	32.1	33.1	34.1	35.1	36.1	37.1	38.1	39.1	40.1	41.1	42.1	43.1	44.1
60	25.0	26.0	27.0	28.0	29.0	30.0	31.0	32.0	33.0	34.0	35.0	36.0	37.0	38.0	39.0	40.0	41.0	42.0	43.0	44.0
61	24.9	25.9	26.9	27.9	28.9	29.9	30.9	31.9	32.9	33.9	34.9	35.9	36.9	37.9	38.9	39.9	40.9	41.9	42.9	43.9
62	24.9	25.9	26.9	27.9	28.9	29.9	30.9	31.9	32.9	33.9	34.9	35.9	36.9	37.8	38.8	39.8	40.8	41.8	42.8	43.8
63	24.8	25.8	26.8	27.8	28.8	29.8	30.8	31.8	32.8	33.8	34.8	35.8	36.8	37.8	38.8	39.8	40.8	41.8	42.8	43.7
64	24.8	25.8	26.7	27.7	28.7	29.7	30.7	31.7	32.7	33.7	34.7	35.7	36.7	37.7	38.7	39.7	40.7	41.7	42.7	43.7
65	24.7	25.7	26.7	27.7	28.7	29.7	30.7	31.7	32.7	33.6	34.6	35.6	36.6	37.6	38.6	39.6	40.6	41.6	42.6	43.6
66	24.6	25.6	26.6	27.6	28.6	29.6	30.6	31.6	32.6	33.6	34.6	35.6	36.6	37.5	38.5	39.5	40.5	41.5	42.5	43.5
67	24.6	25.6	26.6	27.6	28.5	29.5	30.5	31.5	32.5	33.5	34.5	35.5	36.5	37.5	38.5	39.5	40.4	41.4	42.4	43.4
68	24.5	25.5	26.5	27.5	28.5	29.5	30.5	31.5	32.4	33.4	34.4	35.4	36.4	37.4	38.4	39.4	40.4	41.4	42.3	43.3
69	24.4	25.4	26.4	27.4	28.4	29.4	30.4	31.4	32.4	33.4	34.4	35.3	36.3	37.3	38.3	39.3	40.3	41.3	42.3	43.2
70	24.4	25.4	26.4	27.4	28.3	29.3	30.3	31.3	32.3	33.3	34.3	35.3	36.3	37.2	38.2	39.2	40.2	41.2	42.2	43.2
71	24.3	25.3	26.3	27.3	28.3	29.3	30.3	31.3	32.2	33.2	34.2	35.2	36.2	37.2	38.2	39.1	40.1	41.1	42.1	43.1
72	24.3	25.3	26.2	27.2	28.2	29.2	30.2	31.2	32.2	33.2	34.1	35.1	36.1	37.1	38.1	39.1	40.0	41.0	42.0	43.0
73	24.2	25.2	26.2	27.2	28.2	29.1	30.1	31.1	32.1	33.1	34.1	35.1	36.0	37.0	38.0	39.0	40.0	41.0	41.9	42.9
74	24.1	25.1	26.1	27.1	28.1	29.1	30.1	31.0	32.0	33.0	34.0	35.0	36.0	36.9	37.9	38.9	39.9	40.9	41.9	42.8
75	24.1	25.1	26.1	27.0	28.0	29.0	30.0	31.0	32.0	32.9	33.9	34.9	35.9	36.9	37.9	38.8	39.8	40.8	41.8	42.7
76	24.0	25.0	26.0	27.0	28.0	28.9	29.9	30.9	31.9	32.9	33.9	34.8	35.8	36.8	37.8	38.8	39.7	40.7	41.7	42.7
77	24.0	24.9	25.9	26.9	27.9	28.9	29.9	30.8	31.8	32.8	33.8	34.8	35.7	36.7	37.7	38.7	39.7	40.6	41.6	42.6
78	23.9	24.9	25.9	26.9	27.8	28.8	29.8	30.8	31.8	32.7	33.7	34.7	35.7	36.7	37.6	38.6	39.6	40.6	41.5	42.5
79	23.8	24.8	25.8	26.8	27.8	28.8	29.7	30.7	31.7	32.7	33.7	34.6	35.6	36.6	37.6	38.5	39.5	40.5	41.4	42.4
80	23.8	24.8	25.7	26.7	27.7	28.7	29.7	30.6	31.6	32.6	33.6	34.6	35.5	36.5	37.5	38.5	39.4	40.4	41.4	42.3
81	23.7	24.7	25.7	26.7	27.6	28.6	29.6	30.6	31.6	32.5	33.5	34.5	35.5	36.4	37.4	38.4	39.4	40.3	41.3	42.3
82	23.7	24.6	25.6	26.6	27.6	28.6	29.5	30.5	31.5	32.5	33.4	34.4	35.4	36.4	37.3	38.3	39.3	40.2	41.2	42.2
83	23.6	24.6	25.6	26.5	27.5	28.5	29.5	30.4	31.4	32.4	33.4	34.3	35.3	36.3	37.3	38.2	39.2	40.2	41.1	42.1
84	23.5	24.5	25.5	26.5	27.5	28.4	29.4	30.4	31.4	32.3	33.3	34.3	35.2	36.2	37.2	38.2	39.1	40.1	41.1	42.0
85	23.5	24.5	25.4	26.4	27.4	28.4	29.3	30.3	31.3	32.3	33.2	34.2	35.2	36.1	37.1	38.1	39.0	40.0	41.0	41.9
86	23.4	24.4	25.4	26.4	27.3	28.3	29.3	30.3	31.2	32.2	33.2	34.1	35.1	36.1	37.0	38.0	39.0	39.9	40.9	41.9
87	23.4	24.3	25.3	26.3	27.3	28.2	29.2	30.2	31.2	32.1	33.1	34.1	35.0	36.0	37.0	37.9	38.9	39.9	40.8	41.8
88	23.3	24.3	25.3	26.2	27.2	28.2	29.1	30.1	31.1	32.1	33.0	34.0	35.0	35.9	36.9	37.9	38.8	39.8	40.7	41.7
89	23.2	24.2	25.2	26.2	27.1	28.1	29.1	30.1	31.0	32.0	33.0	33.9	34.9	35.9	36.8	37.8	38.7	39.7	40.7	41.6
90	23.2	24.2	25.1	26.1	27.1	28.0	29.0	30.0	31.0	31.9	32.9	33.9	34.8	35.8	36.7	37.7	38.7	39.6	40.6	41.5
91	23.1	24.1	25.1	26.0	27.0	28.0	29.0	29.9	30.9	31.9	32.8	33.8	34.7	35.7	36.7	37.6	38.6	39.5	40.5	41.5
92	23.1	24.0	25.0	26.0	27.0	27.9	28.9	29.9	30.8	31.8	32.8	33.7	34.7	35.6	36.6	37.6	38.5	39.5	40.4	41.4
93	23.0	24.0	25.0	25.9	26.9	27.9	28.8	29.8	30.8	31.7	32.7	33.7	34.6	35.6	36.5	37.5	38.4	39.4	40.3	41.3
94	22.9	23.9	24.9	25.9	26.8	27.8	28.8	29.7	30.7	31.7	32.6	33.6	34.5	35.5	36.5	37.4	38.4	39.3	40.3	41.2
95	22.9	23.9	24.8	25.8	26.8	27.7	28.7	29.7	30.6	31.6	32.5	33.5	34.5	35.4	36.4	37.3	38.3	39.2	40.2	41.1
96	22.8	23.8	24.8	25.7	26.7	27.7	28.6	29.6	30.6	31.5	32.5	33.4	34.4	35.4	36.3	37.3	38.2	39.2	40.1	41.1
97	22.8	23.7	24.7	25.7	26.6	27.6	28.6	29.5	30.5	31.5	32.4	33.4	34.3	35.3	36.2	37.2	38.1	39.1	40.0	41.0
98	22.7	23.7	24.7	25.6	26.6	27.5	28.5	29.5	30.4	31.4	32.3	33.3	34.3	35.2	36.2	37.1	38.1	39.0	40.0	40.9
99	22.7	23.6	24.6	25.6	26.5	27.5	28.4	29.4	30.4	31.3	32.3	33.2	34.2	35.1	36.1	37.0	38.0	38.9	39.9	40.8

Observed Temperature °F	API Gravity at Observed Temperature																			
	25	26	27	28	29	30	31	32	33	34	35	36	37	38	39	40	41	42	43	44
	Corresponding API Gravity at 60°F																			
100	22.6	23.6	24.5	25.5	26.5	27.4	28.4	29.3	30.3	31.3	32.2	33.2	34.1	35.1	36.0	37.0	37.9	38.9	39.8	40.7
101	22.5	23.5	24.5	25.4	26.4	27.4	28.3	29.3	30.2	31.2	32.1	33.1	34.0	35.0	35.9	36.9	37.8	38.8	39.7	40.7
102	22.5	23.4	24.4	25.4	26.3	27.3	28.3	29.2	30.2	31.1	32.1	33.0	34.0	34.9	35.9	36.8	37.8	38.7	39.7	40.6
103	22.4	23.4	24.3	25.3	26.3	27.2	28.2	29.1	30.1	31.1	32.0	33.0	33.9	34.9	35.8	36.7	37.7	38.6	39.6	40.5
104	22.4	23.3	24.3	25.3	26.2	27.2	28.1	29.1	30.0	31.0	31.9	32.9	33.8	34.8	35.7	36.7	37.6	38.6	39.5	40.4
105	22.3	23.3	24.2	25.2	26.1	27.1	28.1	29.0	30.0	30.9	31.9	32.8	33.8	34.7	35.7	36.6	37.5	38.5	39.4	40.4
106	22.2	23.2	24.2	25.1	26.1	27.0	28.0	29.0	29.9	30.9	31.8	32.8	33.7	34.6	35.6	36.5	37.5	38.4	39.4	40.3
107	22.2	23.1	24.1	25.1	26.0	27.0	27.9	28.9	29.8	30.8	31.7	32.7	33.6	34.6	35.5	36.5	37.4	38.3	39.3	40.2
108	22.1	23.1	24.0	25.0	26.0	26.9	27.9	28.8	29.8	30.7	31.7	32.6	33.6	34.5	35.4	36.4	37.3	38.3	39.2	40.1
109	22.1	23.0	24.0	24.9	25.9	26.9	27.8	28.8	29.7	30.7	31.6	32.5	33.5	34.4	35.4	36.3	37.3	38.2	39.1	40.1
110	22.0	23.0	23.9	24.9	25.8	26.8	27.7	28.7	29.6	30.6	31.5	32.5	33.4	34.4	35.3	36.2	37.2	38.1	39.0	40.0
111	22.0	22.9	23.9	24.8	25.8	26.7	27.7	28.6	29.6	30.5	31.5	32.4	33.4	34.3	35.2	36.2	37.1	38.0	39.0	39.9
112	21.9	22.9	23.8	24.8	25.7	26.7	27.6	28.6	29.5	30.5	31.4	32.3	33.3	34.2	35.2	36.1	37.0	38.0	38.9	39.8
113	21.8	22.8	23.8	24.7	25.7	26.6	27.6	28.5	29.5	30.4	31.3	32.3	33.2	34.2	35.1	36.0	37.0	37.9	38.8	39.7
114	21.8	22.7	23.7	24.6	25.6	26.5	27.5	28.4	29.4	30.3	31.3	32.2	33.2	34.1	35.0	36.0	36.9	37.8	38.7	39.7
115	21.7	22.7	23.6	24.6	25.5	26.5	27.4	28.4	29.3	30.3	31.2	32.1	33.1	34.0	35.0	35.9	36.8	37.7	38.7	39.6
116	21.7	22.6	23.6	24.5	25.5	26.4	27.4	28.3	29.3	30.2	31.1	32.1	33.0	34.0	34.9	35.8	36.7	37.7	38.6	39.5
117	21.6	22.6	23.5	24.5	25.4	26.4	27.3	28.3	29.2	30.1	31.1	32.0	32.9	33.9	34.8	35.7	36.7	37.6	38.5	39.4
118	21.6	22.5	23.5	24.4	25.4	26.3	27.2	28.2	29.1	30.1	31.0	31.9	32.9	33.8	34.7	35.7	36.6	37.5	38.5	39.4
119	21.5	22.4	23.4	24.3	25.3	26.2	27.2	28.1	29.1	30.0	30.9	31.9	32.8	33.7	34.7	35.6	36.5	37.5	38.4	39.3
120	21.4	22.4	23.3	24.3	25.2	26.2	27.1	28.1	29.0	29.9	30.9	31.8	32.7	33.7	34.6	35.5	36.5	37.4	38.3	39.2
121	21.4	22.3	23.3	24.2	25.2	26.1	27.1	28.0	28.9	29.9	30.8	31.7	32.7	33.6	34.5	35.5	36.4	37.3	38.2	39.1
122	21.3	22.3	23.2	24.2	25.1	26.1	27.0	27.9	28.9	29.8	30.8	31.7	32.6	33.5	34.5	35.4	36.3	37.2	38.2	39.1
123	21.3	22.2	23.2	24.1	25.1	26.0	26.9	27.9	28.8	29.8	30.7	31.6	32.5	33.5	34.4	35.3	36.2	37.2	38.1	39.0
124	21.2	22.2	23.1	24.1	25.0	25.9	26.9	27.8	28.8	29.7	30.6	31.5	32.5	33.4	34.3	35.3	36.2	37.1	38.0	38.9
125	21.2	22.1	23.0	24.0	24.9	25.9	26.8	27.8	28.7	29.6	30.6	31.5	32.4	33.3	34.3	35.2	36.1	37.0	37.9	38.9
126	21.1	22.0	23.0	23.9	24.9	25.8	26.8	27.7	28.6	29.6	30.5	31.4	32.3	33.3	34.2	35.1	36.0	36.9	37.9	38.8
127	21.0	22.0	22.9	23.9	24.8	25.8	26.7	27.6	28.6	29.5	30.4	31.4	32.3	33.2	34.1	35.0	36.0	36.9	37.8	38.7
128	21.0	21.9	22.9	23.8	24.8	25.7	26.6	27.6	28.5	29.4	30.4	31.3	32.2	33.1	34.1	35.0	35.9	36.8	37.7	38.6
129	20.9	21.9	22.8	23.8	24.7	25.6	26.6	27.5	28.4	29.4	30.3	31.2	32.1	33.1	34.0	34.9	35.8	36.7	37.6	38.6
130	20.9	21.8	22.8	23.7	24.6	25.6	26.5	27.4	28.4	29.3	30.2	31.2	32.1	33.0	33.9	34.8	35.8	36.7	37.6	38.5
131	20.8	21.8	22.7	23.6	24.6	25.5	26.5	27.4	28.3	29.2	30.2	31.1	32.0	32.9	33.8	34.8	35.7	36.6	37.5	38.4
132	20.8	21.7	22.6	23.6	24.5	25.5	26.4	27.3	28.2	29.2	30.1	31.0	31.9	32.9	33.8	34.7	35.6	36.5	37.4	38.3
133	20.7	21.6	22.6	23.5	24.5	25.4	26.3	27.3	28.2	29.1	30.0	31.0	31.9	32.8	33.7	34.6	35.5	36.5	37.4	38.3
134	20.6	21.6	22.5	23.5	24.4	25.3	26.3	27.2	28.1	29.0	30.0	30.9	31.8	32.7	33.6	34.6	35.5	36.4	37.3	38.2
135	20.6	21.5	22.5	23.4	24.3	25.3	26.2	27.1	28.1	29.0	29.9	30.8	31.7	32.7	33.6	34.5	35.4	36.3	37.2	38.1
136	20.5	21.5	22.4	23.3	24.3	25.2	26.1	27.1	28.0	28.9	29.8	30.8	31.7	32.6	33.5	34.4	35.3	36.2	37.1	38.0
137	20.5	21.4	22.4	23.3	24.2	25.2	26.1	27.0	27.9	28.9	29.8	30.7	31.6	32.5	33.4	34.4	35.3	36.2	37.1	38.0
138	20.4	21.4	22.3	23.2	24.2	25.1	26.0	27.0	27.9	28.8	29.7	30.6	31.6	32.5	33.4	34.3	35.2	36.1	37.0	37.9
139	20.4	21.3	22.2	23.2	24.1	25.0	26.0	26.9	27.8	28.7	29.7	30.6	31.5	32.4	33.3	34.2	35.1	36.0	36.9	37.8

APPENDIX 4

Galvanic Series of Metals and Metal Alloys

ANODIC (*Corroded End*)

Magnesium
Zinc
Galvanized Steel
Aluminum
Mild Steel
Cast Iron
50-50 Tin-Lead Solder
Stainless Steel Type 304 (active)
Stainless Steel Type 316 (active)
Lead
Tin
Muntz Metal
Inconel
Hastelloy
Yellow Brass
Admiralty Brass
Aluminum Bronze
Red Brass
Copper
Monel
Stainless Steel Type 304 (passive)
Stainless Steel Type 316 (passive)
Silver
Graphite
Gold
Platinum

CATHODIC (*Protected End*)

APPENDIX 5

Composition of Synthetic Seawater Utilized in ASTM D-665-B

Salt	g/L
NaCl	24.54
$MgCl_2 \cdot 6H_2O$	11.10
Na_2SO_4	4.09
$CaCl_2$	1.16
KCl	0.69
$NaHCO_3$	0.20
KBr	0.10
H_3BO_3	0.03
$SrCl_2 \cdot 6H_2O$	0.04
NaF	0.003

APPENDIX 6

Factors and Metal Densities Needed to Obtain ipy and mdd Corrosion Rates

To obtain mdd

ipy (696 × density) = mdd

To obtain ipy

mdd (0.00144 ÷ density) = ipy
mpy × 1000 = ipy

Metal	Density, g/cm^3
Aluminum	2.72
Brass (red)	8.75
Brass (yellow)	8.47
Copper	8.92
Copper-nickel (70-30)	8.95
Iron	7.87
Lead	11.35
Magnesium	1.74
Nickel	8.89
Monel	8.84
Silver	10.50
Titanium	4.54
Tin	7.29
Zinc	7.14

ipy = inches per year
mpy = mils per year
mdd = milligrams per square decimeter per day

APPENDIX 7

Compatibility of Various Materials with Common Fuels and Solvents

Elastomer or Metal	Hydrocarbon						
	Diesel Fuel	Unleaded Gasoline	Jet Fuel JP-4, JP-5	Kerosene	Mineral Spirits	Heavy Fuel Oil	Xylene
Ryton	OK	OK	OK	OK	OK	OK	OK
Polycarbonate	OK	OK	Poor	Poor	Marginal	OK to marginal	Poor
Polypropylene	OK	Marginal	OK to marginal	OK to marginal	OK to marginal	OK	OK to marginal
PTFE (Teflon)	OK	OK	OK	OK	OK	OK to marginal	OK
PVC	OK	Marginal	OK	OK	OK	OK	Poor
Natural rubber	Poor	Poor	Poor	Poor	Poor	Poor	Poor
Buna-N (nitrile)	OK	OK	OK	OK	OK	Poor	Poor
Neoprene	OK to marginal	OK to marginal	OK	OK	Marginal	OK to marginal	Poor
Silicone	Poor	Poor	Poor	Poor	Poor	Poor	Poor
Tygon	No info.	Marginal	Poor	Poor	OK to marginal	OK	Poor
Viton	OK	OK	OK	OK	OK	OK	Poor
304 Stainless steel	OK	OK	OK	OK	OK	OK	OK to marginal

316 Stainless steel	OK	OK	OK	OK	OK	OK	OK to marginal
Aluminum	OK	OK	OK	OK	OK	Marginal	OK
Copper	OK	OK to marginal	OK to marginal	OK to marginal	No info.	OK	OK
Brass	OK	No info.	OK	OK	No info.	OK to marginal	OK
Cast iron	OK	OK	OK	OK	OK to marginal	OK	OK to marginal
Hastelloy marginal	OK to marginal	OK to marginal	OK to marginal	OK to marginal	OK to marginal	OK	OK
Titanium	OK to marginal	OK	OK	OK	OK to marginal	OK	OK
Carbon graphite	OK	OK	OK	OK	OK	OK	OK

APPENDIX 8

Nomograph for Calculation of Cetane Improver Treatment from CI-0801 Additions

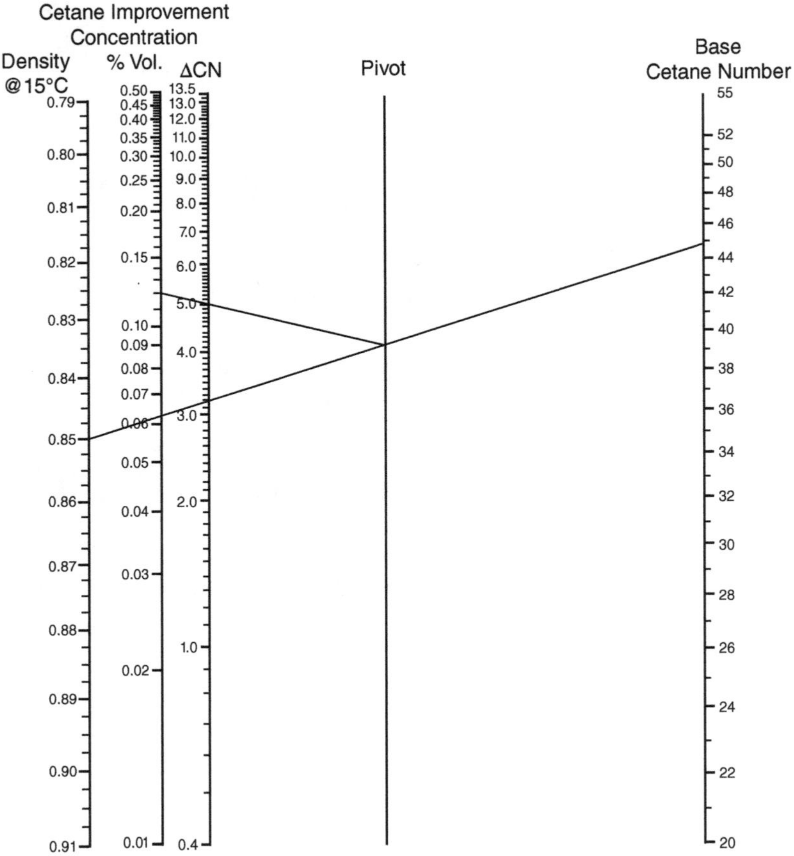

METHOD OF USE

1. Identify the base fuel cetane number and density on the approipriate axes and draw a line joining these two points.
2. Where this line intersects the pivot line, either:
   a) connect the pivot point to the cetane number increase (ΔCN) that is desired and extend the line to the CI-0801 axis to predict the additive concentration, or
   b) connect the pivot point to the chosen CI-0801 concentration and read back along the line to the predicted cetane number increase.

EXAMPLE

Base cetane number 45, density @ 15°C 0.85, desired cetane number increase 5.0. Therefore, the CI-0801 concentration required is 0.12% volume.

(Source: The Associated Octel Company, Ltd.)

APPENDIX 9

U.S. Energy Information, Annual Energy Review, September 2012

Estimated Number of Alternative-Fueled Vehicles in Use in the United States, by Fuel, 2000–2010

Year	Alternative Fuel							
	Liquefied Petroleum Gases (LPG)	Compressed Natural Gas (CNG)	Liquefied Natural Gas (LNG)	Methanol 85 Percent (M85)	Ethanol 85 Percent (E85)*	Electricity	Hydrogen	Total
1995	172,806	50,218	603	386	1,527	2,860	0	246,855
1996	175,585	60,144	663	172	4,536	3,280	0	265,006
1997	175,679	85,571	813	172	9,130	4,453	0	280,205
1998	177,183	78,782	1,172	200	12,788	5,243	0	295,030
1999	178,610	91,267	1,681	198	24,604	6,964	0	322,302
2000	181,994	100,750	2,090	10,426	87,570	11,830	0	394,664
2001	185,053	111,851	2,576	7,872	100,303	17,847	0	425,457
2002	187,680	120,839	2,708	5,873	120,951	33,047	0	471,098
2003	190,369	114,406	2,640	0	179,090	47,485	9	533,999
2004	183,864	118,532	2,717	0	211,800	49,536	43	565,492
2005	173,795	117,699	2,748	0	246,363	51,398	119	592,125
2006	164,846	116,131	2,798	0	297,099	53,526	159	634,562

2007	158,254	114,391	2,781	0	364,384	55,730	223	695,766
2008	151,049	113,973	3,101	0	450,327	56,901	313	775,667
2009	147,030	114,270	3,176	0	504,297	57,185	357	826,318
2010	143,037	115,863	3,354	0	618,505	57,462	421	938,643

Sources: 1992–1994 Science Applications International Corporation, "Alternative Transportation Fuels and vehicles data Development," unpublished final report prepared for the EIA, (McLean, VA, July 1996), and U.S. Department of Energy Efficiency and Renewable Energy Data contents, see EIA, Alternatives to Traditional Transportation Fuels: An Overview (June 1994), Table 22 – 1995–2002-EIA, "Alternatives to Traditional Transportation Fuels 2003 Estimated Data" (February 2004), Tables 1 and 10. Data were revised by using gross instead of net heat contents. – 2003 forward-EIA, Alternative-Fuel Vehicle Interactive Data Viewer (see http://www,eia,gov/renewable/afv/users. Cfm#abs_charts-2 and http://eia.gov/renewable/afv/xls/New%20GEGs.xls; and "Alternatives to Traditional Transportation fuels," annual reports, Table C1.

*Includes only those E85 vehicles to be used as alternative-fuel vehicles (AFVs), primarily fleet-operated vehicles; excludes other vehicles with E85-fueling capability. In 1997, some vehicle manufacturers began including E85-fueling capability in certain model lines of vehicles. For 2010, the U.S. Energy Information (EIA) estimates that the number of E85 vehicles that are capable of operating on E85, motor gasoline, or both, is about 10 million. Many of these AFVs are sold and used as traditional gasoline-powered vehicles.

APPENDIX 10

Rating Scale for the Waring Blender and Multi-Contact Emulsion Tests

Rating	Description of Emulsion
0	Clean break on the interface of oil and water
1	Very slight skin at the oil-water interface that usually does not break on tilting the bottle
2	Skin at oil-water interface, heavier than #1 rating and usually accompanied by dirt and bubbles on the skin. No evidence of any white emulsion
3	First sign of white emulsion. Usually forms at the bottom and in the center of the bottle. It is circular in shape and approximately ¼ to 1 inch in diameter
4	Approximately the same amount of emulsion on the bottom of the bottle as #3 rating. However, emulsion is also beginning to form at the oil-water interface and extends 1/32 to 1/16 inch downward into the water layer. Roughly 15% of water layer occupied by emulsion.
5	Circular emulsion at bottom extends outward and upward resembling spokes. Emulsion at the interface slightly thicker than #4 rating
6	More emulsion than #5 rating. Thin film of emulsion forming on sides of bottle surrounding the water layer. Water is still visible looking through the sides and looking up from the bottom of the bottle.
7	Emulsion on bottom of water layer is almost solid. Emulsion on sides of bottle is broken in a few spots enabling the operator to see the water layer.
8	Semi-solid emulsion with perforations or bubbles similar to honeycomb. No water visible except that seen in the bubbles.
9	Same emulsion as #8 rating but with fewer bubbles. 75–90% is solid.
10	Almost completely solid emulsion with only a few air bubbles visible
11	Completely solid emulsion (mayonnaise type)
	Fuel and Water Layer Ratings C = Clear H = Hazy

Useful Terms and Definitions

A variety of technical terms are used throughout this manuscript. Most of these terms are familiar and known by those who have worked in the petroleum industry for a few years. The purpose of providing this glossary is to focus on the terminology and definitions which are the most useful when describing fuel performance problems.

Absolute Pressure A pressure scale with the baseline zero point at perfect vacuum.

Accumulator A vessel containing fluid stored under pressure which acts to serve as a source of fluid power.

Actuator A mechanism used to convert hydraulic energy into mechanical energy.

Additive A chemical substance added to a product to impart or improve certain properties. Typical fuel additives include antioxidants, cetane improvers, corrosion inhibitors, demulsifiers, detergents, dyes, metal deactivators, octane improvers, and wax crystal modifiers.

Alkylation A refinery process for producing high-octane components consisting mainly of branched chain paraffins. The process involves combining light olefins with isoparaffins, usually butene and isobutane, in the presence of a strong acid catalyst such as hydrofluoric or sulfuric acid.

Aluminum Chloride Processing A refining method using aluminum chloride as a catalyst to improve the appearance and odor of steam cracked naphtha streams. Aluminum chloride functions as a catalyst for the polymerization of olefins into higher-molecular-weight, less-problematic compounds.

Anode The electrode of an electrolytic cell where oxidation occurs. Electrons move away from the anode in an external circuit. At the anode, corrosion occurs and metal ions are solubilized.

Antifoam Agent An additive, usually silicone- or polyglycol-based, used to reduce foaming in petroleum products.

API (American Petroleum Institute) A trade association of firms engaged in all aspects of the U.S. petroleum industry including producers, refiners, marketers, and transporters.

Aromatics Unsaturated hydrocarbons such as benzene, toluene, xylene, and related compounds. Aromatic compounds share electrons equally among carbon atoms.

Ash Content In the petroleum industry, this term usually refers to metals and metal salts contained in crude oil or a petroleum product. These noncombustible substances can form deposits and can impair engine efficiency and power.

Asphaltenes These are complex high-molecular-weight polycyclic aromatic compounds which may contain oxygen, sulfur, or nitrogen heteroatoms. They are found in crude oil and in certain heavy fuel oils in micellar form. Their dispersion throughout oil can be stabilized by asphaltene precursors called resins and maltenes.

Paraffins can agglomerate and crystallize onto asphaltenes. Under high-shear or high-temperature conditions these asphaltenes will tend to coalesce and deposit onto pump parts and other moving components. Also, if blended with other oils, asphaltenes may flocculate and become insoluble in the mixture.

Some of the fuel performance problems which can be associated with asphaltenes include:

- Pump deposit buildup
- Restricted flow in lines and piping
- Tank bottom sludge and deposit accumulation
- Stabilization of existing fuel-water emulsions

Autoignition The spontaneous ignition of a mixture of fuel and air without an ignition source. Diesel and jet engines combust fuel in this manner.

Ball-on-Three-Disks (BOTD) This test method is utilized to measure the lubricity characteristics of distillate fuel. It involves rotating a ball against three disks lubricated by distillate fuel. After testing, the wear scar diameters on three disks are measured and averaged. Wear scar diameters of approximately 0.45 mm or less are presently considered acceptable.

Bar A value equal to 0.987 atm or 29.53 in. Hg.

Barrel A unit of liquid volume utilized by the petroleum industry equal to 42 U.S. gallons or approximately 35 Canadian gallons.

Blow-by Combustion gases which are forced past the piston rings and into the crankcase of an internal combustion engine. Blow-by gases may contain NO_x, SO_x, CO, CO_2, water vapor, hydrocarbon gases, and soot. In combination with water, some of these gases can form corrosive acids.

Borderline Pumping Temperature The lowest temperature at which a lubricating oil can be efficiently pumped. It is also identified as the temperature at which a lubricant's viscosity equals 30,000 cP.

Bosch Number A measurement of diesel exhaust smoke color: clear = 0; black = 10.

Bottoms The high-boiling-point residual liquid from the refining process; also called residuum. Any residue remaining in a distillation unit after the highest-boiling-point material has been removed by distillation. The boiling range will depend upon the feedstock and the amount of material removed.

Brake-Specific Fuel Consumption (BSFC) This is a measure of the pounds of fuel consumed per brake horsepower for a one hour period.

Brake-Specific Horsepower (BSHP) This is a measure of the power output at the end of the crankshaft available for doing work. 1.36 bhp = 1.00 kW.

Bright Stock High-viscosity oil which is highly refined and dewaxed and produced from residual lubestock or bottoms. It is used in finished lubricant blends to provide good bearing film strength, prevent scuffing, and reduce oil consumption.

British Thermal Unit (BTU) A measure of the quantity of heat required to raise the temperature of 1 lb of water 1°F. For fuels such as gasoline, kerosene, fuel oil, and residual fuels, the following formulas can be used to determine the BTU/lb:

Sherman and Kropff Equation:

For gasoline:	18,320 + 40(°API – 10)
For kerosene:	18,440 + 40(°API – 10)
For fuel oil:	18,650 + 40(°API – 10)

Faragher, Morrell, and Essex Equation:

For heavy residual fuels:	17,645 + (54 × °API)
For Bunker "C" fuel:	17,685 + (57.9 × °API)

These equations can be used to calculate the gross heating value or higher heating value of fuel whereby water formed during combustion is condensed and combustion gases are cooled to the initial fuel-air temperature.

Brookfield Viscosity A measurement of the apparent viscosity of a fluid using a Brookfield viscometer. This viscometer is a variable-sheer, variable-rpm viscometer.

BS&W "Bottoms, Sediment, and Water." This centrifuge method is used to measure the approximate amount of suspended solids and water in crude oil and petroleum products.

Buna-N This nitrile rubber material has been used extensively in automatic transmissions. It is compatible with most petroleum products, can be formed into a variety of shapes, and is economical. It is useful at temperatures ranging from –40°F to 230°F (–40°C to 110°C). It has excellent resistance to swelling and softening at higher temperatures.

Camshaft A lobed shaft usually designed to rotate in relation to the speed of an engine. The lobes on this shaft are somewhat elliptical in shape. When rotating, the lobes act to push against a rod or some other mechanism to operate other parts of the valve train.

Carbon Black Oil Usually a viscous, highly aromatic residual oil utilized in the manufacture of carbon black. These oils may also contain polynuclear aromatic compounds, hydrogen sulfide, and ash.

Cathode The portion of a corrosion cell in which reduction occurs. This region is not visibly affected.

Cavitation A condition which exists when gas accumulates within a liquid stream being pumped. This area of low pressure can lead to pump wear and damage.

Centipoise A standard unit of viscosity equal to 0.01 poise (P). At 68°F (20°C), water has a viscosity of 1.002 cP or 0.01002 P.

Cetane Index A calculated value which incorporates fuel distillation properties and °API which can be used to approximate the cetane number of a fuel. This method, however, does not provide information on the performance provided by a cetane improver.

The cetane index of distillate fuel can be related to the aromatic content of the blend. As fuel aromatic content increases, the cetane index will typically decrease. A general relationship exists which relates a cetane index of 40 with a diesel fuel aromatic content of about 35%.

Cetane Number A fuel performance measurement utilizing a standard engine and standard reference fuels. The property of fuel ignition delay is determined by this engine. High-cetane-number fuels have a shorter ignition delay period. High-cetane-number fuels provide for easy starting, quiet running, and more complete combustion.

This number is determined from the percentage of *n*-cetane which must be mixed with heptamethylnonane to give the same ignition performance as the fuel being tested. The cetane number can be obtained by using the following calculation:

$$\text{Cetane number} = \%\ n\text{-cetane} + 0.15(\%\ \text{heptamethylnonane})$$

CFPP (Cold Filter Plugging Point) A measure of the ability of a distillate fuel to be filtered satisfactorily in cold environments. This test measures the temperature at which fuel wax crystals can reduce or halt the flow of fuel through a standardized test filter.

Channeling A term used to describe lubricants, especially gear oils, which fail to readily flow back together when a machine component passes through a volume of cold oil. When channeling occurs, an oil is unable to flow and lubricate.

Check Valve A valve which permits flow of fluid in only one direction. Often a check valve can be nothing more than a spring, ball, and seat separating two openings. Common valve types include in-line check valves and right-angle check valves.

Claus Sulfur Recovery Process The Claus process is a controlled combustion process commonly used for the recovery of sulfur from H_2S. Temperatures >2,000°F (1,093.3°C) are achieved during combustion and yields of about 95% are typical. The basic reaction involves the following:

$$2H_2S + 3O_2 \rightarrow 2SO_2 + 2H_2O$$

$$2SO_2 + 4H_2S \rightarrow S_6 + 4H_2O$$

Clay Treating A process used to improve the color of cracked naphthas and light distillates. It is also used to remove surface active agents which can negatively impact the WSIM rating of jet fuel.

Cloud Point This is the temperature at which a "cloud" or haze of wax crystals appears when a fuel or lubricant is cooled under standard test conditions.

Coefficient of Friction The coefficient of friction between two surfaces is the ratio of the force required to move one surface over the other to the total force pressing the two together.

Coking A refinery process in which fuel oil is converted to lower-boiling-point liquids and coke by a thermal cracking process.

Cold Cranking Simulator (CCS) An intermediate shear rate viscometer that predicts the ability of an oil to permit a satisfactory cranking speed to be developed in a cold engine.

Compression Ignition A fuel combustion process whereby heat generated by the rapid compression of air is utilized as an autoignition source for fuel. This form of fuel combustion is utilized in the diesel engine. No spark plug is required to ignite the fuel.

Compression Ratio In an internal combustion engine it is the ratio of the volume of the combustion space at the bottom of the piston stroke (bottom dead center) to that at the top of the piston stroke (top dead center). High compression ratios usually improve fuel economy but increase NO_x emissions. Compression ratio values for spark-ignition engines typically range from 7:1 to 9:1 to as high as 12:1 for higher-rpm engines. For diesel engines, compression ratios typically range from 14:1 to 18:1 to as high as 24:1 for lighter, higher-rpm engines.

Conductivity The property of a substance which allows the passage of electrical charge. Fuels typically are poor conductors of electrical charge and enable static charge to build. Often, conductivity aids are added to fuel to permit the transfer and release of static charge.

Conjugation In fuel technology, this term typically refers to olefins containing double bonds which exist between alternating carbon atoms. Conjugated olefins are quite unstable and lead to fuel degradation and deposits.

$$CH_2{=}CH{-}CH{=}CH{-}CH{=}CH{-}R$$

A conjugated olefin

Conradson Carbon Number ASTM D-189 Determination of the weight of nonvolatile residue formed after evaporation and atmospheric pyrolysis of fuel or oil. This test method provides some information about the relative coke-forming or deposit-forming tendency of a fuel or oil. Products having a high ash value will have an erroneously high carbon residue value.

Continuous Spray Pump Fuel injection pump which provides a constant rather than an intermittent spray; normally for fuel injection in spark ignition engines.

Cracking A refining process in which high-molecular-weight liquid hydrocarbons are broken down into lighter, lower-boiling-point liquids or gases. This process can be accomplished thermally or catalytically.

Crankcase Dilution Dilution of lubricating oil in the oil pan or sump of an engine by fuel which has entered the crankcase.

Degree-Days The difference between the daily mean temperature and 65°F (18.3°C) when the temperature is below 65°F (18.3°C) for a given day. Degree-day fuel oil delivery systems are based on the principle that the consumption of fuel oil for heating is roughly proportionate to the number of degree-days. The number of degree-days for a given period is arrived at by adding together the number of degree-days for individual days in the period.

Demulsibility The ability of a fuel or oil to separate from water.

Detergent In relationship to fuel technology, a detergent is an oil-soluble surfactant added to fuel aiding in the prevention and removal of deposits. Examples include **anionic** alkyl aryl sulfonates, **cationic** fatty acid amides, or **nonionic** polyol condensates.

Dezincification Some brass alloys are susceptible to pitting corrosion or loss of zinc from the metal matrix. This type of corrosion usually occurs when metal is in contact with high percentages of oxygen and carbon dioxide.

Dimer Acid Produced by the reaction and combination of two unsaturated fatty acids at mid-molecule to form a single molecule.

Direct Injection A type of fuel injection process used in larger commercial and stationary diesel engines. In this process, fuel is injected at high pressure, through a multiple-hole nozzle directly into the combustion chamber. Combustion chamber design, injection pressure, and air flow are optimized to maximize combustion and fuel efficiency. Direct injected engines are typically more fuel efficient than indirect injected engines, but usually generate more noise.

Direct Injection Combustion System

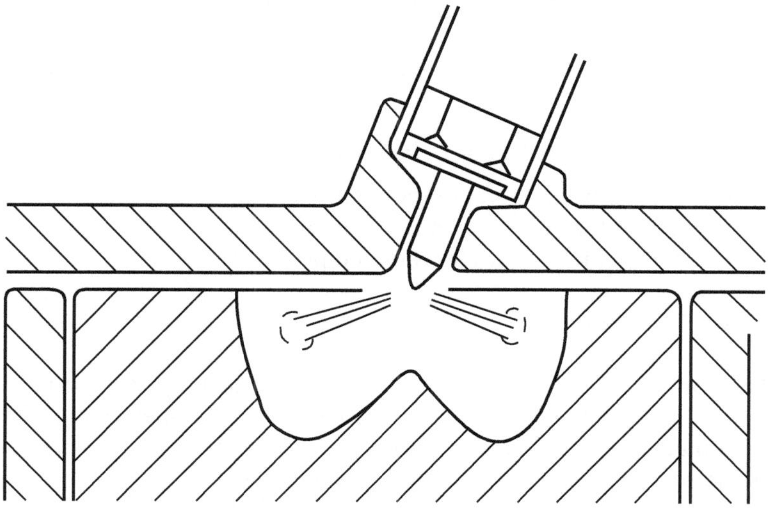

(From Owen, Keith, Editor. 1989. Gasoline and Diesel *Fuel Additives*. Chichester: John Wiley & Sons.)

Displacement The volume of fluid discharged by a pump per revolution or cycle minus loss due to leakage. Terms gal/min or L/min are commonly used to describe the **delivery** of fluid by the pump.To calculate delivery, the following conversion factors can be used:

$$\text{Delivery (gal/min)} = \frac{\text{displacement (in}^3) \times \text{RPM}}{\text{number of revolutions}} \div 231$$

$$\text{Delivery (L/min)} = \frac{\text{displacement (mL)} \times \text{RPM}}{\text{number of revolutions}} \div 1000$$

Doctor Test A method which can be used to determine the **sweet** or **sour** nature of gasoline by treating gasoline with a solution of lead plumbite and sulfur. As an example, the reaction which occurs with methylmercaptan is as follows:

$$2CH_3SH + Na_2PbO_2 \rightarrow CH_3\text{-}S\text{-}Pb\text{-}S\text{-}CH_3 + 2NaOH$$

The addition of sulfur results in the formation of black lead sulfide:

$$(CH_3\text{-}S)_2Pb + S \rightarrow CH_3\text{-}S\text{-}S\text{-}CH_3 + PbS$$

and

$$2CH_3SH + Na_2PbO_2 + S \rightarrow CH_3\text{-}S\text{-}S\text{-}CH_3 + PbS + 2NaOH$$

When used in fuel processing, the PbS and NaOH can be reacted with air and heat. This process results in reformation of the original Doctor Solution:

$$PbS + 4NaOH + 2O_2 \rightarrow Na_2PbO_2 + Na_2SO_4 + 2H_2O$$

Drivability Index This value is calculated by utilizing parameters from the ASTM D-86 distillation profile. The purpose of this measurement is to be able to predict the cold-start drivability of gasoline. It is calculated as follows:

$$DI = 1.5(T\text{-}10) + 3.0(T\text{-}50) + (T\text{-}90)$$

where, T-10 is the 10% evaporation temperature
T-50 is the 50% evaporation temperature
T-90 is the 90% evaporation temperature

Lower drivability index values reflect better cold-start drivability.

Emulsion A dispersed, two-phase system in which one phase is usually water and the other oil. An emulsion in which oil is dispersed in water is termed an oil-in-water emulsion. An emulsion in which water is dispersed in oil is termed a water-in-oil emulsion.

In a dispersed system, it is possible to have both phases in existence at the same time. However, whenever fuels emulsify with water, water-in-oil emulsion typically forms. Agents which comprise the external phase of an emulsion are usually the most soluble in the bulk liquid in which the emulsion exists.

Endothermic Reaction A reaction in which heat is absorbed. To maintain a constant temperature during reaction, heat must be added to reactants and products.

Exhaust Gas Recirculation (EGR) The mixing of exhaust gas with intake air used in fuel combustion. In the diesel engine, mixing of exhaust gas helps in reducing the NO_x emissions. This process lowers the combustion temperature and oxygen concentration, thus lowering the total NO_x. However, excessive EGR leads to the formation of increasing amounts of CO, soot, and hydrocarbon emissions.

Exothermic Reaction A reaction which evolves heat. Refining processes such as alkylation, hydrogenation, and polymerization are exothermic.

Ferric Hydroxide A fragile, reddish-brown product of iron or steel corrosion, $Fe(OH)_3$.

Ferrous Hydroxide A product of iron corrosion which is white in appearance, $Fe(OH)_2$. When ferric ions enter, the product will appear black, brown, or green in color.

Ferrous Oxide (Hydrous) A hydrated, jet black product of iron corrosion, $FeO \cdot nH_2O$.

Fifteen/Five (15/5) Distillation A laboratory distillation which is also referred to as the "true boiling point" distillation. It is performed on a column containing 15 theoretical plates at a reflux ratio of 5:1.

Fines Typically broken or crushed catalyst support media composed of alumina and silica. Smaller fines are approximately 20 microns in diameter.

Flame Front The region between the burning and unburned zone present in gas phase combustion processes or the leading edge of burning fuel within the combustion chamber. In an internal combustion engine, movement of the flame front within the confines of the combustion chamber can influence overall quality of the combustion process.

Good combustion is characterized by a flame burning uniformly across the piston head. Deposits and poor-quality fuel can interfere with the movement of the flame across the combustion chamber. Rough running can result.

Flash Point The lowest temperature at which the vapor above a liquid will ignite when exposed to a flame. The flash point is an important indicator of the fire and explosion hazards associated with a petroleum product.

Floating Roof A special type of storage tank roof which floats on the surface of the oil in storage. This helps to eliminate tank venting and reduces evaporative loss.

Flooding A condition that prevents starting of an engine when more fuel is drawn in than can be ignited.

Fluidizer Oil Compounds such as polyisobutylene or refined naphthenic oils which aid in the removal of deposits from the underside of intake valves.

Fuel Oil A general term applied to petroleum products used for the production of power or heat with a flash point >100°F (37.8°C) . Fuel oils are typically classified as either engine fuels or burner fuels.

A classification system has been developed to describe various fuel oil types. For example #1 fuel oil is similar to kerosene, #2 fuel oil is similar to diesel fuel, and #4 fuel oil is viscous oil at room temperature and is typically used to fuel industrial furnaces. The #5 and #6 fuel oil classifications describe viscous oils which must be heated before burning. These oils are used as bunker fuels in ships and industrial power plants.

Fuel Oil Equivalent (FOE) The heating value of a standard barrel of fuel oil, equal to 6.05×10^6 BTU.

Fungible Fuel which is available from different sources but can be freely exchanged and used as a single source fuel. No interactions will occur between the fuels.

Furfural A colorless liquid which changes to reddish brown upon exposure to light and air. Furfural forms condensation products with many types of compounds including phenols and amines. If present in kerosene or jet fuel, furfural may lead to color degradation.

Galvanic Cell A cell in which a chemical change is the source of electrical energy. The cell is established when two dissimilar metals are in contact or two similar metals in different electrolytic solutions are in contact.

Gas Oil A refinery fraction boiling within a typical temperature range between 330°F and 750°F (165.6°C and 398.9°C). Diesel fuel, kerosene, and heating oil fall into this fraction.

Glow Plug A device used to provide heat to the prechamber of an indirect injected diesel engine. It is used in cold conditions to permit ease of starting. The glow plugs have a preheating time of about seven seconds.

Gum A term used to describe dark-colored deposits which form upon the oxidation of compounds present in fuel. This term has also been used to describe the products of poor combustion deposited within engine cylinders.

Heavy Fuel Oil A residue of crude oil refining. The product remaining after gasoline, kerosene, diesel fuel, lubricating oil, and other distillates have been removed.

Heterocycle A compound which contains a closed ring system in which the atoms are of more than one kind. Sulfur, nitrogen, and oxygen are common heteroatoms found in petroleum heterocycles.

Horsepower The power required to lift 550 pounds 1 ft in height in one second or 33,000 pounds 1 foot in one minute. One horsepower is equal to 746 W or 42.4 BTU/min.

Hydrodynamics The science which deals with liquids in motion and the associated kinetic energy.

Hydrophilic A substance which has an affinity for water. In fuel technology, certain polar organic compounds such as alcohols, organic acids, and amines are hydrophilic.

Hydrophobic A substance which lacks affinity for water. In fuel technology, hydrocarbon species are hydrophobic.

Indirect Injection A type of diesel fuel injection process used primarily in smaller engines to achieve good mixing of injected fuel into air. Fuel is not injected directly into the combustion chamber, but into a smaller, prechamber which feeds into the primary combustion chamber. When compressed air enters the prechamber, fuel is sprayed from the fuel injector at a relatively low pressure to begin the combustion process. The pressure of the burning fuel forces the mixture into the primary combustion chamber where thorough mixing with air occurs. Completion of the combustion process takes place at a lower pressure and at a relatively lower noise level than most direct injected engines. Indirect injected engines are often fitted with a glow plug to aid in cold starting of the engine.

Indirect Injection Combustion System

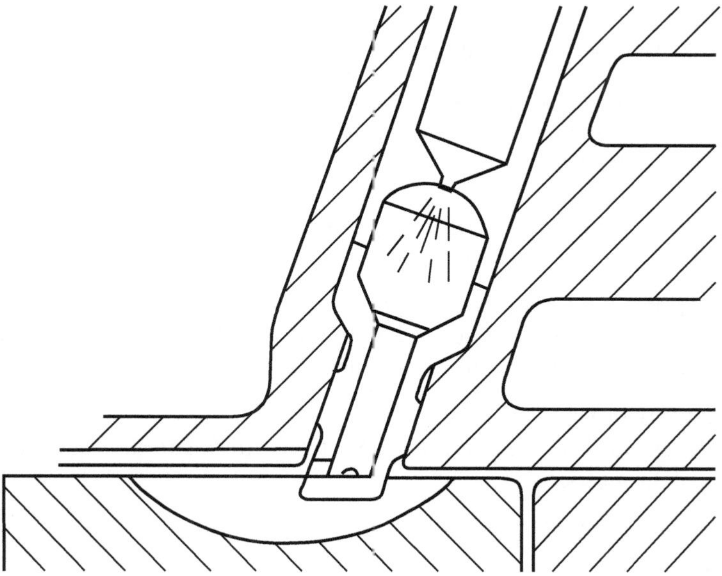

(From Owen, Keith, Editor. 1989. *Gasoline and Diesel Fuel Additives*. Chichester: John Wiley & Sons.)

Induction Time The time period under given test conditions in which a petroleum product does not absorb oxygen at a substantial rate to form gum. It is measured as the time elapsed between the placing of a test bomb containing 50 mL of fuel and oxygen at 100 to 102 psi into a 212°F (100°C) test well and the point in the time-pressure curve preceded by two consecutive 15-minute periods of a pressure drop not less than 2 psi.

Isomerization A refinery process used to produce branched compounds from straight-chain molecules. Butane, pentane, and hexane are isomerized to branched, higher-octane compounds through isomerization. Isomerization reactions are catalyzed through the use of aluminum chloride and other metal catalysts.

JFTOT An acronym for Jet Fuel Thermal Oxidation Test. It is a measurement by ASTM D-3241 to determine the tendency of jet fuel to form deposits when heated.

Kauri-Butanol (KB) Value A measure of the relative solvent power of a hydrocarbon; the higher the KB value, the greater the effect of the hydrocarbon as a solvent.

Kinematic Viscosity A coefficient defined as the ratio of the dynamic viscosity of a fluid to its density. The **centistoke** is the reported value of kinematic viscosity measurement.

Laminar Flow A condition where fluid mass moves in a continuous, streamlined parallel path.

Light-ends A term used to describe the low-boiling-point fractions in a fuel or refined petroleum fraction.

Liquefied Petroleum Gas (LPG) Compressed and liquefied light-end petroleum fractions. LPG is primarily propane with low concentrations of ethane and butane.

Manifold A common conducting unit for fluids and gases into which multiple ports are connected.

Mercaptan A thiol compound containing the structure R-SH where R is carbon. These compounds are liquids and possess strong, unpleasant odors. They are quite common in crude oil and can be found in finished fuels.

Merox Sweetening A process used in fuel refining to remove primarily hydrogen sulfide and mercaptans from fuel. This is a patented, proprietary process.

Meter-in To regulate the amount of flow into a system.

Meter-out To regulate the amount of flow out of a system.

mpy Weight Loss An acronym for mils per year weight loss. This rating is used to describe the loss in the thickness of a metal surface often due to corrosion.

Naphtha A term used to describe a light crude oil distillate fraction boiling in the gasoline range. The typical boiling range is from about 100°F to 420°F (37.8°C to 215.6°C). Naphthas can be subdivided into the following fractions:

IBP – 160°F (71.1°C)		Light naphtha
160°F (71.1°C) – 280°F (137.8°C)		Intermediate naphtha
280°F (137.8°C) – 330°F (165.6°C)		Heavy naphtha
330°F (165.6°C) – 420°F (215.6°C)		High-boiling-point naphtha

Naphthenes Saturated hydrocarbons found in fuels which contain at least one closed ring system.

Naphthenic Acids Acids derived from crude oil which are usually monocarboxylic, monocyclic, and completely saturated. Many are derivatives of cyclopentane and more complex alicyclic ring systems.

These acids can also be produced by oxidation of cycloparaffins during distillation or other refining processes. The boiling range of these compounds is usually between about 400°F to 575°F (204.4°C to 301.7°C) with molecular weights ranging from 180 to 350. They are frequently removed from petroleum product by caustic washing. The sodium salts of these acids are water soluble and act as detergents and emulsifying agents.

Natural Gasoline Hydrocarbons condensed from natural gas which consist primarily of pentanes and heavier-molecular-weight compounds.

Neoprene Neoprene, an elastomer, is compatible with most petroleum products and is used in the manufacture of seals and gaskets. However, above temperatures of 150°F (66°C), neoprene becomes unstable and less useful.

Neutralization Number The amount of acid or base required to neutralize all components present.

Newtonian Fluid A fluid with a constant viscosity at a given temperature regardless of the shear rate.

Octane A hydrocarbon with the formula C_8H_{18} having 18 different isomers. These isomers have typical boiling points between 210.2°F and 257°F (99°C and 125°C). The most important of these isomers is 2,2,4-trimethyl pentane or isooctane. It is a colorless liquid with a boiling point of 210.2°F (99°C) and is used as a standard for rating the antiknock properties of gasoline.

O-ring An O-ring is a seal or gasket formed in the shape of a circle or letter "O." It can be made from a variety of elastomeric materials and can be used to provide a very effective seal in high-pressure static and dynamic systems. It is "squeezed" between two opposing grooved fittings or mating parts. Also, O-rings are forced against the walls of the seal by internal pressure making for a very tight seal.

Oxidation Loss of electrons from a chemical species. Any process which increases the proportion of electronegative constituents in a compound such as increasing the proportion of oxygen in organic compounds.

Passivation Changing of the chemically active surface of a metal to a much less active state.

Performance Number Primarily used to rate the antiknock values of aviation gasolines with octane numbers over 100. This value is expressed as the maximum knock-free power output obtained from fuel expressed as a percentage of the power obtainable from isooctane. The relationship between octane number and performance number is listed as follows:

$$\text{Octane number} = 100 + \frac{\text{performance number} - 100}{3}$$

Pilot Pressure Auxiliary pressure used to actuate or control hydraulic system components.

Pipestill A furnace containing a series of pipes through which oil is pumped and heated. During heating, the oil is vaporized prior to introduction into the distillation unit or thermal cracking unit.

Polymerization The process of producing higher-molecular-weight compounds by reacting two or more unsaturated molecules together. In the refinery, propylene and butylene are the primary polymerization feedstocks.

Poppet A part of certain valves which prevents flow when the valve closes against a seat.

Positive Displacement A characteristic of a pump or motor which has the inlet port sealed from the outlet so that fluid cannot recirculate within the system.

Potentiometer A component of a servo system which measures and controls electrical potential.

Power The time rate at which work is done.

Pressure Force applied to, or distributed over, a surface and measured as force per unit area.

Absolute pressure. Pressure measured with respect to zero pressure.

Gauge pressure. Pressure measured with respect to that of the atmosphere.

Pyrrole A nitrogen-containing, heterocyclic compound found at low concentrations in crude. Pyrrole can readily polymerize in the presence of light to form dark-brown compounds. Pyrrole derivatives, such as indole, have been identified as sources which can lead to darkening of fuel color and eventual deposit formation.

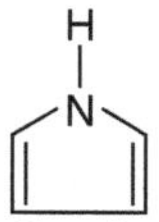

Pyrrole

Raffinate A term used to describe the material recovered from an extraction process. Examples of refinery raffinates include lubestock recovered from furfural or methylpyrrolidone extraction and kerosene recovered from SO_2 extraction. Furfural, methylpyrrolidone, and SO_2 extraction processes are employed to remove aromatic compounds.

Range Oil A term sometimes used to describe kerosene or #1 fuel oil.

Rankine Temperature Scale A temperature scale with the size of degree equal to that of the Fahrenheit scale and zero at absolute zero. Therefore 0°R = −459°F (−273.2°C) and the normal boiling point of water is 671.67°R.

Red Dye These dyes are dilutions of C.I. Solvent Red 164 utilized throughout the petroleum industry to identify specific fuels and oils. The dyes are termed "2-Naphthalenol (phenylazo) azo alkyl derivatives."

Reduced Crude This term is used to describe the residual material remaining after crude oil has been distilled to remove lighter, more volatile fractions.

Refining Catalyst Most catalysts used in the refining industry are solid heterogeneous catalysts. The chemical reactions they enhance would not proceed at all, or would do so quite slowly, in the absence of the catalyst. Reactions of this nature are believed to occur due to a dramatic disruption of the existing chemical bonds of an absorbed molecule. Molecules or molecule fragments may enter into reactions much different from those which occur in uncatalyzed reactions.

Examples of heterogeneous catalysts used in the refining industry are shown in the following table:

Catalyst	Use in Refining
Silica alumina gel	Cracking of heavy petroleum fractions
Chromic oxide gel, chromium on aluminum, nickel-aluminum oxide	Hydrogenation; dehydrogenation of hydrocarbons
Phosphoric acid on kieselguhr	Polymerization of olefins
Co, ThO_2, MgO on kieselguhr (Fischer-Tropsch catalyst)	Synthesis of hydrocarbons from H_2 and CO
Platinum	Isomerization of hydrocarbons
$Al(C_2H_5)_3$; $TiCl_4$	Polymerization of olefins

Note: Kieselguhr is a fine, variously colored earth derived from the accumulated deposits of cell walls of diatoms.

Reforming Both thermal and catalytic processes are utilized to convert naphtha fractions into high-octane aromatic compounds. Thermal reforming is utilized to convert heavy naphthas into gasoline-quality aromatics. Catalytic reforming is utilized to convert straight-run naphtha fractions into aromatics. Catalysts utilized include oxides of aluminum, chromium, cobalt, and molybdenum as well as platinum-based catalysts.

Relief Valve A relief valve is used in pumps, reactors, and equipment which create positive pressure. These valves are usually connected between the pump outlet and a vent or a large-volume accumulation vessel. The valve is normally closed, but will open whenever the system pressure exceeds the desired maximum operating pressure.

Rotary Actuator A device for converting hydraulic energy into rotary motion.

Ryton A polyphenylene sulfide elastomer.

Sensitivity The difference between the research octane number (RON) and the motor octane number (MON) of a gasoline. Fuels blended with alkylate material have low sensitivity values.

Sequence Valve A sequence valve will permit flow to a secondary system only after an action has been completed in a primary system.

Servo Mechanism A mechanism which responds to the action of a controlling device and operates as if it were directly actuated by the controlling device. However, it is capable of supplying power output much greater than the original controlling device. The power is derived from an external and independent source. Both mechanical and electrical servo units exist.

Silicone Seal This elastomer is one of the original materials used as sealing material. It has a wider useful temperature range than Buna-N, but unlike Buna-N, it is not tough and cannot be used in reciprocating systems. Silicone seals tend to swell and absorb oil when operating at high temperatures. Silicone seals are useful in applications where temperatures range from –60°F to 400°F (–51°C to 204°C).

Smoke Point A measurement of the burning quality of jet fuel, kerosene, and lamp oils. This value is determined by ASTM D-1322.

Space Velocity The volume or weight of a gas or liquid which flows through a catalyst bed or reaction zone per unit time divided by the volume or weight of catalyst. The terms **liquid hourly space velocity** (LHSV), and **weight hourly space velocity** (WHSV), are used to describe this type of flow measurement. High values correspond to short reaction times.

Specific Gravity Defined as the ratio of the mass of a given volume of liquid at 60°F (15°C) to the mass of an equal volume of pure water at the same temperature. Typical specific-gravity values for some common compounds are listed below:

Compound	Specific Gravity
Water	1.0
Gasoline	0.730 – 0.760
Jet fuel (aviation turbine kerosene)	0.755 – 0.840
#1 Diesel fuel and fuel oil	0.790 – 0.820
#2 Diesel fuel and fuel oil	0.830 – 0.860
#6 Fuel oil and residual fuels	0.890 – 0.930

Specific Heat Three different values for specific heat can be determined and are described as the **specific heat capacity, specific heat of fusion,** and **specific heat of evaporation**. These values are described as follows:

Specific heat capacity. Defined as the quantity of heat in joules required to raise the temperature of 1 kg of a substance by 1°K. For gases, it is necessary to differentiate between specific heat capacity at constant pressure and constant volume.

Specific heat of fusion. Defined as the quantity of heat in joules required to transform 1 kg of a substance at fusion temperature from the solid to the liquid state.

Specific heat of evaporation. Defined as the quantity of heat in joules required to evaporate 1 kg of a liquid at boiling temperature. The specific heat of evaporation is dependent upon pressure.

Petroleum Product	Specific Heat Capacity @ 68°F (20°C), kJ/kg·K
Gasoline	2.02
Diesel fuel	2.05
Heating oil	2.07
Kerosene	2.16
Lubricating oil	2.09
Ethanol	2.43
Water	4.18

Spool A term applied to most any moving cylindrical part of a hydraulic component which directs flow through the system.

Static Dissipater During transport and flow, static charge can build up in fuel and lead to electrical discharge which can ignite fuel/air vapors. This phenomenon can occur in low-conductivity fuels such as jet fuel and low sulfur diesel fuel when pumped and filtered at high flow rates. Polar additives known as static dissipaters or electrical conductivity improvers are added to these fuels to promote the transfer of static charge through the fuel to a ground source.

Straight-Run Gasoline This stream is the gasoline fraction which is collected from the distillation tower. It is primarily paraffinic in nature and typically has a lower octane number rating than cracked gasoline from the same crude oil feed.

Stripping The separation of the more-volatile components of a liquid mixture from the less-volatile components. The less-volatile components are collected in a relatively pure state, but the more-volatile components collected may not be as pure. Stripping may be performed through fractional distillation or by contacting a liquid mixture with a vapor or gas into which the lighter material will diffuse.

Surface Viscosity Molecular films on liquid surfaces may be either readily mobile or slow to flow under the action of a two-dimensional stress. Some surface films show surface plasticity and behave as solids until a critical stress is applied. These films exhibit surface viscosity.

Surfactants Soluble compounds which possess groups of opposite polarity and solubilizing tendencies. They can form oriented monolayers at phase interfaces, form micelles, and possess detergency, foaming, wetting, emulsifying, and dispersing properties.

Tall Oil Fatty acids and some unsaponifiable material obtained as a major by-product from the sulfite-pulping process of wood.

Theoretical Plate In a distillation column, it is a plate onto which perfect liquid-vapor contact occurs so that the two streams leaving are in equilibrium. It is used to measure and rate the efficiency of a column at separating compounds. The ratio of the number of theoretical plates to the actual number of plates required to perform a separation is used to rate the efficiency of a distillation column. Actual separation trays in refinery distillation units are usually less effective than theoretical plates.

Thermal Expansion Coefficient for Liquids The coefficient of volume expansion for liquids is the ratio of the change in volume per degree to the volume at –17.8°F (0°C). The value of the coefficient varies with temperature.

Three-Way Catalytic Converter A container in line within the gasoline engine exhaust system which contains platinum, palladium, and/or rhodium catalysts. These catalysts convert CO, NO_x, and unburned fuel to CO_2, H_2O, and N_2.

Topping A process of removing light products from crude oil by distillation. Heavy products remain in the column.

Turbocharging This process utilizes the energy supplied by the combustion exhaust gases to power a turbine which compresses the intake air. The compressed air is more dense and, consequently, higher in oxygen content. When delivered to the combustion chamber, compressed air improves fuel economy and engine performance.

Turbine A rotary device that is actuated by the impact of a moving fluid or gas against blades or vanes.

Turbulent Flow A condition where the fluid mass moves in random paths rather than in continuous parallel paths.

Turpentine Light, volatile oils obtained as exudates or from the distillation of rosin oils. It is a mixture of cyclic terpenes composed mostly of α-pinene and has a typical boiling point of about 302°F (150°C).

Two-Stroke Cycle Engine An engine which produces one power stroke for each crankshaft revolution. Fuel combustion proceeds as follows:

1. Air Intake. With air transfer and exhaust ports open, air under slight pressure in the crankcase flows into the engine cylinder.
2. Compression. The rising piston covers ports, compresses air in the cylinder and creates suction in the crankcase. Fuel injection and combustion is initiated.
3. Expansion. Burning fuel expands, pushing the piston down. Air flows into the crankcase to be compressed as the piston descends.

4. Exhaust. The descending piston uncovers the exhaust port. A slight pressure builds up in the crankcase enough to move air into the cylinders.

Vapor Pressure The pressure created in a closed vessel when a solid or liquid is in equilibrium with its own vapor. The vapor pressure is a function of the substance and the temperature.

Volatility Factor A rating of gasoline quality under environmental and operating conditions. It is a function of the Reid vapor pressure, the percent distilled at 158°F (70°C), and the percent distilled at 212°F (100°C). It can be used to help predict the vapor locking tendency of gasoline.

Water White A description of the color of petroleum products which is equivalent to a rating of +21 or lighter in the Saybolt color scale. Water-white fuel components appear as clear liquids to the eye.

Wax Tailings A pitchlike substance which is the last volatile product distilling from vacuum charge prior to coking.

Weathered Crude A term applied to crude oil which has lost an appreciable amount of volatile components due to evaporation and other conditions of storage and handling.

Wetting Agent Surface active compounds such as detergents usually consisting of molecules possessing an oil-attracting hydrophobic group and a water-attracting hydrophilic group such as a carboxylic or sulfonic acid group. These compounds act to decrease the contact angle between water and another surface, thereby lowering the surface tension to permit the wetting of the surface.

Wick Char The weight of deposit which forms on a wick after burning kerosene or lamp oil under specified test conditions. It is a rating of burning quality.

Wiese Formula An empirical formula developed by General Motors Research for expressing motor fuel antiknock values above 100. This value is obtained by measuring the number of milliliters of tetraethyl lead added per gallon to achieve the same knocking tendency as the fuel being tested. The determined value is then expressed as a **performance number**.

Zeolites Aluminosilicates containing a $(SiAl)_nO_{2n}$ matrix with a negative charge which is balanced by the presence of cations in the cavities. The cations are easily exchanged, and water and gases can be selectively absorbed into the cavities. Zeolites are also used to remove molecules of specific sizes by absorption into the pores.

Useful Calculations, Conversions, and Equations

Determination of specific gravity from °API gravity

$$\text{Specific gravity} = \frac{141.5}{131.5 + \text{API @ } 60°\text{F}}$$

Approximation of parts per million (ppm) value from lb/1000 barrel (bbl) pounds per thousand barrels (ptb) value

For gasoline	1 ptb = 4.0 ppm
For kerosene/diesel	1 ptb = 3.3 ppm
For residual fuel	1 ptb = 3.0 ppm

Conversion Factors

1 cu ft = 28.317 L
1 cu ft = 7.4805 gal
1 gallon = 3785.4 mL
0°F = 459.67°R
Density of water @ 60°F = 8.337 lb/gal

1 kg·cal = 1.8 BTU/lb

Calculated Values

Density of liquid @ 60°F and atmospheric pressure

1 lb/gal = specific gravity × 8.3372 lb/gal

Density of gas @ 60°F and atmospheric pressure

$$\text{Specific gravity} = \frac{\text{molecular weight}}{28.964}$$

$$\text{Cubic foot vapor/gallon liquid} = \frac{\text{lb/gal} \times 379.49}{\text{molecular weight}}$$

References

Chapter 1

Kepner, C. H. and B. B. Tregoe. 1981. *The New Rational Manager*. Princeton, NJ: Princeton Research Press.

Chapter 2

Barnett, J. W. 1996. New salt product dries distillate streams, eliminates haze. *Oil & Gas Journal*, Vol. 94, Issue 16.

Giles, H. N. 2010. Crude oils. In: *Significance of Tests for Petroleum Products*. ed. S. J. Rand, pp. 106–122. West Conshohocken, PA: ASTM International.

Hengstebeck, R. J. 1959. *Petroleum Processing: Principles and Applications*. New York: McGraw-Hill.

McPherson, L. J. and M. F. Olive. 1992. Alkylation, isomerization, polymerization, hydrotreatment and sulfur production. In: *Modern Petroleum Technology*, ed. G. D. Hobson, pp. 517–575, London: Institute of Petroleum.

McPherson, L. J. and M. F. Olive. 1992. Cracking and reforming. In: *Modern Petroleum Technology*, ed. G. D. Hobson, pp. 517–575, London: Institute of Petroleum.

Nalco Chemical Company. 1994. *Refining Technology Manual*. Sugar Land, TX.

Scott, L. 1996. CG/MS analysis of polar compounds. *Report No. 965391*. Houston, TX: Core Laboratories.

Suchanek, A. J. 1990. Catalytic routes to low-aromatics diesel look promising. *Oil & Gas Journal*, May 7, pp. 109–119.

Chapter 3

Allenson, S. J. 1996. (*Internal report*) *Asphaltenes*. Sugar Land, TX: Nalco/Exxon Energy Chemicals, L.P.

American Society for Testing and Materials. 1996. *Marine Fuels: Specifications, Testing, Purchase and Use*. May 19–21, New Orleans, LA.

Banks, R. E. and P. J. King. 1992. Chemistry and physics of petroleum. In: *Modern Petroleum Technology*, ed. G. D. Hobson, pp. 517–575, London: Institute of Petroleum.

Bluth, M. J. 1996. (*Internal report*) *Carbon Number Characterization of Atmospheric Tower Residual*. Sugar Land, TX: Nalco/Exxon Energy Chemicals, L.P.

Boren, D. S. 1994. (*Internal report*) *Diesel Fuel Cetane Number Testing*. Sugar Land, TX: Nalco/Exxon Energy Chemicals, L.P.

Chevron Research and Technology Company. 1990. *Motor Gasolines*. Richmond, CA: Chevron.

Gibbs, L. M., B. R. Bonazza, and R. L. Furey. 2010. Automotive spark-ignition fuels. In: *Significance of Tests for Petroleum Products*. ed. S. J. Rand, pp. 3–15. West Conshohocken, PA: ASTM International.

Griffith, M. G. and C. W. Sigmund. 1985. Controlling compatibility of residual fuel oils. In: *Marine Fuels*, ed. C. H. Jones, pp. 227–247, Philadelphia, PA: American Society for Testing and Materials.

Kueter, K. E. 1985. *Asphaltene Precipitation: Cause and Effects within Oxy-Peru Jungle Operation*. Bakersfield, CA: Occidental Oil and Gas Corporation.
Martin, C. J. and L. Hicks. 2010. Burner, heating, and lighting fuels. In: *Significance of Tests for Petroleum Products*. ed. S. J. Rand, pp. 65–79. West Conshohocken, PA: ASTM International.
Nalco Company. 2003. Magnesium carboxylate products, properties, and uses. *Technical Bulletin*. Naperville, IL.
United States Department of Defense. 2014. Detail specification, naval distillate. MIL-DTL-16884N. Washington.
Waite, R. 1989. Filtration equipment. In: *Manual of Aviation Fuel Quality Control Procedures*, ed. R. Waite, pp. 56–79, Philadelphia, PA: American Society for Testing and Materials.
Westbrook, S. R. and R. T. LeCren. 2010. Fuels for land and marine diesel engines and for nonaviation gas turbines. In: *Significance of Tests for Petroleum Products*. ed. S. J. Rand, pp. 33–52. West Conshohocken, PA: ASTM International.
Winkler, M. F. 1985. Shipboard fuel handling and treatment for diesel engines. In: *Marine Fuels*, ed. C. H. Jones, pp. 154–173, Philadelphia, PA: American Society for Testing and Materials.
Yen, T. F., W. H. Wu, and G. V. Chilingar. 1984. A study of the structure of petroleum asphaltenes and related substances by infrared spectroscopy. *Energy Sources*, 7(3):203–235.

Chapter 4

Allinger, N. L., et al. 1971. *Organic Chemistry*. New York: Worth Publishers.
Annual Book of ASTM Standards. 2016. American Society for Testing and Materials. 2016. Volume 5, Issue 2. ASTM D5853 Note 5-The effect of 24-hour aging in D 97 pour point. West Conshohocken, PA: ASTM International.
Chandler, J. E., F. G. Horneck, and G. I. Brown. 1992. The effect of cold flow additives on low temperature operability of diesel fuels. *Report No. 922186*. Warrendale, PA: Society of Automotive Engineers.
Coley, T. 1981. Diesel fuel systems for low temperature operability. Paper presented at the Coordinating European Council International Symposium on the Performance of Automotive Fuels and Lubricants, Rome, Italy.
Ethyl Corporation. Diesel fuel additives. *Report No. PCD 417872*. Richmond, VA.
Exxon Corporation. 1973. *Tables of Useful Information*. Houston, TX.
Gary, J. H. and G. E. Handwerk. 1975. *Petroleum Refining: Technology and Economics*. New York: Marcel Dekker.
Jones, C. H., ed. 1985. *Marine Fuels*. Philadelphia, PA: American Society for Testing and Materials.
Kates, E. J., and a team of POWER editors. 1964. Oil and Gas Engines: A POWER Special Report, pp. 2–32.
Nalco Champion Diagnostic Solutions. 2015. TK 150-Bottom Water. *Report 1403847*. Sugar Land, TX: Nalco Champion.
Nalco Chemical Company. 1996. *Water Chemistry*: Part 1. Naperville, IL.
Stavinoha, L. L., and C. P. Henry, eds. 1981. *Distillate Fuel Stability and Cleanliness*. Philadelphia, PA: American Society for Testing and Materials.
Sutton, D. L. 1986. Investigation into diesel operation with changing fuel property. *Report No. 860222*. Warrendale, PA: Society of Automotive Engineers.
University of California. College of Agricultural and Environmental Sciences. 2010. Gluconacetobacter liquefaciens. Davis, CA.
http://wineswever.ucdavis.edu/industry/enology/winemicro/winebacteria/gluconacetobacter_liquefaciens.html.

Chapter 5

American Society for Testing and Materials. 1996. Marine fuels: specifications, testing, purchase and use. *Proceedings from ASTM Technical Course May 19–21*, New Orleans, LA.

Barry, E. G., L. J. McCabe, D. H. Gerke, and J. M. Perez. 1985. Heavy-duty diesel engine/fuels combustion performance and emissions: A cooperative research program. *Report No. 852078*. Warrendale, PA: Society of Automotive Engineers.

Berridge, S. A. 1992. Finishing processes. In: *Modern Petroleum Technology*, ed. G. D. Hobson, pp. 517–575, London: Institute of Petroleum.

Boren, D. S. 1994. (*Internal report*) *Diesel Cetane Engine Testing*. Sugar Land, TX: Nalco/Exxon Energy Chemicals, L.P.

Lacey, P. I. and S. R. Westbrook. 1995. Diesel fuel lubricity. *Report No. 950248*. Warrendale, PA: Society of Automotive Engineers.

Mushrush, G. W. 1992. Fuel instability 2: Organo-sulfur hydroperoxide reactions. *Fuel Science and Technology International*, 10(10):1563–1600.

Owen, K. and T. Coley. 1995. *Automotive Fuels Reference Book*. 2nd ed. Warrendale, PA: Society of Automotive Engineers.

Paramins. 1996. *Proceedings from Paramins lubricant seminar May 3–5*, New Brunswick, NJ.

Pyburn, C. M., F. P. Cahill, and R. K. Lennox. 1978. Sulfur compound interactions on copper corrosion test in propane. *Energy Processing/Canada*. March–April, pp. 60–64.

Russell, L. A. 1996. (*Internal report*) *Carbon Number Increase*. Sugar Land, TX: Nalco/Exxon Energy Chemicals, L.P.

Schweitzer, P. A. 1986. *Corrosion Resistance Tables*. 3rd ed. New York: Marcel Dekker.

Stinson, K. W. 1971. *Diesel Engineering Handbook*. 11th ed. Stamford, CT: Diesel Publications.

Unzleman, G. H. 1993. Cetane index: EPA's control for aromatics. *Octane Week*, August 1, p. 7.

Yon-Hin, P. 1996. (*Internal report*) *Corrosion Inhibitors*. Sugar Land, TX. Nalco/Exxon Energy Chemicals, L.P.

Chapter 6

Annual Book of ASTM Standards. 2016. ASTM D5001 Standard test method for measurement of lubricity of aviation turbine fuels by the ball-on-cylinder lubricity evaluator (BOCLE). West Conshohocken, PA: ASTM International.

Annual Book of ASTM Standards. 2016. ASTM D6079 Standard test method for evaluating lubricity of diesel fuels by the high frequency reciprocating rig (HFRR). West Conshohocken, PA: ASTM International.

Coordinating Research Council. 2009. Silver fuel level sensor corrosion program. *CRC Report No. 653*. Alpharetta, GA.

Deen, H. E., A. M. Kaestner, and C. M. Stendahl. 1968. Additives to improve quality and low temperature handling of middle distillate and residual fuels. Published paper presented at World Petroleum Congress, Paris, France.

Dorris, M. M. and D. Pitcher. 1988. Effective treatment of microbiologically contaminated fuel storage tanks. In: *Distillate Fuel: Contamination, Storage and Handling*. eds. H. L. Chesneau and M. M. Dorris, pp. 146–156. Philadelphia, PA: American Society for Testing and Materials.

Garcia, J. M. and S. A. Lordo. 2007. Chemistry and impacts of commonly used amine-based H_2S scavengers on crude unit towers and overheads. *Paper No. 07571*. Houston: National Association of Corrosion Engineers.

Garcia, J. M., S. S. Tyler, and H. E. McFarlin. 1994. Knowing and controlling the risks associated with hydrogen sulfide. *Report ENV-94-167*. Washington, D.C.: National Petroleum Refiners Association.

Herbstman, S. 1990. Diesel additives to improve fuel quality. *Lubrication*. Vol. 76, No. 2. White Plains, NY: Texaco.

Johnson, J. C. 1975. *Antioxidants: Syntheses and Applications*. Park Ridge, NJ: Noyes Data Corporation.

Jones, I., K. B. Peyton, and J. L. Sorrells. 2015. Low sulfur dehazers for fuels. IASH Newsletter, Issue No. 54. Atlanta, GA: International Association for the Stability and Handling of Liquid Fuels.

Leonard, E. C. ed. 1975. *The Dimer Acids*. Memphis, TN: Humko Sheffield Chemical Operation of Kraftco Corp.
Mikkonen, S., A. Niemi, M. Niemi, and J. Niskala. 1989. Effect of engine oil on intake valve deposits. *Report No. 89211*. Warrendale, PA: Society of Automotive Engineers.
Nalco Champion. 2014. *Pipeline Batch Data*. Sugar Land, TX.
Owen, K., and T. Coley. 1995. *Automotive Fuels Reference Book*. 2nd ed. Warrendale, PA: Society of Automotive Engineers.
Peyla, R. J. 1991. Additives to have key role in new gasoline era. *Oil & Gas Journal*, February 11, pp. 53–57.
Richards, P. 2014. *Automotive Fuels Reference Book*, 3rd ed. Warrendale, PA: Society of Automotive Engineers.
Ryznar, J. W. 1954. Role of chemical additives in the use of lower cost diesel fuel oils. *Proceedings from presentation at the Southeastern Railway Diesel Club*, Chicago, IL: National Aluminate Corporation. August.
Vardi, J. and B. J. Kraus. 1992. Peroxide formation in low sulfur automotive diesel fuels. *Report No. 920826*. Warrendale, PA: Society of Automotive Engineers.

Chapter 7

Amoco Chemicals Corporation. Pumpability test for distillate fuels. Technical bulletin. Chicago.
Annual Book of ASTM Standards. 2016. Vol. 05.01. West Conshohocken, PA: ASTM International.
Annual Book of ASTM Standards. 2016. Vol. 05.02. West Conshohocken, PA: ASTM International.
Annual Book of ASTM Standards. 2016. Vol. 05.03. West Conshohocken, PA: ASTM International.
Annual Book of ASTM Standards. 2016. Vol. 05.04. West Conshohocken, PA: ASTM International.
Annual Book of ASTM Standards. 2016. Vol. 05.05. West Conshohocken, PA: ASTM International.
British Standard. BS EN 14112. 2003. Fat and oil derivatives—fatty acid methyl esters (FAME)—determination of oxidation stability (accelerated oxidation test). London.
British Standard BS EN 15751. 2009. Automotive fuels—fatty acid methyl ester (FAME) fuel and blends with diesel fuel—determination of oxidation stability by accelerated oxidation method. London.
Diesel Cold Flow Test Development Group. 1991. Method for determining the simulated filter plugging point of middle distillate fuels. Abingdon, England.
Enjay. Enjay fluidity test. *Report No. ELD-59848*. Linden, NJ.
Institute of Petroleum. 1990. Standard method for analysis and testing of petroleum and related products. In: *Cold Filter Plugging of Distillate Fuels IP309*. London.
NACE International. 2001. Standard test method: Determining the corrosive properties of cargoes in petroleum product pipelines. In: *Book of Standards*. NACE Standard TM0172-2001. Houston.
STMCS. 1963. Method for determination of Amsterdam maximum pour point. Amsterdam. Shell Technology Centre.
Texaco. 1988. Instructions for performing the Waring blender test procedure. Beacon, NY.
Universal Oil Products Company. 1974. Doctor test for petroleum distillates. *Report No. 41-74*.
Williams Brothers Pipeline Company. 1982. Williams Brothers fuel oil stability test. Tulsa, OK.: The Williams Companies.

Chapter 8

Allen, A. 2015. (Internal report) Determination of additive effect on JFTOT. *TSR-3403*. Sugar Land, TX: Nalco Champion.
Allen, A. 2015. (Internal report) Optimization of wax crystal modifiers—pour and cloud point testing. *TSR-4661*. Sugar Land, TX: Nalco Champion.

Allen, A. and K. R. Trejo. 2015. (Internal report) Determination of additives' effect on fuel stability. *Harms Testing Report*. Sugar Land, TX: Nalco Champion.
Allen, A. and K. R. Trejo. 2015. (Internal report) EN 590 product performance testing for cetane, conductivity, lubricity, and CFPP. *TSR-5396*. Sugar Land, TX: Nalco Champion.
Buggs, R. N. 1992. (Internal report) Dehazer for gasoline package. 92/496. Sugar Land, TX: Nalco/Exxon Energy Chemicals, L.P.
Ekoue-Kovi, K. 2016. (Internal report) Antioxidant screening for AGO fuel. *Report S-11037*. Sugar Land, TX: Nalco Champion.
Ekoue-Kovi, K. 2016. (Internal report) CFPP for GTL diesel. *TSR-5260*. Sugar Land, TX: Nalco Champion.
Ekoue-Kovi, K. 2016. (Internal report) Gasoline ASTM D525 antioxidant testing. *TSR-5324*. Sugar Land, TX: Nalco Champion.
Garcia, J. M. 2005. (Internal report) H_2S abatement of sour naphtha. *TSR-110*. Sugar Land, TX: Nalco Company.
Garcia, J. M. 2011. (Internal report) Reduction of mercaptans in gas condensate. *TSR-1296*. Sugar Land, TX: Nalco Champion.
Garcia, J. M. 2011. (Internal report) Testing alternate H_2S scavengers. *S-8104*. Sugar Land, TX: Nalco Champion.
Garcia, J. M. 2013. (Internal report) H_2S scavenger efficiency in reformer feed. *S-9811*. Sugar Land, TX: Nalco Champion.
Jones, I. 2014 (Internal report) 84 octane gasoline NACE testing. *TSR-2712*. Sugar Land, TX: Nalco Champion.
Jones, I. 2015 (Internal report) Gasoline detergent package multiple-contact testing. *TSR-2012*. Sugar Land, TX: Nalco Champion.
Mata, Z. 1994. (Internal report) Tank bottom and fuel filter analysis. 94/877. Sugar Land, TX: Nalco/Exxon Energy Chemicals, L.P.
Nord, R. F. 1995. (Internal report) Source of ASTM D-2276 failure. 95/693. Sugar Land, TX: Nalco/Exxon Energy Chemicals, L.P.
Sheets, S. O. 1993. (Internal report) Effect of antifoulants on JP-5 WSIM. 93/285. Sugar Land, TX: Nalco/Exxon Energy Chemicals, L.P.
Soriano, N. 2015. (Internal report) Wax dispersant testing. *TSR-4514*. Sugar Land, TX: Nalco Champion.
Soriano, N. 2016. (Internal report) Ester-based lubricity improver evaluation. *TSR-6143*. Sugar Land, TX: Nalco Champion.
Soriano, N. 2016. (Internal report) Export diesel lubricity testing. *TSR-6539*. Sugar Land, TX: Nalco Champion.
Tong, N. 2016. (Internal report) Analysis of precipitate in stored product blend. *TSR-6388*. Sugar Land, TX: Nalco Champion.
Tong, N. 2016. (Internal report) Lubricant dewaxing aid evaluation. *TSR-5929*. Sugar Land, TX: Nalco Champion.
Tong, N. 2016. (Internal report) NACE corrosion testing. *TSR-6216*. Sugar Land, TX: Nalco Champion.
Wang, S. L. 1994. (Internal report) Crude oil pour point reversion. *94/720*. Sugar Land, TX: Nalco/Exxon Energy Chemicals, L.P.
Woodson, T. 2010. (Internal report) Effect of lubricity improver on fuel electrical conductivity performance. *Report LLCX-86QJCA*. Sugar Land, TX: Nalco Company.
Woodson, T. 2016. (Internal report) FCC slurry oil characterization. *TSR-6373*. Sugar Land, TX: Nalco Champion.
Woodson, T. 2016. (Internal report) Low sulfur crude pour point testing. *TSR-5784*. Sugar Land, TX: Nalco Champion.

Chapter 9

Allen, D. K. 1979. *Metallurgy Theory and Practice*. Chicago, IL: American Technical Society.
Garrett, T. K. 1994. *Automotive Fuels and Fuel Systems*. Vol. 1: Diesel. London: Pentech Press.

Herro, H. M. and R. D. Port. 1993. *The Nalco Guide to Cooling Water System Failure Analysis*. New York: McGraw-Hill.
Russell, J. H. 1990. (Internal report) Chemical storage and injection systems. Sugar Land, TX: Nalco Chemical Company.
Schweitzer, P. A. 1994. *Corrosion Resistant Piping Systems*. New York: Marcel Dekker.
Sperry Vickers. 1983. *Mobile Hydraulics Manual*. Troy, MI: Sperry Corporation.
Tschoke, H. 1994. Distributor injection pumps: VE. In: *Diesel Fuel Injection*. eds. U. Adler, H. Bauer, and A. Beer., pp. 148–161. Stuttgart: Robert Bosch GmbH.

Chapter 10

International Air Transportation Association. 2015. *Dangerous Goods Regulations*. Montreal.
Lewis, R. J. 1993. *Hazardous Chemicals Desk Reference*. 3rd ed. New York: Van Nostrand Reinhold.
Nalco/Exxon Energy Chemicals, L.P. 1997. *Shipping Samples Safely*. Sugar Land, TX.

Chapter 12

Amarasekara, A. S. 2013. *Handbook of Cellulosic Ethanol*. Hoboken, NJ: John Wiley & Sons, and Salem, MA: Scrivenger Publishing.
Bechtold, R. L. 1997. *Alternative Fuels Guidebook: Properties, Storage, Dispensing and Vehicle Facility Modifications*. Warrendale, PA: Society of Automotive Engineers.
Brinkman, N. D., E. E. Ecklund, and R. J. Nichols, eds. 1989. *Fuel Methanol: A Decade of Progress*. Warrendale, PA: Society of Automotive Engineers.
Bureman, P. E. and K. B. Peyton. 2009. Use of ethanol and biodiesel in fuels today. Reliability and Maintenance Conference, Ft. Worth, TX: National Petroleum Refiners Association.
Cheng, W-H. and H. H. Kung, eds. 1994. *Methanol Production and Use*. New York: Marcel Dekker.
Clark, N., M. Gautam, D. Lyons, C. Atchinson, W. Xie, P. Norton, K. Vertin, et al. 1999. On road use of Fischer-Tropsch diesel blends. *Report No. 1999-01-2251*. Warrendale, PA: Society of Automotive Engineers.
Enjay Chemical Company. 1962. *Ethyl Alcohol*. New York: Enjay Laboratories.
Garrett, T. K. 1994. *Automotive Fuels and Fuel Systems*. Vol.2: Diesel. London: Pentech Press.
Johnson, R. W. and E. Fritz, eds. 1989. *Fatty Acids in Industry*. New York: Marcel Dekker.
Kirk-Othmer Encyclopedia of Chemical Technology. 4th ed. 1997. Soybeans and other oilseeds. New York: John Wiley & Sons.
Klass, D. L., ed. 1981. *Biomass as a Nonfossil Fuel Source*. Washington, D.C.: American Chemical Society.
Li, X., W. L. Chippior, Ö. L. Gülder, J. Cooley, E. K. Richardson, and K. Mitchell. 1998. *Report No. 982487*. Warrendale, PA: Society of Automotive Engineers.
McGinnis, T. P. and K. B. Peyton. 2010. Analytical characterization of sediment formed in soy oil-based B20 biodiesel. *Paper ID JAI102584*, West Conshohocken, PA: ASTM International.
Moore, R. M. (Bob). 2000. Direct methanol fuel cells for automotive power systems. *Report No. 2000-01-0012*. Warrendale, PA: Society of Automotive Engineers.
Nylund, N-O. and P. Aakko. 2000. Characterization of new fuel qualities. *Report No. 2000-01-2009*. Warrendale, PA: Society of Automotive Engineers.
Ofner, H., D. W. Gill, and C. Krotscheck. 1998. Dimethyl ether as fuel for CI engines—a new technology and its environmental potential. *Report No. 981158*. Warrendale, PA: Society of Automotive Engineers.
Pickard, A. 2010. Gaseous fuels and light hydrocarbons (methane through butanes, natural gasoline, and light olefins. In: *Significance of Tests for Petroleum Products*. ed. S. J. Rand, pp. 164–172. West Conshocken, PA: ASTM International.
Probstein, R. F. and E. R. Hicks. 1990. *Synthetic Fuels*. Cambridge, MA: pH Press.
Renewable Fuels Association. Washington D.C. July 16, 2016. http://www.ethanolrfa.org/resources/publications/pocket.
Richards, P. 2014. *Automotive Fuels Reference Book*, 3rd ed. Warrendale, PA: Society of Automotive Engineers.

Senatore, A., M. Cardone, V. Rocco, and M. V. Prati. 2000. A comparative analysis of combustion process in D.I. diesel engine fueled with biodiesel and diesel fuel. *Report No. 2001-01-0691*. Warrendale, PA: Society of Automotive Engineers.

Society of Automotive Engineers. 2000 Surface vehicle information report—alternative automotive fuels. Warrendale, PA: Society of Automotive Engineers.

Speight, J. G., 2008. *Synthetic Fuels Handbook: Properties, Processes, and Performance.* New York: McGraw-Hill.

Stavinoha, L. L., E. S. Alfaro, H. H. Dobbs, Jr., and L. A. Villahermosa. 2000. *Alternative Fuels: Development of a Biodiesel B20 Purchase Description*. Warrendale, PA: Society of Automotive Engineers.

Suppes, G. J., T. J. Fox, K. R. Gerdes, H. Jin, M. L. Burkhart, and D. N. Koert. 2000. Cold flow and ignition properties of Fischer-Tropsch fuels. *Report No. 2000-01-2014*. Warrendale, PA: Society of Automotive Engineers.

Tesfaw, A. and F. Assefa. 2014. Current trends in bioethanol production by *Saccharomyces cerevisiae*: Substrate, inhibitor reduction, growth variables, coculture, and immobilization. International Scholarly Research Notes. Volume 2014. Addis Ababa, Ethiopia.

U.S. Energy Information Administration. *Annual Energy Review*. 2012. Table 10.5. Estimated Number of Alternative-Fueled Vehicles in Use and Fuel Consumption, 1992–2010. (2012): 1, July 17, 2016 http://www.eia.gov/totalenergy/data/annual/showtext.cfm?t=ptb

U.S. Energy Information Administration. *Today in Energy*. 2013. Cellulosic biofuels begin flow but in lower volumes than forseen by statutory targets. (2013): 1, July 17, 2016 http://www.eia.gov/todayinenergy/detail.cfm?id=10131.

Index

B

C

G

N

O

P

Q

R

Printed by Libri Plureos GmbH in Hamburg,
Germany

9 780071 848091